Biotechnological Innovations in Energy and Environmental Management

BOOKS IN THE BIOTOL SERIES

BIOTECHNOLOGY BY OPEN LEARNING

Biotechnological Innovations in Energy and Environmental Management

PUBLISHED ON BEHALF OF :

Open universiteit   and   **University of Greenwich
(formerly Thames Polytechnic)**

Valkenburgerweg 167   Avery Hill Road
6401 DL Heerlen   Eltham, London SE9 2HB
Nederland   United Kingdom

Butterworth-Heinemann Ltd
Linacre House, Jordan Hill, Oxford OX2 8DP

A member of the Reed Elsevier plc group

OXFORD LONDON   BOSTON
MUNICH NEW DELHI SINGAPORE SYDNEY
TOKYO TORONTO WELLINGTON

First published 1994

British Library Cataloguing in Publication Data
A catalogue record for this book is
available from the British Library

Library of Congress Cataloguing in Publication Data
A catalogue record for this book is
available from the Library of Congress

ISBN 0 7506 0615 0

Composition by University of Greenwich
(formerly Thames Polytechnic)
Printed and Bound in Great Britain

The Biotol Project

The BIOTOL team

OPEN UNIVERSITEIT, THE NETHERLANDS
Prof M. C. E. van Dam-Mieras
Prof W. H. de Jeu
Prof J. de Vries

UNIVERSITY OF GREENWICH (FORMERLY THAMES POLYTECHNIC), UK
Prof B. R. Currell
Dr J. W. James
Dr C. K. Leach
Mr R. A. Patmore

This series of books has been developed through a collaboration between the Open universiteit of the Netherlands and University of Greenwich (formerly Thames Polytechnic) to provide a whole library of advanced level flexible learning materials including books, computer and video programmes. The series will be of particular value to those working in the chemical, pharmaceutical, health care, food and drinks, agriculture, and environmental, manufacturing and service industries. These industries will be increasingly faced with training problems as the use of biologically based techniques replaces or enhances chemical ones or indeed allows the development of products previously impossible.

The BIOTOL books may be studied privately, but specifically they provide a cost-effective major resource for in-house company training and are the basis for a wider range of courses (open, distance or traditional) from universities which, with practical and tutorial support, lead to recognised qualifications. There is a developing network of institutions throughout Europe to offer tutorial and practical support and courses based on BIOTOL both for those newly entering the field of biotechnology and for graduates looking for more advanced training. BIOTOL is for any one wishing to know about and use the principles and techniques of modern biotechnology whether they are technicians needing further education, new graduates wishing to extend their knowledge, mature staff faced with changing work or a new career, managers unfamiliar with the new technology or those returning to work after a career break.

Our learning texts, written in an informal and friendly style, embody the best characteristics of both open and distance learning to provide a flexible resource for individuals, training organisations, polytechnics and universities, and professional bodies. The content of each book has been carefully worked out between teachers and industry to lead students through a programme of work so that they may achieve clearly stated learning objectives. There are activities and exercises throughout the books, and self assessment questions that allow students to check their own progress and receive any necessary remedial help.

The books, within the series, are modular allowing students to select their own entry point depending on their knowledge and previous experience. These texts therefore remove the necessity for students to attend institution based lectures at specific times and places, bringing a new freedom to study their chosen subject at the time they need and a pace and place to suit them. This same freedom is highly beneficial to industry since staff can receive training without spending significant periods away from the workplace attending lectures and courses, and without altering work patterns.

SOFTWARE IN THE BIOTOL SERIES

BIOcalm interactive computer programmes provide experience in decision making in many of the techniques used in Biotechnology. They simulate the practical problems and decisions that need to be addressed in planning, setting up and carrying out research or development experiments and production processes. Each programme has an extensive library including basic concepts, experimental techniques, data and units. Also included with each programme are the relevant BIOTOL books which cover the necessary theoretical background.

The programmes and supporting BIOTOL books are listed below.

Isolation and Growth of Micro-organisms
Book: *In vitro* Cultivation of Micro-organisms
        Energy Sources for Cells

Elucidation and Manipulation of Metabolic Pathways
Books: *In vitro* Cultivation of Micro-organisms
        Energy Sources for Cells

Gene Isolation and Characterisation
Books: Techniques for Engineering Genes
        Strategies for Engineering Organisms

Applications of Genetic Manipulation
Books: Techniques for Engineering Genes
        Strategies for Engineering Organisms

Extraction, Purification and Characterisation of an Enzyme
Books: Analysis of Amino Acids, Proteins and Nucleic Acids
        Techniques used in Bioproduct Analysis

Enzyme Engineering
Books: Principles of Enzymology for Technological Applications
        Molecular Fabric of Cells

Bioprocess Technology
Books: Bioreactor Design and Product Yield
        Product Recovery in Bioprocess Technology
        Bioprocess Technology: Modelling and Transport Phenomena
        Operational Modes of Bioreactors

Further information: Greenwich University Press,
University of Greenwich, Avery Hill Road, London, SE9 2HB.

Contributors

AUTHORS

Dr D.J. Hill, University of Wolverhampton

Dr S.M. Keith, University of Teesside, Middlesbrough, UK

Dr C. K. Leach, De Montfort University, Leicester, UK

Dr E. Middelbeek, Open universiteit, Heerlen, The Netherlands

Prof M. C. E. van Dam-Mieras, Open universiteit, Heerlen, The Netherlands

Dr G.W.G. Montgomery, University of Teesside, Middlesbrough, UK

Dr S.E. Montgomery, University of Teesside, Middlesbrough, UK

Dr S.W. Shales, University of the West of England, Bristol

Dr G. Mijnbeek, Bird Engineering bv, Schiedam, The Netherlands

EDITORS

Dr C. K. Leach, De Montfort University, Leicester, UK

Prof M. C. E. van Dam-Mieras, Open universiteit, Heerlen, The Netherlands

SCIENTIFIC AND COURSE ADVISORS

Prof M. C. E. van Dam-Mieras, Open universiteit, Heerlen, The Netherlands

Dr C. K. Leach, De Montfort University, Leicester, UK

ACKNOWLEDGEMENTS

Grateful thanks are extended, not only to the authors, editors and course advisors, but to all those who have contributed to the development and production of this book. They include Mrs A. Allwright, Miss K. Brown, Ms H. Leather and Miss J. Skelton.

The development of this BIOTOL text has been funded by **COMETT, The European Community Action Programme for Education and Training for Technology**. Additional support was received from the Open universiteit of The Netherlands and from the University of Greenwich (formerly Thames Polytechnic).

Contents

How to use an open learning text

An open learning text presents to you a very carefully thought out programme of study to achieve stated learning objectives, just as a lecturer does. Rather than just listening to a lecture once, and trying to make notes at the same time, you can with a BIOTOL text study it at your own pace, go back over bits you are unsure about and study wherever you choose. Of great importance are the self assessment questions (SAQs) which challenge your understanding and progress and the responses which provide some help if you have had difficulty. These SAQs are carefully thought out to check that you are indeed achieving the set objectives and therefore are a very important part of your study. Every so often in the text you will find the symbol Π, our open door to learning, which indicates an activity for you to do. You will probably find that this participation is a great help to learning so it is important not to skip it.

Whilst you can, as an open learner, study where and when you want, do try to find a place where you can work without disturbance. Most students aim to study a certain number of hours each day or each weekend. If you decide to study for several hours at once, take short breaks of five to ten minutes regularly as it helps to maintain a higher level of overall concentration.

Before you begin a detailed reading of the text, familiarise yourself with the general layout of the material. Have a look at the contents of the various chapters and flip through the pages to get a general impression of the way the subject is dealt with. Forget the old taboo of not writing in books. There is room for your comments, notes and answers; use it and make the book your own personal study record for future revision and reference.

At intervals you will find a summary and list of objectives. The summary will emphasise the important points covered by the material that you have read and the objectives will give you a check list of the things you should then be able to achieve. There are notes in the left hand margin, to help orientate you and emphasise new and important messages.

BIOTOL will be used by universities, polytechnics and colleges as well as industrial training organisations and professional bodies. The texts will form a basis for flexible courses of all types leading to certificates, diplomas and degrees often through credit accumulation and transfer arrangements. In future there will be additional resources available including videos and computer based training programmes.

Preface

'What is environmental biotechnology?' is not a simple question to answer. Virtually all biotechnological endeavours can be claimed, justifiably, to have environmental consequences. We can cite for example, that biotechnologically mediated improvements in health care will influence population growth and individual longevity both of which will have an environmental impact. Similarly, the development of new crop plants and crop protectants will influence the use of land and our ability, to feed a growing population and the replacement of the 'dirty' manufacturing processes by cleaner biotechnological strategies can be regarded as environmentally important. The objective of this text is not to examine all of the environmental ramifications of biotechnology but to provide a perspective on the application of biotechnology specifically targeted at removing, circumventing or reducing the damaging consequences that human activities have on the environment. Thus, in the context of this text, environmental biotechnology means the specific application of biotechnology to the management of environmental problems. Even with this more restricted perspective, the potential topic range is enormous and it is impossible to review the whole field comprehensively in a single text. Instead, a careful selection of the main activities has been made to provide readers with both an indepth study and a broad perspective.

The text has been written on the assumption that the reader is familiar, at least in principle, with the science and technology which underpin contemporary biotechnology. The text, however, has many helpful reminders and explanations. It begins by providing a natural sciences perspective on the environment and explains that the solutions to environmental problems depends not solely upon technological development but is also greatly influenced by social, political and economic factors.

Environmental problems may be tackled in two fundamentally different ways. One is to adopt processes that do not create the problem in the first place. We might call this the 'front of pipe' approach. The other is to treat the problem once it has been created, sometimes referred to as the 'end of pipe' approach. Both strategies are examined although the balance of the text reflects current practices of attempting to find 'end of pipe' solutions. Thus we consider the treatment of solid and liquid wastes by examining the use of landfill, aerobic waste water treatment, anaerobic waste water treatment and the treatment of recalcitrant organic materials.

The text is not, however, confined to a discussion of 'end of pipe' processes. We also examine the application of biotechnology to achieve desired ends currently fulfilled by environmentally more damaging technologies. Important in this area are improvements in soil fertility and crop protection. We use the production of biofuels to illustrate how biotechnology may serve as a bridge between the end of pipe and the front of pipe strategies. In this case, biotechnology may be used to reduce the problems caused by the wastes derived from 'dirty' technologies to produce a usable product normally derived by the consumption of non-renewable resources.

This book should be of value to a wide range of readers with interests in biotechnology, the environment or both. It should benefit students, researchers, industrialists and planners who require information concerning the potential of biotechnology to provide solutions to environmental concerns.

Scientific and Course Advisors: Professor M.C.E. van Dam-Mieras
Dr C.K. Leach

An introduction to biotechnology innovations in environmental and energy management

An introduction to biotechnology innovations in environmental and energy management

Societies generally become concerned with the long-term environmental consequences of their actions only after industrialisation has provided them with substantial economic and material affluence. Recently, it has become more widely recognised and accepted that environmental matters very often have a global dimension and that environmental problems are not restrained by national borders. The actions of one community often impinge on many others. The international conscience has, therefore, been aroused and concern about Man's use and abuse of the environment has been widely expressed.

The catalogue of Mankind's misdemeanours on soil, sea and air is both squalid and long. Solving these problems will be complex and depends on the development of appropriate technologies, social attitudes and awareness, economic consensus and political will. Set in a historical background of the diverse ethnic cultures, economic fortunes, social structures, political and legal systems and religious beliefs of the world's nation states, it is not surprising that few solutions command universal

strategies in solving environmental concerns

support. From issues surrounding the conservation of whales to the generation of electricity using nuclear energy, there are a multitude of attitudes and perspectives which are adopted. Nevertheless we can identify a number of major strategies that are commonly described as providing solutions to environmental concerns. Some people, for example, place their faith in the attainment of more scientific knowledge and better technological control. Others see the solution as mainly dependent upon development of a more environmentally-sensitive, socio-economic morality, while still others see the cultivation of spiritual values as providing the key. The growing trend is, however, to recognise that they are all important elements in the evolution of globally-accepted approaches to sound environmental management.

The development of environmentally sustainable practices will undoubtedly depend upon a wide variety of inputs. Solutions will not arise simply from advances in technology but will call upon changes in social attitudes and the development of civic acceptance and political will, on an international scale, to meet the environmental

central role of the natural sciences in generating environment- ally-sound practices

challenges which confront us. Central to these developments will be the contribution of the natural sciences in general, and biotechnology in particular, to the development of new, environmentally acceptable, practices. It is after all, the natural sciences that provide much of the data which gave rise to our awareness and concerns. Much is expected of biotechnology in helping us in responding to these concerns. The fulfilment of the potential of biotechnology to alleviate some of these issues is, however, dependent upon the social acceptability and economic viability of biotechnological processes and products. Biotechnology will not, by itself, solve all environmental problems.

The purpose of this text is to provide a description of the processes and potential of what might be called, 'environmental biotechnology'. In its widest sense, environmental biotechnology is essentially the application of biotechnology to the management of the environment. It encompasses aspects of natural resource management, the treatment of waste and the control of pollution. It is, however, impossible to provide an in depth discussion of all of the current and potential applications of biotechnology to the development of environmentally sustainable strategies and practices in a single text. A deliberate policy has been adopted here to predominantly confine discussion to aspects of waste treatment, pollution control and energy management. Biotechnology will also have important contributions to make to agriculture and these, too, will have a major impact on the way we use our environment. For example, it should be realised that through the development of biological agents to replace chemically-produced crop protectants and by the generation of higher yielding, easier to harvest crops, biotchnology will have a major impact on pollution levels and land use. Similarly, the application of biotechnology to animal production, fish farming and food manufacture will have an impact on the provision of food and the use of land and water resources. We can also cite the use of biotechnological procedures to preserve germplasm and its consequences on biological conservation.

The exclusion of some of these aspects from this text does not mean that they are not of extreme environmental importance, it merely reflects our desire to keep the text down to manageable proportions. Many of the topics excluded from detailed discussion here are dealt with elsewhere in the BIOTOL series, especially in the BIOTOL 'Innovation' texts. In this text we have included a balanced selection of the main types of activities falling within 'environmental biotechnology' to which the themes energy, waste and pollution may be directly applied. You will learn as you progress through this text that, the application of biotechnology, in the short term, is predominantly focused on rectifying problems that arise from the use of conventional, non-biological, environmentally-damaging technologies. This type of application might be conveniently thought of as an 'end of pipe' technology. In other words, a technology which is used to treat the output from other technologies. Increasingly, however, it is

biotechnology as a cleaning and a cleaner technology

becoming apparent that biotechnology offers completely novel solutions to human needs and to be able to supply these needs in a more environmentally acceptable and sustainable way. Biotechnology is not only a 'cleaning' technology but also a 'cleaner' technology.

structure of the text

In order to provide a context in which you may study the application of biotechnology to solving or circumventing environmental problems, we have provided an overview of the contribution of the natural sciences, biotechnology, sociology and politics to the formulation of strategies for managing the environment in Chapter 2. The perspectives and views expressed in this chapter are those held by the authors. They have attempted to be non-controversial but recognise that their opinions and philosophy are expressed from the perspective of a long-term study in the natural sciences. Others, with different professional occupational backgrounds, may see environmental issues in a different light. Chapter 2 should, therefore, be regarded as the personal philosophy of the authors rather than as a catalogue of hard, factual materials that needs to be learnt by the prospective environmental biotechnologist.

In Chapter 3 we described the origins and nature of wastes and pollution especially drawing attention to the distinction between biologically-derived and degradable wastes and non-biodegradable wastes and their consequences on environmental health. We also consider the use of landfill as a mechanism for handling solid wastes and briefly discuss the exploitation of landfill to recover derelict land and to generate methane.

In Chapters 4 and 5 we turn our attention to the treatment of water-borne wastes. In Chapter 4 we provide an overview of water treatment processes and describe the use of aerobic processes to remove biodegradable wastes. This theme is developed further in Chapter 5 where we describe the use of anaerobic bioprocesses to treat water-borne wastes and the concomitant generation of methane.

Many of the organic chemicals produced by industry are rather recalcitrant to biodegradation and cannot be successfully treated by conventional anaerobic and aerobic water treatment processes. The application of biotechnology to treating and/or transforming these so-called xenobiotics is described in Chapter 6.

In Chapter 7, we consider the quality standards that are applied to the environment. The focus here is on the analysis and control of water quality. In this chapter, we include a discussion of the potential of biotechnology to provide analytical devices of value in water quality control.

The generation of methane as a by-product of landfill and anaerobic waste water-treatment may be regarded as a potential asset since methane is a valuable energy source. This theme of biologically-generated energy sources is taken up in Chapter 8 where we discuss the generation and use of biomass as fuel, biogas, bio-ethanol and biohydrogen. Energy is also an important element in Chapter 9 where we examine the application of biotechnology to agriculture. We draw attention to the high energy cost and environmentally-polluting practices of agriculture in many Western communities. We examine some of the ways in which biotechnology may be applied to reduce or circumvent these environmentally-damaging factors. In particular, we consider the maintenance of soil fertility and the application of environmentally-safe crop protectants.

In the final chapter we turn our attention to the application of biotechnology to mineral extraction. Mining and mineral extraction are of tremendous importance to modern economies and have frequently been associated with environmental damage and degradation.

The text has been written on the assumption that readers have some previous experience of the main enabling techniques of biotechnology namely; genetic engineering, protein engineering, *in vitro* cell cultivation and process technology. Although this assumption has been made, authors have provided many helpful reminders and reference is made to other texts in the BIOTOL series which provide additional support.

A natural sciences perspective on environmental management

A natural sciences perspective on environmental management

2.1 Introduction

This chapter is designed to provide a context in which to study the contribution of biotechnology to environmental and energy management. Although subsequent chapters are mainly confined to the discussion of 'end-of-pipe' applications, in this chapter we take a rather wider view.

We begin by considering the natural order of life on Earth and the interaction that each organism has with other organisms and with its abiotic environment. We then briefly review the evolution of human society(s) and provide an overview of the consequences this evolution has had on the natural environment. We use the consideration of energy as the underpinning theme of this chapter using it to both illustrate the general flow of matter within the biosphere and to discuss the environmental consequences of human activities. Included in this discussion will be a description of Man's use of fossil fuels and the possibility of using renewable (biologically generated) energy sources. Inevitably this necessitates discussion of agriculture and the impact biotechnology may have on agricultural practices. The use of fossil fuels as chemical - as well as energy - resources also poses concerns about resource depletion and pollution. We outline briefly the potential of biotechnology to produce alternative approaches to the generation of chemicals and the removal of pollutants.

In the final part of the chapter, we discuss the concept of integral life cycle analysis and management which may enable human societies to achieve a sustainable life-style and we indicate how biotechnology is an essential component of such an achievement.

2.2 The natural order - food webs and energy

interactions between organisms and their surroundings

It is axiomatic to say that the environment is of vital importance to living organisms. Each organism lives in continuous interaction with its surroundings. It imports the energy and raw materials it needs for growth, development and reproduction from its locality. At the same time, its surroundings also form a threat. Abiotic (for example, wind, temperature fluctuations, acidity etc) and biotic (other organisms) factors within the vicinity of the organism may threaten its survival. For example, through its interaction with its environment, the organism may be mechanically or chemically damaged, infected by pathogens or eaten by predators. Organisms have, of course, evolved nutritional and defence strategies to maintain themselves in this 'struggle for life'. Nevertheless, the success of these strategies is largely dependent upon the quality of the environment and, as we shall see, in many cases Man-induced changes to the environment may pose problems which either alter the types of organisms that can survive or create such severe problems that there are no satisfactory biological solutions.

Within the biosphere, organisms are arranged into food webs. In such food webs, plants are the primary producers which are 'eaten' by animals and other organisms which, in turn, are devoured by other predators. This type of nutritional relationship forms the energy basis of the natural struggle for life. It is, therefore, important that we look at it in more detail.

Π   You have probably met the concept of food webs before. Before reading the next section see if you can identify the energy source which allows a food web to be maintained and see if you can explain why it is possible for one organism to be a source of nutrition for another.

2.2.1 Food webs and energy

the key role of photosynthesis

For most plants, their only energy source is solar radiation and they derive their carbon from carbon dioxide. During photosynthesis, solar energy is converted to chemical energy and stored in the bonding energy of sugar molecules. These sugar molecules can be considered as the reservoir of both raw materials and energy of the plant. The sugar molecules are converted to other compounds by the non-photosynthetic reactions of plant metabolism. These compounds make up the components of plant cells. The biomolecules produced consist mainly of carbon, hydrogen, oxygen, nitrogen, phosphorus and sulphur. In addition to these, small amounts of other elements also occur in these cells. These mainly play a catalytic role in cellular metabolism and are taken up from soil by the roots of the plant. Nitrogen, phosphorus and sulphur-containing materials and water are also absorbed from the soil.

Plants therefore absorb a variety of nutrients from their environment and trap the energy of sunlight in the form of chemical energy. This energy, initially stored as sugar molecules, can be released in times of need into more mobile energy 'carriers'. Predominant amongst these is the cellular energy 'carrier' adenosine triphosphate (ATP). The chemical energy within these 'carriers' may be used by the organism to drive a wide range of activities including biosynthesis and the maintenance of osmotic balance. Comparative biochemistry reveals that the same energy 'carriers' are found in all living organisms. This similarity between plants and other organisms is not, however, confined to their energy carriers. All organisms, in spite of their tremendous diversity, are mainly composed of the same four main groups of biomolecules; carbohydrates, lipids, proteins and nucleic acids. The major metabolic pathways found in diverse organisms are also remarkably similar. However, only organisms capable of photosynthesis can take up energy directly from solar radiation. These universal

universal principles of metabolism

principles of metabolism imply that for growth, development, homeostasis and reproduction, all organisms draw from the same energy reservoir. This reservoir is, of course, the sun.

There is, however, another important consequence arising from the similarity between molecules which make up diverse organisms and the universality of metabolism. It means that the constituents and products of one organism may easily be taken up and metabolised by another. Therefore, with the exception of plants that can mostly only be eaten, all organisms can potentially use each other as a source of energy and raw materials. We have represented these underpinning principles of food webs in Figure 2.1.

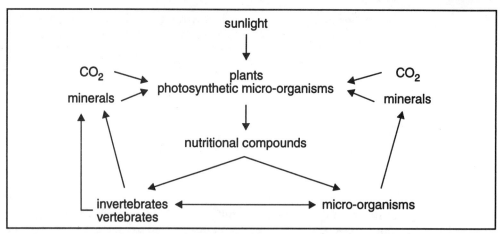

Figure 2.1 Highly simplified schematic representation of the flow of energy and matter in living nature.

We need, however, to elaborate a little on this unified scheme. Some micro-organisms such as the green algae and the cyanobacteria are also capable of photosynthesis. Just like plants, they can directly use solar energy as their metabolic driving force. Other micro-organisms, the so called chemoautotrophs (chemolithothrophs) can form ATP and other energy 'carriers' during the oxidation of simple inorganic compounds. They also use carbon dioxide as their source of carbon. However, with the exception of plants and the micro-organisms mentioned above, all other organisms need an organic energy source, which implies that, for their energy supply, they are completely dependent on the photosynthetic organisms that are able to convert solar energy to chemical energy. The non-photosynthetic organisms thus draw their energy from the chemical bonds of the products of photosynthesis. Not all organisms use the same nutritional compounds and micro-organisms especially show a quite impressive diversity in this respect. Such nutritional habits determine in which environment a micro-organism will be able to grow. Because of their diversity and flexibility (nutritional adaption), micro-organisms play an important role in degrading dead materials. It should be, therefore, of no surprise to learn that micro-organisms are often employed by Man to transform 'waste' organic matter.

chemo-autotrophs

nutritional needs

the relationship between catabolism and biosynthesis

When a micro-organism is provided with the necessary nutritional compounds, the organism can make all the cellular biomolecules it needs. To that purpose, the nutritional compounds are degraded intracellularly to smaller molecules which are subsequently used in biosynthetic processes. During the degradation of nutritional compounds, part of the energy present in chemical bonds is converted to usable forms of cellular energy (for example, ATP, reduced pyridine nucleotides and concentration gradients) which are subsequently used during biosynthetic reactions and processes.

∏ Will all of the biomolecules be synthesised at the same time?

The capability to synthesise all the biomolecules needed does not, necessarily imply that the micro-organism will carry out all such synthetic processes simultaneously. The micro-organism will continuously monitor its environment and adapt its biosynthetic status to the local supply. Thus, given the option, the micro-organism will preferentially absorb pre-formed biomolecules from its environment rather than consume energy to synthesise such molecules *de novo*. This behaviour implies that, in the presence of large numbers of micro-organism, there is competition between them for these nutritional materials.

absorption and
use of nutrients
in higher
heterotrophic
organisms
Higher, non-photosynthetic organisms (heterotrophs) must also obtain raw materials and energy from the environment. They do so by devouring plants or other animals or products derived from these sources. During passage through the alimentary tract, the nutritional compounds are degraded to smaller molecules (for example, monosaccharides, amino acids, fatty acids) and these are taken up by the surrounding cells. In turn, these molecules may be further transported and imported into cells in other organs of the body. Within these cells, the molecules derived from the nutritional source are degraded further to produce smaller units which are suitable for the biosynthesis of cellular compounds. At the same time, some of the nutritionally-derived compounds are used to generate utilisable forms of energy. We have illustrated the generalised flow of cellular metabolism in Figure 2.2.

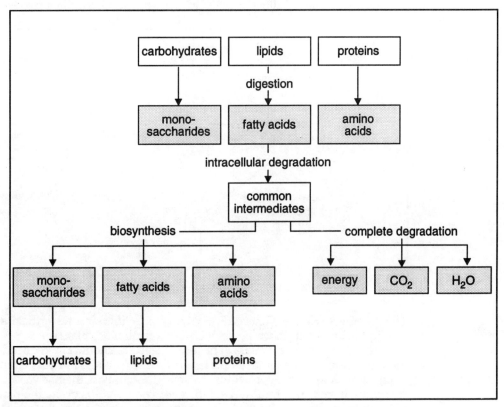

Figure 2.2 Flow chart of metabolic activity.

SAQ 2.1

Briefly outline the functions of carbohydrates, lipids and proteins in living systems.

essential
nutrients
Heterotrophic, higher organisms differ from the majority of micro-organisms in that they cannot synthesise all cellular components. Some biomolecules must, therefore, be taken up with the nutrients. Such components are called essential nutritional components and include, amongst other things, nutrients we refer to as vitamins and essential amino acids. The occurrence of essential components further increases the dependence of organisms on their environment. The reason behind this 'choice' for dependency is probably an energetic one; very often the biosynthetic route for essential

components is highly energy-demanding and, therefore, import of these molecules is more favourable from an energetic point of view.

Π   In the flow of energy in the biosphere described above, it would appear that there is a unified flow of energy from plants (and photosynthetic micro-organisms) to heterotrophs. But are plants dependent in any way on heterotrophs?

The answer is that plants are dependent upon heterotrophs to regenerate the carbon dioxide and minerals they need in order to carry out photosynthesis and to build their own cellular constituents. If we consider the biosphere in terms of the turnover of elements, rather than in terms of energy flow, then cyclical rather than linear patterns are observed. We will illustrate this using carbon as an example (see Figure 2.3).

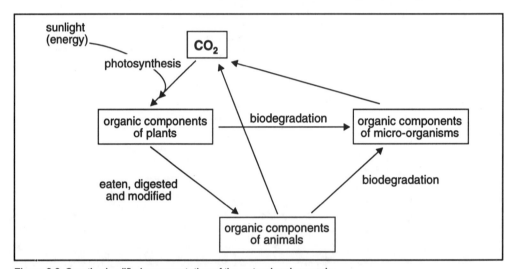

Figure 2.3 Greatly simplified representation of the natural carbon cycle.

The scheme shown in Figure 2.3 is greatly simplified but nevertheless illustrates the principles involved. If we begin with carbon dioxide, plants using the processes of photosynthesis convert this form of carbon into the chemically reduced form found in the biomolecules which make up plant cells. This plant material is either devoured by animals or degraded by micro-organisms. The animals themselves may be eaten by predators in complex food webs. However complex this cycle, the biomolecules will ultimately be taken up by degrading micro-organisms. At each stage of the food web, some of the biomolecules will be oxidised releasing carbon dioxide. Thus over a period of time, all (or virtually all) of the carbon dioxide reduced by photosynthesis will be re-oxidised and return to the system as carbon dioxide.

cyclic flow of carbon

geochemical cycles

The cyclic flow of carbon and other elements within the biosphere is well documented and the cyclical flows of these elements are often referred to as geochemical cycles. We will not elaborate on them further here except to emphasise a particular point. The cyclical flow of elements is driven by the energy input derived from photosynthesis. In other words, solar energy provides the energy input which drives these cycles.

Π From this general overview of the natural order, can we conclude that a natural system is stable and sustainable?

the sustainable but dynamic nature of biocycles

The answer is not quite as straight forward as at first it might seem. In simple terms, it might appear that such a system could be both stable and sustainable providing the sun continues to shine! With a continued energy input and a cyclical flow of elements, in principle this would appear to fulfil the necessary criteria for stability and sustainability. However, this would imply that the metabolic activities within the cycles were not capable of change. We know, however, that evolution continually modifies the metabolic capabilities of organisms. Thus over an evolutionary time scale, we might anticipate changes in the biological components which participate within the geocycle. The evidence is that changes in for example, plants lead to evolutionary responses in animals. Thus in terms of geochemical cycles little change might be anticipated over many millions of years even whilst there may have been many changes to the form and behaviour of the organisms which participate in these cycles. We can conclude, therefore, that although the situation is highly dynamic with respect to changes in the participants, the system is sustainable. In the absence of humans, we could imagine that such a system of slow evolutionary change would continue to occur for many aeons to come. In such a condition only after some major disruption due to, for example, the energy input from the sun, would this system be under threat.

Before we leave this section, we make one final point. The development of technology has, in evolutionary terms, taken place in a very short space of time. The natural world, reliant upon evolutionary mechanisms, requires significant periods of time to adapt to changing circumstances. Thus changes within the biosphere caused by human activities may pose threats not only to specific organisms but also to the stability and sustainability of the biosphere as a whole; threats for which there is no natural mechanism capable of responding in a timescale which will enable stability to be retained. In the next section, we will examine briefly these threats by considering the impact of human activities on the biosphere.

SAQ 2.2

In Figure 2.1 we have indicated that the only source of energy within the biosphere is the sun. Is the generation of biologically usable energy in chemoautotrophs an exception to this rule?

2.3 Two revolutions

hunting and gathering society

The schematic model of the interactions between living systems described above is consistent with a hunting and gathering human society. It has been calculated that if Mankind had continued to live in this way, this would be compatible with a human population of perhaps, about 10 million people. In 1990, however, the human population was estimated to be 5321 millions. Such a large population immediately poses two main questions:

- what enabled such a population growth?

- are there limits to the human population the Earth can support and if so, what are they?

We will need to go back in time to find answers to these questions and examine the development of human society. More than 10 000 years ago Man changed from a hunting and gathering lifestyle to agriculture. The driving force behind this development is largely unknown.

Some archaeological sources suggest that the fact that large animals had become scarce may have contributed. Also climatological changes occurring after the end of the last ice age (18 000 years ago) were probably of importance. These developments took place in the Neolithicum and, therefore, the transition from a hunting and gathering life style

Neolithic Revolution

to an agricultural life style is sometimes described as the Neolithic Revolution. This revolution probably continued for thousands of years and did not take place everywhere on Earth at the same time.

The oldest farms have been found in the Middle East, in a region where wild wheat grew naturally. Probably people started to collect grains of corn and to use part of the harvest as sowing-seed for the next growing season. In Mexico, beans and maize were already cultivated 8000 years ago. Man also started to develop cattle-breeding by letting tamed, wild animals reproduce in captivity. Together with agriculture and cattle-breeding, the need to develop techniques for food conservation also arose and this most probably stimulated pottery. In the same period, Man also invented the wheel, which facilitated transportation, and started to use metals like copper and bronze for making tools. The use of these new tools of course further improved, among other things, agricultural techniques.

This revolution is not usually regarded as a threat to the stability of the biosphere. Within such human communities, sunlight still provided the primary energy source and humans, and their crops, were merely components of the natural geocycles. The human population was still relatively small and human-directed activities still represented only a modest proportion of the total global activity. Thus, although human activities resulted in some changes in land use, these activities could, it is generally believed, be accommodated for by the biosphere's inherent abilities to adapt and restore.

Now, however, let us take a big step in time and go to Europe in the middle of the eighteenth century, the stage of a second and, environmentally, perhaps more important revolution.

population growth and improvements in agriculture

In the beginning of the Christian era about 37 million people were living in Europe. By about 1500, Europe had 70 million inhabitants and after 1650 the population rapidly increased. The reason for this rapid population growth was that, due to improved agricultural techniques, a relatively large amount of food was produced. Therefore, a product like wheat was cheap, people were better nourished and thus had an increased resistance to disease. As a result, population growth increased rapidly. This implies that a growing number of people needed food, clothing and housing. Also a growing number of people needed work to earn their living. However, agriculture did not offer sufficient employment because of low prices and trade also only gave work to a limited number of people. Thus the perspective in those days was not rosy. According to the

the views of Malthus

English vicar and economist, Thomas Robert Malthus, (1766-1834) it would be impossible to provide the ever increasing population with food. In his Essay on the 'Principles of Population and its Effects on Human Happiness' (1798) he states there was a tendency for the population to double each 25 years while the means of support always grew slower. War, famine and catastrophes, however, would continuously reduce the size of the population to a level which could be supported.

minimum of
existence

According to Malthus, due to this mechanism, the standard of living of the majority of individuals would never be higher than the minimum of existence. A remedy would be a restriction of reproduction especially for those not capable of maintaining a family. This theory obviously met a lot of criticism from both Christians and 'socialists'. It is a pity we must conclude that, at least at first sight, this pessimistic eighteenth century view does not seem to have lost much of its actuality.

the views of
Condorcet and
Godwin

A much more optimistic eighteenth century view was that of the French mathematician, pedagogue and politician, Marie-Jean Nicholas Caritat Condorcet, (1743-1794) and the English writer and political theorist, William Godwin, (1756-1836) who explained in, respectively, their 'Esquisse d'un Tableau Historique des Progrès de l'Esprit Humain' (1794) and 'Enquiry Concerning Political Justice and its Influence on General Virtue and Happiness' (1793) their belief that the advance of science and technology would bring Mankind prosperity and possibilities for personal development.

the Industrial
Revolution

In retrospect, we could say that the rapid changes that have taken place in technology, economics and society since 1750 and which are commonly referred to as the Industrial Revolution, have made it possible for far more people to live on Earth than may have been anticipated from the relationships described in Figure 2.1. Of course, we have to realise that for some parts of the world this merely means staying alive while for others it does, indeed, mean prosperity and chances for personal development. We can pose two important questions:

- are the prospects for Mankind better now than those expressed in the middle of the eighteenth century?

- what changes have taken place between the beginning of the Industrial Revolution and now and do these changes present threats to the viability of planet Earth?

To answer these questions we need to think a little more about the Industrial Revolution.

Ⅱ Before reading on, see if you can list some changes in life style that have been brought about as a result of the Industrial Revolution. You will be able to check your ideas against ours in the next section.

2.4 The Industrial Revolution in focus

Let us go back once more to the cradle of the Industrial Revolution, English society of the middle of the eighteenth century. Around that time, the increasing population growth caused an increasing demand for food, clothing and housing. Spinning and weaving were, until that time, mainly a matter of a cottage industry, carried out in the countryside during the winter season. The products made by spinning and weaving were sold to travelling merchants. The increasing demand for textiles resulted in an

the origins of
the Industrial
Revolution

interference with the textile cottage industry by these merchants. First the merchants started to provide the cottage workers with raw materials. The products made from these materials were collected after an agreed period. Later the degree of organisation of their activities increased still further. The merchants (entrepreneurs) started to hire buildings and made textile labourers work under supervision and during fixed periods of time in these so-called manufactories. The most important source of energy in these manufactories was muscular strength, sometimes supplemented with water- and wind-energy. This situation dramatically changed after the development of the steam engine by the Scottish engineer, James Watt, (1736-1819).

The principle of the steam engine was known already in antiquity but the first practical application was realised by Thoman Newcomen (1663-1729) who, around 1708, developed a pump driven by steam for pumping ground water out of mines. Coal was becoming increasingly important as a fuel because of the growth in population. Until then wood was commonly used for that purpose, but as wood could also be used for the construction of houses, vehicles and ships it became too precious to be used as a fuel. Thus production of the alternative fuel, coal, became increasingly important.

coal as an alternative to wood

As deeper coal layers were exploited and shafts thus became deeper, ground water became more and more of a problem. Primitive steam-driven pumps were used to pump this water out of the shafts. Increasing technical demands were the driving force behind further development of these pumps. Therefore, the development of mining directly caused further development of steam engines. The improved engines designed by James Watt could be used for other purposes as well. They could be used, for instance, to drive the apparatus in the textile manufactories. Energy derived from fuel increasingly replaced muscular strength.

The development of the steam engine resulted in an increasing demand for machinery, coal and iron. Industrial production necessitated transport and stimulated both the development of means of transport and the associated infrastructure (such as roads, warehouses etc).

the conglomerisation of mining factories and human habitation

To be sure of the supply of fuel and raw materials and to keep the price of transportation as low as possible, factories were built close to mines and blast furnace plants. The houses for the workmen were built close to the factories and in this way extensive industrial areas with bad housing and working conditions were created. The Industrial Revolution not only brought Mankind factories, mass production and the accompanying economic developments, but at first, also serious social abuse. The idea of human equality progressively found its way into the industrialised society and the striving for social rights resulted in a beginning of social legislation in the nineteenth century.

We thus could conclude that, in the long run, the Industrial Revolution brought prosperity to the industrialised parts of the world and created possibilities for individual development.

the acquisition of material wealth

A feature of individual development is the desire to own material things; fine clothes, grand homes, devices of transport and so on. In the early phases of the Industrial Revolution, Europe was short of 'consumer goods' to supply the newly prosperous. Ships were sent to acquire goods such as silks and spices. Unwittingly, the national states of Western Europe created a worldwide market. Europe's maritime powers (Holland, Britain, Portugal and France) first sought, and then protected these on a global scale. The world began, therefore, to experience the changes which led to the industrial and social circumstances of the twentieth century, the development of a worldwide trading network in which individuals seek to protect their apparent trading interests.

creation of worldwide markets

We can conclude that through the development of technology, the primary necessities of life such as nutrition, clothing, housing and medical care are increasingly provided for. In addition, 'luxury' products, free time and the possibilities to travel are increasingly being made available. Thus lifestyle is now completely different from that of the eighteenth century European citizens and hardly shows any resemblance to that of our hunting and gathering ancestors. The key question is, is it sustainable?

2.5 The crucial importance of energy

Three questions may be obvious:

- how could system Earth which, according to the biological model shown in Figure 2.1, was able to maintain perhaps 10 million people, provide for this tremendous population growth?

- in what respect do the present boundary conditions of the system differ from those in the middle of the eighteenth century?

- what are the consequences of this population growth and what are the limits of system Earth?

The first and second questions can be answered together and these answers can help us in our search for an answer to the third question. In finding an answer to the latter question, Mankind will have to call on its full scientific, technological, economical and social expertise and judgement.

the Earth is a closed system

Let us begin with the answers to questions one and two and return once more to the biological model of the closed system Earth shown in Figure 2.1. A closed system is defined as a system that exchanges energy but not matter with its environment. We can take Earth as a closed system because the Earth exchanges energy with its surroundings but the total amount of matter on Earth remains more-or-less constant. In our description of the biological model of living nature, we described how solar energy can be trapped and converted to chemical energy by the process of photosynthesis in plants and certain micro-organisms. Solar energy is 'stored' in biomolecules and released if necessary. Release can take place not only in the plant itself, but also in an organism that has directly or indirectly fed on the plant. Thus, all living organisms use directly or indirectly solar energy. Living organisms are very skilful in converting energy from one form to the other.

biological systems use energy efficiently

The efficiency of energy conversions taking place in nature far exceeds that of Man-designed processes. But not only the efficiency of natural energy conversions deserves our attention. We should also think for a moment about the fact that a living organism only produces the materials it needs to function and in doing this uses energy in a highly efficient way. In times of a positive energy balance, energy is stored in compounds such as starch, glycogen and lipids. Furthermore, each living organism degrades biomolecules that have fulfilled their biological function to smaller units and subsequently uses these for the production of new biomolecules or as a cellular fuel. Moreover, the biomolecules present in an organism can, after the death of this organism, be used by other organisms, especially micro-organisms.

the contrast in the elements used by living systems and Man-made products

A second characteristic difference between natural and Man-designed production processes is, in addition to the efficiency of energy conversion and the built-in integrated recycling, the type of 'raw materials' used. In natural production processes, materials containing mainly the elements carbon, hydrogen, oxygen, nitrogen, phosphorous and sulphur are used. They may also contain very small amounts of other elements. In Man-designed processes targetted at fulfilling social 'desires', practically all elements present on Earth are used. This often implies that, when the product is no longer needed, degradation of dumped material by micro-organisms is much less easy than for natural products or even impossible despite the very diverse metabolic potential of micro-organisms.

the non-biodegradable nature of many manufactured products

Furthermore, even with the production of materials involving the elements commonly found in biomolecules, Man often uses processes which make the products virtually non-biodegradable. For example, the polymerisation of molecules using free radicals to produce plastics and the halogenation of organics to produce pesticides involve processes alien to biological systems. Free-radicals are very reactive and, if generated within living systems, may be highly damaging. Thus the use of a new reactive species and of unnatural reaction conditions (for example, very high temperatures and pressures), leads to the production of recalcitrant products. These are, unlike bio-products, effectively removed from geocycling and may accumulate in the environment posing various problems.

energy released during the oxidation of carbon compounds

Let us, however, linger a little longer on the products of bioprocesses. During their production, solar energy was directly or indirectly converted to chemical energy. Biomolecules thus contain a certain amount of energy stored in a huge variety of compounds of the element carbon. The carbon atoms in biomolecules are, on average, in a more reduced form than in carbon dioxide. During oxidation of carbon compounds, the energy can be released again. This principle is used by living organisms during the biologically controlled energy-yielding reactions of metabolism. Mankind applies this oxidation principle on a large scale during the combustion of fuel in Man-designed production processes. However, the efficiency of the latter is mostly far less than that of natural ones. Moreover, during industrial combustion, unwanted side reactions take place that do not occur in living nature. These side products often have negative effects on the environment.

When a living organism dies, it stops converting and storing energy, but the reduced carbon compounds still exist. They can be used by other organisms. This does not, however, always happen. Under favourable geological conditions, this organic material can be converted into gas, oil or coal. In this way huge amounts of solar energy have been stored in the fossil energy sources, gas, oil and coal. What Mankind began doing on a large scale during the middle of the eighteenth century was maintaining human society at the expense of the energy reserves of system Earth. The large-scale change to using fossil energy enabled us, until now, to produce our social *desiderata*. We are, however, rapidly exhausting our energy resources and environmental effects are becoming more and more evident.

the biological origins of fossil fuels

the non-recyclable nature of fossil energy resources

Thus, in satisfying its needs, Mankind practically uses all elements present on Earth and the Earth's energy resources. In doing so, matter is converted from one form into another often at the expense of much energy. Matter is not lost during conversion and can, in principle be recycled, again at the expense of energy. As far as the energy resources are concerned the situation is completely different; fossil energy is simply being exhausted.

We can conclude, therefore, that we can keep 5321 million individuals alive on Earth at the expense of the fossil energy resources. It will be obvious that this situation cannot last for ever. Therefore, a resemblance between the present situation and the situation at the beginning of the Industrial Revolution is apparent. At present, a rapidly growing population needing food, clothing, housing, medical care and work to earn a living, is a matter of major concern, just as in the middle of the eighteenth century. From an energetic point of view the perspective is, however, less optimistic.

We can foresee two fundamental problem areas. Given that the total amount of matter on Earth is constant, the production of materials which are not easily recyclable (either biologically or by human activity) means that the pool of resources is being depleted. If, at the same time, the accumulated products have deleterious properties (for example, are toxic or cause climatic change), then their production, use and dispersal become crucial issues. Important as these issues are, a much more crucial question arises from the need to achieve a situation in which the energy balance of the system Earth is in equilibrium. This means that we should not use more energy than is entering the system. We will have to find ways to fulfil our energy needs using renewable energy. An important question, therefore, is how we can trap, convert and store the energy that is entering system Earth. In concluding this we have not answered the question as to what the limits of system Earth are, but we have formulated a firm boundary condition for maintaining system Earth.

the need to achieve an energy balance

The energy supply of industrialised societies is strongly based on non-renewable energy (fossil fuels, nuclear energy). Moreover, the energy sources used are a threat to the environment. We should also consider that the energy use per capita in areas like South East Asia and Middle Africa only amounts to 7 GJ annum^{-1}, which is far less than the 35 GJ per capita used by the United Nations Industrial Development Organisation (UNIDO) as a necessary minimum to fulfil basic human needs. (Note GJ = J x 10^9). In industrialised societies, like in Europe and North America, the energy use is much higher. We, therefore, will be confronted with a justifiable make-up for energy use in developing countries, an ever increasing energy need in industrialised countries (if no policy changes occur) and an urgent necessity to limit the use of fossil energy.

comparison of energy consumption in developing and developed nations

According to forecasts, in 2030, the energy use in the industrialised parts of the world will be 207% of that in 1990. In this scenario, technological energy savings and a population growth limited to 1.4 x 10^9 -1.6 x 10^9 individuals is taken into account. In the developing parts of the world, the energy use will have increased by the year 2030 to 350% of that in 1990 according to a scenario in which the populations is estimated to grow to 6.4 x 10^9 and the energy use to 40 GJ per person per year, which is still a small amount compared to the 350 GJ per person per year in the USA in 1990.

It is still far from clear how we can combine the forecast of global energy needs given above with the necessity to limit the use of non-renewable energy. Some would, for example, anticipate a significant increase in the use of nuclear fuel, at least over the short to medium term. However, ultimately we will face depletion of these fuel sources and the problems of accumulations of radioactive wastes are immense. The extensive use of such fuels poses enormous problems. How then, can we combine the forecasts of global energy need to the necessity to use renewable energy sources?

nuclear fuels and the problems of radioactive wastes

The complexity of the problem suggests that, in tackling it, the full potential of sciences and technology will have to be integrated and used. Well known concepts in this respect are 'sustainability' and 'sustainable development'. In the UN report, *Our Common Future*, also known as the Brundtland report, the objective of sustainable development is described as follows:

the Brundtland report

> 'Sustainable development is a process of change in which the exploitation of resources, the direction of investments, the orientation of technological development, and institutional change are all in harmony and enhance both current and future potential to meet human needs and aspirations. ... (It is) meeting the needs of the present without compromising the ability of future generations to meet their own needs.'

the link
between
environment
and
development is
essential

We can, a bit cynically, note that satisfying needs can have different meanings, depending on the place on Earth; in 1990 it will mean enough food to survive for one part of the world and not renouncing a rather luxurious lifestyle for the other. Therefore, it is not surprising that one of the conclusions of the United Nations Conference on Environment and Development (UNCED) held in Rio de Janeiro in 1992 was that environment and development can no longer be separated. They both are important aspects in our search for a sustainable society.

SAQ 2.3

1) Each year about 2.4×10^{12} GJ solar energy is stored in biomass. (Note 1 GJ = 10^9 J). According to the FAO (Food and Agriculture Organisation) each person needs an intake of about 10 500 kJ day^{-1}. Assuming that the world population is 5321 million, what proportion of the world's biomass needs to be eaten each year to supply human nutritional needs?

2) Does the value you have determined in question 1) indicate that there is sufficient energy stored in biomass each year to, at least theoretically, fulfil non-nutritional energy requirements? (Justify your answer and read our response carefully)

2.6 Agriculture as a contributor to sustainability

Green
Revolution

In addition to our concerns on the availability and use of energy, the supply of food for the growing world population is an important issue of concern. Therefore, agricultural developments are a focus of interest. Important here is the so called Green Revolution, a term introduced in the 1960s to describe changes in agricultural practices. Essentially the Green Revolution centred on the development and cultivation of new, high-yielding hybrid crop varieties. There are many such new varieties including the major crops of Europe, the rice grown in South East Asia and maize in Latin America. Cultivation of these new varieties goes hand-in-hand with the use of fertilisers, irrigation, herbicides, fungicides and pesticides. For these reasons, the Green Revolution was more successful in delivering food in areas where an agricultural infrastructure already existed or could be stimulated, than in areas where these favouring conditions were absent. The increase in yields resulted in economic and social developments within these agricultural societies and created employment in the food industry and distribution services.

positive effects
of the Green
Revolution

negative
effects of the
Green
Revolution

The Green Revolution also has negative effects. As agriculture becomes less labour intensive there is a loss of jobs in rural communities and this, in turn, leads to an increase in the population flow to the cities. The development of new high-yielding varieties is expensive and will, in the absence of international aid, stay out of reach for many poor farmers in developing countries. This, in turn, can also increase further the gap between rich and poor countries in the world. Furthermore, the use of fertilisers, herbicides, fungicides and pesticides can also give rise to negative environmental effects. Insects and pathogens, through evolution and natural genetic exchange, may develop resistance against crop protectants which makes the combat against them more difficult and, often, more expensive. The trend of agriculture to produce ever larger monocultures has the tendency to reduce genetic diversity and to increase the vulnerability of crops to disease and pests. The use of non-renewable energy to generate crop 'aids' (fertilisers and protectants) is also an important issue. We will return to this again a little later.

Let us consider agriculture as an energy supply/energy use system in a little more detail.

Plants are, as described above, capable of converting solar energy to chemical energy stored in sugar molecules. These sugar molecules can subsequently be used as a source of raw materials and energy for growth, development, homeostasis and reproduction of the plant itself or for other organisms in the food web. Historically, plants have also been used by humans as an energy source to supply heat. For example, our ancestors were wood burners to supply heat for cooking and warmth. In principle, therefore, we could use the primary energy harvesting capacity of plants to provide not only 'nutritional energy' but also energy for other functions. We need, however, to think very carefully about the energetics of agriculture.

agricultural energetics

Plants not only 'harvest' sunlight but require additional compounds from the soil as well as water and carbon dioxide. Thus production by plants depends on three factors: solar energy, nutritional compounds and water (taken up from the soil). Obviously the final growth rate can be determined by a limited availability of these three factors. The limiting factor is usually not solar energy and, therefore, the use of fertilisers, together with irrigation and mechanic treatment of soil can improve crop yield. Yield may be further improved by the use of crop protectants (for example, herbicides, fungicides and pesticides). But this increase in yield of biomass costs energy because the production of herbicides, fungicides and pesticides consumes energy, as do irrigation and the mechanical treatment of soil. All this added external energy nowadays is derived from the non-renewable sources.

solar energy is not usually the limiting factor

The manufacture of the machines used to plough and break down soils, deliver fertilisers and crop protectants and so on also cost energy. Presently, agriculture is responsible for about 5% of the world energy use.

Investing energy in the form of fertilisers, irrigation, crop protection agents and mechanical treatment of soil, however, results in an increased crop growth and thus in a higher biomass yield, or, in other words, in an increase in conversion and storage of solar energy in biomolecules.

It will be self-evident that in agricultural production we have to analyse which factor is the rate-limiting one; it is, after all, no use adding fertilisers to soil when water is the growth-determining compound.

Table 2.1 provides data concerning an energy analysis of agricultural production using grain yields as an example.

Π Table 2.1 reports data from four areas with different levels of intensification of farming practices. From this data what can be concluded about: 1) the amount of product produced per Man h^{-1} in respect to the level of intensification; 2) the ratio of solar energy collected to energy expended with increasing mechanisation.

1) With low intensity farming (level 1), the yield per Man h^{-1} only 0.8 kg h^{-1} whilst in high intensity farming (level 4) the yield is 700 kg h^{-1}. Obviously the high intensity farming gave much greater efficiency in terms of yield Man h^{-1}. 2) In contrast it can be concluded that the ratio of solar energy collected to energy expended, falls with increasing mechanisation. However, the total energy output increases with increasing mechanisation. Also it should be realised that in the analysis given in Table 2.1, the energy supplied by human muscular strength is considered to be free. This is, of course,

IL country	Resource input*				Labour time (h ha^{-1})	Energy input (GJ ha^{-1})	Grain yield (kg ha^{-1})	Energy output (GJ ha^{-1})	Energy use efficiency (out/in)	Labour productivity (kg h^{-1})
	H	A	F	M						
1 Guatemala	*	-	-	-	1415	0.2	1066	15.9	79.5	0.8
2 Guatemala	*	*	-	-	700	0.2	1066	15.9	79.5	1.5
3 USA	*	*	*	-	120	25.6 **	7000	102.6	4.0	58.3
4 USA	*	*	*	*	20	46.2 ***	7000	102.6	2.2	700.0

* H = human labour; A = animal traction; F = fertiliser (150 kg N ha^{-1}); M = mechanisation

** in a relatively wet climate additional energy input required for drying (= 6J ha^{-1})

*** in a dry climate additional energy input required for irrigation (= 9 GJ ha^{-1})

Table 2.1 The energy yield of grain in relation to time and energy applied at different locations, IL = level of intensification. (Data derived from Pimentel D. and Hall, C.W. Food and Natural Resources Academic Press, New York, 1989 and World Food Production - Biophysical Factors of Agricultural Production, Open universitiet, Heerlen, The Netherlands, 1992).

not completely true. Human muscular activity is, after all, a component of global energy consumption. Furthermore, it should also be realised that transport and further treatment of agricultural products must be taken up in an energy analysis. Remember crops can be used directly for nutrition, they can be processed in the food industry or they can be used as fuel. In the first two cases, smaller or larger amounts of energy will be invested depending on the nature of the crop, its distribution and the extent of its processing and conditions of storage. For example, canning and freezing impose energy demands. The question we need to address is, could crops be used as a renewable energy source and meet the current and future demands of human societies?

It has been calculated that currently each year 2.4×10^{12} GJ solar energy is stored in biomass while 2×10^{11} GJ energy are used from fossil sources by human societies. (You might recall the calculation we did in SAQ 2.3). If, as was stated above, agricultural production is responsible for 5% of the world energy use we can conclude that under the right agricultural conditions more solar energy is converted to biomass than is invested in it as fossil energy. This implies that energy production from biomass is a realistic option. As long as the sun is shining, energy conversion and storage in organisms can be considered as a renewable energy source. Therefore, technological crop improvement resulting in an increased efficiency in energy conversion or exploiting the possibilities potentially offered by photosynthetic organisms in oceans should be considered in energy forecasting.

Another important point in considering crop production in the context of renewable energy is that, in principle, the same amount of carbon dioxide is liberated during combustion as is used during photosynthesis, and, therefore no negative environmental carbon dioxide effects should be expected.

Theoretically, crops can store about 7.5% of solar energy during photosynthesis, but as plants subsequently use about 40% of the stored energy in the non-photosynthetic reactions of plant metabolism, the net energy efficiency of crop production is about 4.5%. Of course differences exist between different types of crop in this respect, and not all crops will be suitable as energy crops. We also have to consider the conversion of biomass to a suitable fuel, for instance, alcohol from wheat or sugar cane, or bio-diesel from rape-seed.

ratios of energy yield/energy invested of potential energy crops

Generally it is agreed that, in order for an energy crop to be used efficiently, it must produce five times the amount of energy that was invested in it. In temperate zones the ratio of energy yield/invested energy is 6 for hemp, 5 for wood, 4.5 for beet, 2.5 for potatoes and wheat and about 2 for grass. In these zones, hemp would thus have a chance as an energy crop. In (sub) tropical zones, large areas of arable land are available. In these zones climatic conditions enables agricultural production to take place all the year round. In these circumstances energy crops could be an economically viable option. A well known example in this respect is the production of ethanol from sugar cane in Brazil. The question marks presently put by the Brazilian government on this successful alternative fuel programme are of a political and infrastructural character rather than of a technological one.

Apart from cultivating special energy crops we must also consider the small-scale energy production from plant, human and animal wastes. Mostly only part of the energy stored in biomass is used, and thus part of the biomass energy is still present in wastes. Wastes can be degraded by micro-organisms. When these degradation processes are carried out under controlled conditions in a bioreactor, biogas can be produced. Often the biogas is used on the farm itself. We will discuss these processes and related issues in more detail in later chapters.

other products derived from agricultural crops

Finally we must realise that crops can not only be used as raw materials for the food industry and as a potential energy crop, but can also be used as raw materials in other branches of industry. In this context, we can think both of micro-biological conversion of plant materials to basic chemicals and the production of special compounds by plants, plant tissues or -cells. Examples are gums, glues, starch, dyes, ink, detergents, cosmetics, flavouring compounds, preservatives, emulsifiers, oils, textile fibres and fibres in composites. An additional advantage of using raw materials derived from plants is that such compounds can easily be degraded in nature and thus can contribute to maintaining cycles.

Thus, we could conclude that agro-industrial production could be promising from both an energetical and an environmental point of view. However, much research remains to be done, not only in the field of natural sciences and technology, but also in economics and social sciences to turn these possibilities into realities.

SAQ 2.4

Below we have provided a graphical representation of the energy input per unit of energy recovered in crops produced using a variety of farming practices. (Note that the energy ratios reported are averages).

1) From these data, which of the farming practices produce less energy than is consumed in their production?

2) Which of the modern farming practices look most promising as a potential energy crop?

Ratio of energy input/energy output for a variety of ancient and modern agricultural practices.

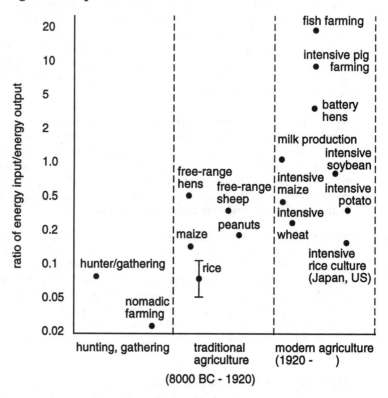

2.7 Integral life cycle management

Let us return once more to the boundary conditions of system Earth in relation to the products society needs or wants to have.

Mankind is withdrawing fossil energy and raw materials from the Earth's reserves to make products for fulfilling social needs. Products that have been used for shorter or longer times and are not needed any more are brought back into the environment. Also, during production processes, unwanted products are released into the environment. The composition of the compounds that are returned to the environment via discharge, emissions or wastes is often very different from the composition of the raw materials.

In principle, recycling is possible, but most of the time recycling is rather energy-demanding.

product-orientated and process-orientated choices

A first, obvious, consequence of these considerations is that we should not only critically ask ourselves if we really need a specific product, but we must also consider what the costs of the production process are in a broad sense. Thus, we should not make our choice for a specific product in a product-oriented way, but in a process-oriented way. We must ask ourselves questions such as:

- which raw materials have been used?

- how much energy has been invested in the production process?

- can the invested energy be recovered?

- is it possible to recycle the product and how much energy will be needed to do so?

We, therefore, have to consider the total life cycle of the product. The design of a production process taking into account these aspects is often referred to as integral life cycle management.

The concept of integral life cycle analysis and management sounds logical and promising. It will be obvious to almost everybody that we will have to use energy and raw materials far more economically than we do now. However, the development of new products based on cleaner production processes and on alternative raw materials is not only a question of technology. It is not only the technical production that matters, products also have to be sold, which implies calling in an extensive transport network and a complex tertiary sector. All these social considerations have to be taken into account. The social behaviour of both producer and consumer will have to change. Acceptance of a new product must not be founded simply on considerations of price; energy and environmental factors must also play a role.

need to provide social information

The approach requires that there is some freedom of choice. We may hope that integral life cycle management can indeed be introduced in industrial societies where this freedom of choice exists. Of course a deliberate policy will have to be followed and all means of providing social information, such as teaching and using the modern media, will have to be brought into place. We have to realise also that, in spite of the tremendous tasks that need to be undertaken, the starting point in industrialised societies is a rather good one relative to that in developing countries. In these latter areas, large parts of the population are living at the minimum level of existence and the freedom of choice is, therefore, extremely limited. It is self-evident that environment and development have an intimate relationship and, therefore, in addition to scientific and technological research, the consequences of introducing new technologies must also be considered from social, economical and cultural points of view.

2.8 The new revolution

The arguments given for integral life cycle management described above demand not only that the social and political will is present to press for the necessary changes but also that the technological means to achieve these changes are available. The recent developments in biotechnology suggest that the technological means are, potentially, within our grasp.

a definition of
biotechnology

Biotechnology has been defined in many different ways but is generally regarded as the application of organisms, biological systems or biological processes to manufacturing and service industries. This technology is underpinned by our knowledge of the mechanisms and processes involved in living organisms. It is, therefore, founded upon systems which utilise the processes and products which occur in nature and which are components of the sustainable geochemical cycling of materials described in Figures 2.1 and 2.2. Biotechnology itself is not new, it is a technology that has been used by human societies for thousands of years in such processes as brewing and baking. Thus far, this technology has not created the environmental problems associated with many of the processes and products of the chemically-based and physically-based technologies. The development of the techniques of genetic engineering, protein technology, *in vitro* cell cultivation and process technology in recent decades has resulted in a rapid expansion of biotechnologically-based enterprises. Is, however, this expansion likely to lead to environmentally more satisfactory processes for fulfilling human needs? If so, how will this be achieved?

If we ignore, for the moment, the potential for biotechnology to contribute to human (and animal) health care and focus on the applications of biotechnology in agriculture, chemical synthesis and energy management, then we can divide the application of biotechnology into two types of strategies. These are:

• the use of biotechnology to remove or reduce the environmental problems arising from conventional technology;

• the use of biotechnology to replace existing, environmentally-damaging, technology.

Let us deal with each of these in turn.

recalcitrant and
inorganic
pollutants

Apart from the underpinning considerations concerning the use of non-renewable energy resources described earlier, existing chemical and physical technologies generate products and by-products which create many environmental problems. The generation of recalcitrant organic materials and inorganic pollutants are universally regarded as environmental disasters. Despite the increasing stringency of legislative controls on emissions and disposal of these products, contamination of soil, water and air continues. Although some of the more obvious consequences of this contamination have been identified, international action is slow to materialise. Moreover, the dangers from the long-term, more subtle effects on sustainability are less easily quantifiable. As the developing world aspires to the living standards of the West, we can anticipate that these problems which arise from abiotic products will become even more acute. Some would maintain that they are already chronic.

treatment of
biologically-
derived
materials

The ability of biotechnology to alleviate these problems by treating (transforming) the products of conventional technologies is somewhat limited. This is because many of these products are not compatible with biological activities. In other words, the metabolisms of organisms are unable to cope with them. All is not, however, lost. Not all of the products of human activities present unsurmountable biological problems. For example, human excreta and many of the products derived from biological sources (for example, paper and certain textiles) are biodegradable. Biotechnology has been used for quite some time to alleviate the environmental problems which would arise from simply dumping these materials in waterways and landfill sites. We should anticipate that as biotechnological expertise develops and is extended, by education and training, to new geographical areas, that environmental degradation from the dumping of these materials will diminish. The newer processes not only provide routes for the degradation of these materials regarded as wastes, but also provide opportunities to generate useful products, especially energy.

The treatment of abiotic materials, however, presents greater difficulties. Generally, these cannot be metabolised by organisms. Their dispersal in low concentrations can exacerbate these problems. For example, the use of halogenated, chemically synthesised, compounds and pesticides pose, not only metabolic problems, but also difficulties because they are spread thinly over large areas. There are, however, circumstances in which such materials may be converted into non-noxious forms. Most of these processes, but not all, depend upon the use of micro-organisms. The rapid growth of micro-organisms and their large numbers are reflected in their abilities to evolve rapidly. Typically, one gram of soil may contain, for example, 10^9 bacteria. Mutations arise spontaneously at a rate of perhaps 1 in 10^6 organisms. Many of these mutations are harmful to the host cells and, under natural conditions, are rapidly selected against. Nevertheless, under appropriate selection conditions, mutants with new metabolic capabilities may be cultivated. This, together with the natural gene transfer systems of micro-organisms and the genetic engineer's ability to create novel gene combinations and even new genes, means that there are some prospects of creating biotechnological processes for treating and removing specific environmentally-hazardous materials.

treatment of abiotic materials

However, there are, as we have indicated, many constraints and there are many environmental pollutants arising from current practices for which we can see no real prospect of finding biotechnological solutions.

One could argue that attempting to detoxify/remove existing environmental problems using biotechnology is the wrong solution. It would be much better to prevent the 'environmental disease' than to curve it. This does not mean that we should abandon the development of 'end-of-pipe' technology. In the shorter term, we will be stuck with existing technologies and their undesirable consequences and it is difficult to envisage processes for treating human and animal excrements that do not involve biotechnology. Thus biotechnology will continue to be of great importance in the management of degradative processes. Much more radical, but longer-term, is the desirability of replacing existing technology by biotechnological processes. In this way, the current 'end-of-pipe' problems arising from non-biocompatable products and processes will be reduced.

The situation we envisage is as follows. There could be a movement away from non-renewable energy and chemical resources (for example, coal, oil etc) to renewable, biological resources. In other words, the primary production system would become agriculture which would not only be called upon to supply food, and raw materials for textiles but also for chemicals used for a wide variety of functions. Here, biotechnology has much to contribute. Through biotechnological manipulations, we may envisage that the harvesting of light by plants may be made more efficient and yields may be improved. We may also speculate on the prospects of plant biotechnologists creating crops that are easier (less energy-demanding) to harvest and process. Similarly we should also anticipate the generation of biological control agents (viruses, bacteria, fungi) to replace the recalcitrant xenobiotics currently used in crop protection. The overall effects of such changes would be to improve crop yields, lower the demand for the use of fossil fuels and reduce the use and accumulation of harmful xenobiotics.

The technological achievement of these objectives is, as you might well have anticipated, not entirely straightforward and much needs still to be done to realise the potential of biotechnology. Also, care has to be taken not to create new environmental problems such as the generation of new weeds and pathogens. Nevertheless, for the moment let us assume these technical achievements could be achieved. How then could these products be used to replace the existing fossil-fuel consuming, environmentally polluting technologies?

In principle this does not present insurmountable technical problems. The products of the 'new' agriculture would be biological in character and would thus be susceptible to biotransformation. In other words, we could use agricultural products as the raw feed-stocks for a wide range of products and processes. For example, starch could be converted to ethanol, a transportable energy commodity; we could resort to using cellulose fibres to replace the synthetic polymers derived from petroleum used for clothing and so on. The key questions are:

- will these new materials display all of the characteristics and cover the range of products we desire? (The production of plastic may be environmentally undesirable but it is, after all, a very useful material);

- how do we encourage the move away from the fossil-fuel consuming, environmentally-polluting techniques to a sustainable biologically-based technology?

Here we will not deal with the techniques of biotechnology and their application to agriculture and chemical synthesis in any great depth since these are dealt with in the BIOTOL texts, 'Biotechnological Innovations in Crop Improvement' and Biotechnological Innovations in Chemical Synthesis'. Nevertheless, there are some important points that we need to make. Currently the adoption of new technology and new products is mainly dependent upon economic considerations. If the process or product will make a profit or extend the profit margin, then there is a strong chance that it will be adopted. If, on the other hand, it is more expensive or will not make a profit, then its adaption is unlikely. Until recently environmental considerations were not included in this equation and as a consequence we are facing current environmental difficulties. We must ensure that society learns to calculate the 'price' of a product by taking into account both economic and environment factors.

The arguments we described earlier indicated that the implementation of new technologies must be considered in a much wider context. The main problem is to find ways in which individual companies or, individual states, or group of states, can buck the current straight jacket of responding predominantly to economic considerations.

There are, however, increasing signs that the world is beginning to act co-operatively on these issues. Although progress is slow, an increasing number of international accords and agreements are being signed and acted upon. For example, the agreement to reduce the emission of 'global warming' gases and the reduction in the emission of 'ozone-depleting' gases are hopeful signs. There are, however, still enormous social, cultural and religious hurdles to be overcome. Perhaps the international realisation that Mankind has no option but to adopt a more sustainable technology in order to achieve and maintain a life style of higher quality than that predicted by Malthus in the mid-eighteenth century will be the greatest motivator for the adoption of biotechnology.

SAQ 2.5

1) Explain the difference between a clean technology and a sustainable technology.

2) Does biotechnology have the potential to be a clean or a sustainable technology or both? (Justify your answer)

Summary and objectives

In this chapter, we have given what we might call a natural scientist's perspective on environmental management. Although some may disagree on specific points we have made here, most would concur with our central tenet that we will have to find ways to use raw materials in a more efficient way and to bring the energy balance of the Earth into equilibrium. This is the only way in which we can bring a sustainable society within our reach. The manner in which this may be achieved is, however, much more conjectural, but of fundamental importance will be the technology that is available to us to meet the challenges of sustainability.

In the short term, we will have to mainly limit ourselves to finding ways of saving fossil energy resources and reducing the production and/or release of environmentally damaging pollutants. Using the potential for energy conversion and storage available from biological sources may, in this context, be extremely helpful. The use of biotechnology for crop improvement and process optimisation offer potentially great opportunities.

We must also become more process-orientated and consider, in our demands, the environmental consequences of these demands. Here sociological and cultural factors will play as important a role as developing the technology to fulfil these demands. Whatever perspective we place on the future development of human society, the application of the principles and practices of biotechnology are seen as critical in developing a new relationship with system Earth.

Now that you have completed this chapter you should be able to:

- explain why the energy entering the system Earth should be sufficient to supply the energy needs of human society;

- explain why products based on biotechnological processes are more likely to be biodegradable and less likely to accumulate as recalcitrant pollutants in the environment;

- explain, in outline, how, by changing human needs together with the exploitation of biotechnology, it may be possible to develop a sustainable development of human society;

- accept the argument that to meet the challenges of achieving sustainable development requires more than simply the application of technological innovation. It calls upon the need to evolve a new socio-economic morality which accommodates and responds to the cultural and religious traditions of Mankind;

- recognise that the development of human society based on the consumption of fossil fuels is, in the longer term, unsustainable;

- explain why the use of conventional technology based on chemical and physical processes leads to the production of materials that are non-biologically compatible;

- explain, in general terms, why the production of materials by chemical and physical processes may be directly, or indirectly, harmful to organisms.

Waste, pollution and the use of landfill

Waste, pollution and the use of landfill

3.1 Introduction

In the previous chapter we explained that the consumption of fossil fuels to provide the energy to drive human-based activities is, in the long term, unsustainable and that there is a need to move towards a greater use of renewable energy sources and a reduction in dependence on non-renewable sources. Using biomass as a source of energy can make important contribution in this respect. However, the use of biologically-derived energy sources does not, in itself, guarantee sustainability. Human activities create other problems of considerable environmental importance. These problems especially relate to the generation of waste and pollution.

definition of
waste and
pollution

For our purposes, we may define the term waste as the generation of products which are not profitably utilised. These may be biological or chemical entities such as the residues and by-products from commercial or domestic processes or physical entities such as heat. The term pollution is usually used to mean the process of contaminating the environment with entities that are deemed harmful. Pollutants may be biological, chemical or physical in nature. Waste and pollution are frequently, but not always, intricately linked.

The aim of this chapter is to provide a context in which to discuss the ways that biotechnology may alleviate or circumvent the environmental consequences of waste production and pollution. We will begin by providing an overview of Man's interaction with his environment especially in terms of generating wastes and pollutants. This will enable us to examine, in general terms, the components of wastes and the nature of pollutants. We will then move on to consider the potential targets for the application of biotechnology providing an outline discussion of the practice of landfill. In subsequent chapters, we will examine some of these applications in greater depth.

3.2 The interactions of Man and the environment

The interactions of Man with the environment are complex. The environment provides the support systems for human activities. As a result of these activities, these support systems may be modified. Thus we can visualise a rather cyclical set of interactions in which the environment affects Man and Man affects the environment.

Humans have always been exposed to hazards arising within the environment.

Π   Attempt to make a list of some of these. Think in terms of physical, chemical and biological hazards.

The sorts of items we anticipate you might have listed are:

• Physical: earthquake, floods, drought, lightening, landslips;

- Chemicals: poisons, allogens and irritants (mainly derived from plants and animals);
- Biological: infections, predators, starvation.

As the human population has grown and developed, the activities of Man have significantly influenced the nature and extent of many of these hazards. In some instances the hazards have been substantially reduced. For example, the hazards from predatory animals have, for large areas, become minimal and the threats from infectious diseases have been greatly reduced. However, as human activities have been extended and diversified, new hazards have been created or intensified. Many of these hazards are attributed to the generation of wastes and pollutants. In some instances, the introduction of pollutants and the generation of wastes are a direct and immediate threat to the well-being, in other cases the threat is more insidious and indirect.

Biotechnology has an important role in reducing a wide variety of hazards. Biotechnology may be seen as offering opportunities and processes to achieve particular ends; reducing the prospect of generating hazards is a very important one. It is not surprising, therefore, that biotechnology is seen as a potentially 'environmentally-friendly' technology.

In this and the following three chapters, we are predominantly concerned with the use of biotechnology to remove or reduce the hazards that arise from the generation of wastes and pollutants. Therefore, it is important that we focus on the generation of these through the activities of Man.

In Figure 3.1, we provide a diagrammatic representation of Man's interaction with the environment.

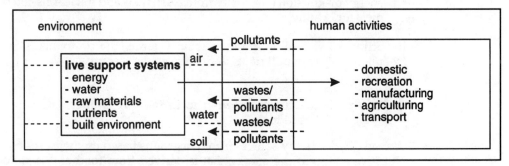

Figure 3.1 The interactions between Man and the environment.

On the left we have represented the environment in which humans live. The environment provides a range of life support systems which enable Man (and other biological systems to live). These life support systems provide:

- sources of energy;

- sources of water;

- sources of nutrition;

- sources of raw materials;

- a place to live (built environment).

Providing these support systems are suitable, life is sustained and mankind can carry out a range of activities. These we have divided, on the right in Figure 3.1, into domestic activities (feeding, bathing), recreation, manufacturing, agriculture and transport. These activities consume resources from the environment and produce wastes and pollutants which ultimately find their way back into the environment. These residues and wastes may be in the form of solids, liquids, gases or energy and will be distributed through water, land and atmosphere. In turn, these contaminants will influence the life support systems.

⫫ Which activities of Man do you think to contribute most to the production of wastes and the introduction of pollutants in the environment?

The answer to this question depends very much on the kind of society we live in. In most highly industrialised West European countries the main social activities acting as a source of wastes and pollution (see Figure 3.1) are:

- manufacturing (chemical industries, oil refinery);

- agriculturing and intensive cattle-breeding;

- housekeeping and house-building;

- transport and traffic;

- energy production (electricity from fossil fuels and nuclear fission burning of fossil fuels for heating).

In Section 3.3 we will examine, in general terms, the components of wastes and the (chemical) nature of pollutants derived from these sources.

⫫ What is the energy referred to in the box entitled 'life support systems' in Figure 3.1 used for?

nutritive and non-nutritive energy

We can divide energy sources into two groups based on their uses by human society. Nutritive energy is the energy that is derived from food and is biological in origin. Non-nutritive energy is the energy used to carry out the many different activities by Man. This energy is used in transport, manufacture, heating, lighting, communication systems and so on. This requirement is seen as essential to maintaining life especially in large, sophisticated modern societies.

⫫ What sources of energy are used to supply the non-nutritive energy?

use of fossil fuels, nuclear energy and renewable energy sources not used evenly

The answer to this is not straightforward. The availability of fossil fuels and the technology associated with the generation of usable energy from, for example, nuclear fuels is not uniformly distributed through human societies. As a consequence, the sources of energy used by different societies show some differences. In general, so called developed countries use fossil fuels to provide up to 99% of their non-nutrition energy. Renewable energy sources (mainly biomass) contribute only about 1% to the energy used. In contrast biomass contributes over 40% to the energy used in developing countries (Figure 3.2). The general trend is that the energy supplied from biomass in developing countries, expressed as a proportion of the total energy used in these countries, will decrease as they aspire to achieve the living standards of developed countries.

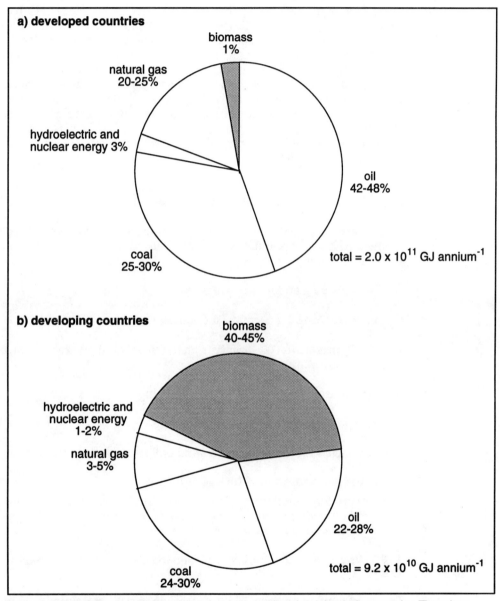

Figure 3.2 The distribution of energy sources used in developed and developing countries. The values reported in this figure are the means taken from a variety of estimates. (Data derived from Parikh, J.K. Energy Systems and Development, Oxford University Press, Oxford, 1980).

∏   In Figure 3.1, we refer to the 'built environment'. What do we mean by this term?

Man's impact
on the
environment
extends well
outside areas
of habitation

You probably think of the 'built environment' as areas inhabited by people including various forms of housing and buildings which shelter manufacturing processes and services. We should include transport facilities (road, rail, canals), water supply and disposal and recreational facilities. We should, therefore, see the built environment as extending from urban areas into the surrounding rural areas. Frequently these rural areas are managed and provide food and other materials to support urban communities. This man-made (or man-modified) environment extend well outside the range of areas of inhabitation and relatively few areas of the world can be properly regarded as natural environments. We must anticipate that Man's activities will have greatest impact on the environment in areas close to regions of intense habitation. However, this does not mean that so called 'natural' environments entirely escape all the consequences of human activities. Many pollutants are highly motile and quickly pervade the biosphere and changes in one region inevitably have consequences elsewhere. We are led to the conclusion, therefore, that the wastes and pollutants which arise from human activities may add to the environmental hazards and may become so widespread as to affect life support systems.

3.3 The nature of wastes and pollutants

In this section we will examine, in general terms, the components of wastes and the (chemical) nature of pollutants. In industrialised countries, the main sources of wastes and pollution are:

* housekeeping and house-building;

* manufacturing (eg chemical industries, oil refinery);

* energy production (electricity from fossil fuels and nuclear fission, burning of fossil fuels for heating);

* agriculturing and intensive cattle-breeding;

* transport and traffic.

We will discuss the nature of wastes and pollutants derived from these sources shortly.

3.3.1 Housekeeping and housebuilding

bodily
discharges

All humans, whether they live in a sophisticated modern society or in a primitive one, produce bodily discharges, mainly faeces and urine. These discharges are considered to be very hazardous to mankind.

∏   Why are faecal materials regarded to be hazardous to mankind?

intestinal
diseases

The most obvious reason is that faecal materials are sources of human intestinal diseases and may readily enable the transmission of these disease organisms to food and water. One typically thinks of diseases such as cholera, typhoid and dysentery in this context. Thus it is essential to have proper treatment processes to reduce or remove this hazard. Much of Chapters 4 and 5 deal with the treatment of sewage. The importance of these processes cannot be over estimated. When sewage treatment processes fail, for example after a natural disaster such as an earthquake or as a result or war, the outcome is often an outbreak of intestinal diseases.

effects of
sewage on the
amenity value
at water

The hazards arising from sewage are not, however, confined to sewage acting as a source and vector of diseases. Faeces are composed largely of bio-molecules which may be degraded (oxidised) by micro-organisms. During the catabolism of these bio-molecules, oxygen is consumed. If the faeces are discharged to rivers or impounded water, the activities of the faecal-degraders will reduce the oxygen tension within the water and may even induce anoxia. Under these circumstances fish and other aquatic organisms die and the water loses its amenity value as a source of potable water or as a place of recreation.

recalcitrant
domestic
wastes

composition of
garbage

In addition to the bodily discharges, human habitation is also accompanied by the production of food scraps and wastes which make up what we call garbage. It is in the production of garbage, however, that primitive and modern sophisticated societies diverge. Although garbage (refuse) from all societies contains food scraps, many other wastes are added to the garbage in sophisticated societies. Although these other components vary considerably from community to community and from household to household, you can get some idea of the complexity of this mixture by considering what materials you place in your garbage container. Typically these include such items as tins, plastic wrappings and containers, cardboard boxes and glass vessels.

∏   What is the fundamental difference between food scraps and many of the other components that make up the garbage (refuse) produced by societies in developed countries?

Whereas food scraps are biological products and are composed of biomolecules, many of the other components are man-made and are not bio-products. For example, tins are made by smelting processes and are composed of metals and many plastics are produced by polymerising petrochemicals-derived molecules.

∏   Can you anticipate what consequence this has for the biodegradability of garbage?

If materials have been biologically produced (for example food scraps), it is implicit that biological systems will be able to metabolise (degrade) these materials. The rate at which this may be achieved is very variable. Some materials such as soft plant tissues and animal carcases are degraded quite rapidly while others such as bones and lignified plant materials are more recalcitrant. Nevertheless we might anticipate that over a period of time, the material would be degraded and the constituents returned to the biosphere. In contrast, many of the man-made materials are not readily biodegradable and will persist in the environment in a more-or-less unchanged state for a considerable time.

We can represent this situation as shown in Figure 3.3.

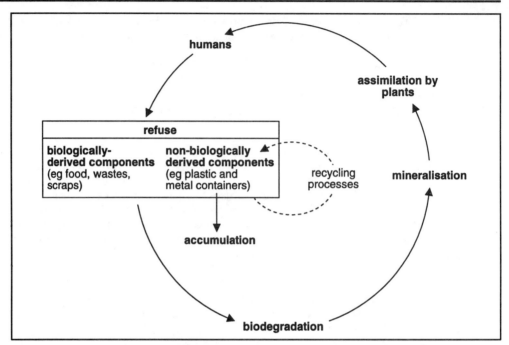

Figure 3.3 The potential recycling of biologically-derived components of refuse and the single use of non-biologically derived products. Note that, only in a few instances, the non-biologically-derived components of refuse are collected and recycled.

In this figure, we show that the biologically-derived constituents of refuse may be recycled through the degradative activities of organisms, especially micro-organisms. The products of biodegradation may be re-assimilated by plants which in turn may serve as a resource for humans and other animals. Many of the non-biologically manufactured products use natural, non-renewable resources such as oil and ore deposits, but the products are not biodegradable and will, if not processed by human activity, remain more-or-less in their manufactured form.

Π In Figure 3.3, we have indicated that only a very small portion of the non-biologically-derived components are cycled. What do you think mainly governs the extent to which this recycling takes place?

the difficulties and environmental costs of recycling non-biologically degradable wastes in refuse

The main factor is cost. Refuse is a complex mixture and particular components (such as tin or glass containers) make up only a small portion of the total. This, together with the fact that refuse is generated at widely different points, means that it is very costly to collect the material for recycling. The situation is improved by organisations which encourage the separation of garbage into separate components. For example, many Western European Communities encourage the separation of glass from general refuse. Even so the costs are still high. Public-spirited individuals who deposit used glass bottles in centralised 'bottle banks' may be encouraging the recycling of the glass, but if they have to travel considerable distances to do so, the environmental advantages of the recycling process may be out-weighed by the environmentally-damaging consumption of fossil fuels used during transport. Thus although the economic cost of the local authority may be low, the environmental costs are high.

After collection, the major part of the domestic garbage is transported to landfill sites (Section 3.5.1) or is incinerated. Processing domestic wastes in these ways, however,

After collection, the major part of the domestic garbage is transported to landfill sites (Section 3.5.1) or is incinerated. Processing domestic wastes in these ways, however, may cause severe environmental problems because of the production of highly toxic components (eg dioxines). Table 3.1 summarises wastes and pollutants generated during housekeeping, housing and processing of domestic wastes.

	Source/compartment	Components
wastes	housekeeping	sewage (faeces, urine, wash-water) food scraps, garbage, scrapped domestic appliances
	housing	building and demolition refuse (wood, stone, asbestos, synthetics)
	processing	incineration residues (fly ash, scoriae) effluent sewage treatment, sewage sludge
pollutants	atmosphere	CO_2, CO, SO_x, NO_x (burning fossil fuels) hydrochloric acid, polychlorinated biphenyls, heavy metals, dioxines (refuse incineration) volatile organic solvents (painting) chlorofluorohydrocarbons (refrigerator, insulating materials) methane (landfill)
	surface waters	oxygen binding components eutrophicating compounds (nitrate, phosphate) percolation water (landfill sites)
	soil	solid residues, heavy metals, dioxines (refuse incineration) percolation waters (landfill sites)

Table 3.1 Wastes and pollutants derived from housekeeping, housing and processing of domestic wastes.

3.3.2 Manufacturing

The compositions of the wastes generated from manufacturing activities are quite variable. They depend not only on the nature of the product but also on the design and operation of the processes involved. We do not intend here to catalogue all of the possibilities of waste composition. It is, however, important to establish some general principles.

Manufacturing processes involving biologically-derived materials (for example food processing and brewing) will mainly produce waste products which are biological in nature. Thus, just as with domestic sewage and the bio-components of garbage, these waste products may be recycled through the action of micro-organisms and plants. Thus the wastes (solids, water borne or gaseous) from these manufacturing processes may, at least in principle, be recycled through the biosphere.

On the other hand, manufacturing processes using non-biological materials and processes generally generate wastes that are not biodegradable and which may persist in the environment in unmodified form.

Some examples of wastes and pollutants derived from industrial sources are listed in Table 3.2. In this table, we have not attempted to give a comprehensive list.

Source	Wastes	Atmosphere	Surface water	Soil/ground water
chemical industry	chemical refuse (solids/liquids) sewage sludge	volatile organic compounds SO_x NO_x	miscellaneous (in)organic compounds heavy metals cyanides	miscellaneous (in)organic compounds heavy metals cyanides
oil refinery	waste water	volatile hydrocarbons SO_x, NO_x heavy metals polycyclic aromatic hydrocarbons aerosols	oil acids phenolics	oil
steel industry	heavy metal containing scoriae and sewage sludge	SO_x, NO_x, CO hydrocarbons aerosols	heavy metals (in)organic chemicals oil	heavy metals
fertiliser industry	contaminated gypsum	SO_x, NO_x, NH_3	inorganic chemicals	

Table 3.2 Some examples of wastes and pollutants derived from industrial sources.

SAQ 3.1

Match the following contaminants with their most likely occurrence.

Contaminants	Occurrence
1) asbestos	a) coal-fired power station
2) copper	b) domestic sewage
3) *Salmonella typhi*	c) building demolition site
4) methane	d) out-flow from an electroplating process
5) soot	e) land fill site

3.3.3 Energy production

The high consumption of energy by sophisticated societies coupled to the use of fossil fuels leads to the production of vast quantities of by-products. The generation of electricity from fossil fuels inevitably leads to the production of enormous quantities of CO_2. Although CO_2 can be recycled by plants through photosynthesis, it is anticipated that the atmospheric concentration of CO_2 will be increased with subsequent consequences on global warming and climate. What the use of fossil fuels effectively does is to increase the pool of carbon being recycled within the biosphere.

∏ Make a list of other by-products of electricity generation using fossil fuels.

<table>
<tr><td>acid rain and
summer smogs</td><td>The sorts of items we anticipate you would have listed would be various oxides of sulphur and nitrogen produced during the combustion of oil and coal. These have been of major concern since they cause acid rain which may have profound effects on natural eco-systems especially on conifer forests and on impounded water. They are also responsible for many lung disorders (note that the consequences of summer smogs are described in the BIOTOL text 'Defence Mechanisms').</td></tr>
</table>

acid rain and summer smogs

The sorts of items we anticipate you would have listed would be various oxides of sulphur and nitrogen produced during the combustion of oil and coal. These have been of major concern since they cause acid rain which may have profound effects on natural eco-systems especially on conifer forests and on impounded water. They are also responsible for many lung disorders (note that the consequences of summer smogs are described in the BIOTOL text 'Defence Mechanisms').

ash, soot and heat

Not all of the by-products of energy generation are gaseous. The combustion of coal and coke is accompanied by the production of ash and soots. You may have also included waste heat as another by-product of energy generation. Usually the waste heat exchange is to water (rivers and cooling towers) or directly to the atmosphere. The change in temperature may profoundly effect the eco-system especially if the system is held at an elevated temperature for a long time or if there are sudden and frequent changes in temperature.

Another product of energy generation is water (usually as vapour) and its discharge may cause localised meteorological changes such as increased fogging.

deposit of radioactive wastes

You should also remember that electricity generation using nuclear energy also produces particularly harmful waste products and strenuous efforts have to be undertaken to ensure that the hazards to environmental and human health are minimised. As a result of the long half-lives of some radioisotopes, no really satisfactory method of disposing of these wastes have been developed until now especially when it is recognised that even very small amounts of radiation can have profound consequences on biological health.

We have summarised wastes and pollutants derived from energy production in Table 3.3.

	Compartment	Components
wastes		(fly) ash, soot, residues from coal burning (heavy metals) gypsum from flue gas desulphurisation radio-active waste
pollutants	atmosphere	carbon monoxide (CO), carbon dioxide (CO_2), sulphur- (SO_x) and nitrogen oxides (NO_x) aerosols
	surface waters soil	thermal pollution (heating) coal and cokes residue (scoriae)

Table 3.3 Wastes and pollutants derived from energy production.

Π Is energy generation the only producer of atmospheric pollution?

The answer is clearly no. Transport (especially automobiles), manufacturing of many products and the disposal of wastes (for example by incineration) all may contribute to atmospheric pollution. Atmospheric pollution has been the cause of death and disease. Although regulations are in force that govern the disposal of materials to the atmosphere, there is still enormous concern about this aspect of pollution.

3.3.4 Agriculture and intensive cattle-breeding

In some Western European countries (eg the Netherlands and the UK) agriculture and cattle-breeding contribute considerably to environmental pollution. Mechanisation and the use of pesticides, fertilizers and fodder additives have all contributed to intensification and scaling-up of agricultural practices.

∏ Are all of the wastes generated by agricultural practices in Western Europe biologically-degradable? If not, give some examples of non-biodegradable wastes.

The answer to this question is no. Although most agricultural wastes, such as vegetable crop residues and manure, are biodegradable, agricultural practices in Western Europe involve, for example, the use of recalcitrant (non-biodegradable) pesticides and herbicides. Also the consumption of fossil fuels to drive tractors and other machinery leads to the production of compounds of limited biodegradability especially if leaded fuels are used. Other examples are plastics and fodder additives, such as copper.

In Table 3.4, we have summarised examples of wastes and pollutants generated by agriculture and intensive cattle-breeding.

	Compartment	Agriculture	Intensive cattle-breeding
wastes		vegetable wastes (crop residues) pesticides plastics	manure
pollutants	atmosphere	pesticides	ammonia (NH_3)
	surface waters	pesticides fertilisers (nitrate, phosphate)	fertilisers from manure
	soil/ground water	fertilisers pesticides	fertilisers from manure fodder residues and additives (eg copper)

Table 3.4 Wastes and pollutants derived from agriculturing and intensive cattle-breeding.

∏ What kind of environment problems can arise from the emission of ammonia to the atmosphere and the leaching of natural and artificial fertilisers from agricultural soils to surface and ground waters?

acidification

algal blooms

Emission of ammonia from manure to the atmosphere contributes to acid deposition which leads to acidification of sandy soils and lakes, and destruction of forest ecosystems. Leaching of fertilisers, such as phosphate and nitrate, from agricultural soils to surface and ground waters causes an accumulation of N- and P- nutrients in these waters. In surface waters, the result can be a massive bloom of algae leading to death of fish due to anoxia. Pollution of ground water with nitrate is a potential threat for drinking water supply, since high concentrations of this compound in drinking water pose serious health risks, eg methaemoglobinaemia in children and increased risk for cancer due to the formation of nitrite and nitrosamines in the gastrointestinal tract.

3.3.5 Transport and traffic

Road traffic plays a major role in transport of passengers and goods in densely populated parts of Europe. Increased prosperity has resulted in a strong increase of the possession of cars, increased mobility and growth of the carriage of goods. Because of this, the contribution of traffic to environmental pollution is substantial.

∏ What are the major wastes and pollutants produced by road traffic?

The use of leaded and sulphur containing fuels derived from fossil sources contributes considerably to atmospheric pollution. Major pollutants are: carbon monoxide, sulphur- and nitrogenoxides, volatile hydrocarbons, soot and lead. In many big towns exhaust gases are responsible for what is called 'summer smog'. Other wastes and pollutants derived from transport activities are summarised in Table 3.5.

	Compartment	Components
wastes		scrapped transport vehicles (cars, trucks, planes, ships) finished oil products
pollutants	atmosphere	carbon monoxide (CO), carbon dioxide (CO_2), sulphur- (SO_x) and nitrogenoxides (NO_x) hydrocarbons, aerosols lead
	surface waters	oil products
	soil	oil and fuel spoilage

Table 3.5 Wastes and pollutants derived from traffic and transport.

∏ Can you see opportunities for biotechnology in reducing environmental pollution caused by transport activities?

Biotechnology can play a role in reduction of atmospheric pollution and in a more sustainable use of energy resources, if renewable biofuels (Chapter 8) were used. In Brazil, there is a long experience with a partial substitution of petrol by bioethanol from fermented sugar cane. Also microbial removal of sulphur from fuels could be helpful in reducing this pollution.

3.4 The hazards that arise from the generation of wastes and pollutants

In this section we will predominantly consider the hazards to human well being that arise through the generation of wastes and pollutants. It should be realised, however, that many threats to human health are also threats to other organisms and ecosystems. Lead, for example, is not only toxic to Man, it is also toxic to almost all plants and animals. In this section we will briefly review the hazards under three sub-headings:

* biological;

- chemical;

- physical.

We should, however, realise that wastes and pollution also influence the sociological and psychological well-being of communities and individuals. The sociological and psychological consequences of pollution are rather poorly defined but we do know that noise, overcrowding, odour and poor health all have profound sociological and psychological consequences. This aspect falls largely outside of the scope of this text but, however, they are of vital importance in deciding how we want to use (and design) our environment.

3.4.1 Hazards arising from biological entities

sources of infectious diseases

The principle biological hazards to human health are infectious diseases. These may be transmitted directly from human to human or be transmitted by vectors such as water, food and insects. The major source of human diseases are humans (think of how you are infected when you catch colds, influenza and measles). Some diseases may be transmitted from animals (for example *psittacosis* from birds, brucellosis from domestic animals). In some cases the transmission of diseases from human to human or from animal to human may involve vectors such as flies and mosquitoes. In Table 3.6 we have listed just a small fraction of the many communicable diseases. Contamination of the environment by wastes derived from human or animal sources (for example by human or animal faeces) increases the risk of disease outbreaks. Conditions which favour the vectors of diseases (for example flies and rodents such as rats) will also increase the risks of disease outbreaks.

Disease	Sources	Main means of transmission
amoebiasis	human excreta	contamination of water/food
anthrax	cattle, sheep, goats, horses	hair, wool, hides, soil
brucellosis	domestic animals	milk, urine
cholera	human excreta	faecal contamination of water/food
diarrhoea	human excreta	faecal contamination of water/food
leptospirosis (Weil's disease)	rats and other wild animals	water contaminated by animal excreta
paratyphoid	animal and human excreta	contamination of food and water
Q fever	domestic animals	excreta, milk
shigellosis	domestic animal/human excreta	contamination of food/water
typhoid	humans, rats, pets, poultry	contaminated food/water

Table 3.6 Some examples of communicable diseases of human.

∏ What is the most common means of transmission amongst the diseases listed in Table 3.6?

You should have identified transmission by contamination of food and water by human or animal excreta.

importance of
hygiene
standards

Generally, the incidence of disease organisms and their biological vectors is increased by increased production and accumulations of biological wastes. A considerable effort must be made to reduce the chances of food and water being contaminated by infected materials and this is achieved through the imposition of hygiene standards. Critical to the success of these steps is the reduction or removal of biologically-derived wastes. We can broadly divide the treatment of these biologically-derived wastes into four main types of processes (Table 3.7). These are landfill, incineration, anaerobic digestion and aerobic digestion.

Physical form	Type of processing
solid	landfill, incineration
liquid	aerobic digestion, anaerobic digestion

Table 3.7 The main types of treatment of biologically-derived wastes.

The actual process used depends upon the physical state of the waste, its quantity and upon regio-specific factors such as the availability of the associated technology. We will examine landfill processes in a little more detail later in this chapter and aerobic and anaerobic digestion are described in greater detail in Chapters 4 and 5.

Π Bearing in mind what we said about the congestion of residential refuse and the fact that industrial wastes are often disposed of through the same drains as domestic wastes, can residential refuse and sewage be successfully treated using microbially-mediated degradation?

the effects of
non-
biodegradable
components on
the
degradation of
biodegradable
materials

The answer is not straightforward. We might for example anticipate that the biologically-derived material in refuse would be biodegradable. Some other components would not. Thus if refuse is used in landfill, we would anticipate that the biologically-derived components would be degraded but recalcitrant materials such as tins, plastics and glass would remain. This is in generally what happens but in some instances the non-biologically degraded portion may interfere with the efficacy of the biodegradation process. This is especially true in liquid-based systems such as sewage. The presence of toxic compounds such as heavy metal ions and some man-made organics may greatly restrict the metabolism and growth of degrading organisms in treatment processes. Also the residence times of these wastes in treatment processes may be insufficient to allow for the degradation of molecules that are only slowly degraded. The members of the latter group are usually regarded as recalcitrants.

3.4.2 Hazards arising from chemical entities

Modern sophisticated societies generate an enormous variety and very large quantities of chemical wastes. These wastes may be deposited in the environment in a wide variety of forms. They may be released as solids, as liquids or as gases. Although some of these may be readily biodegradable, many are not. The latter are called recalcitrant and may persist for a considerable time. Those that are biodegradable can, generally, be returned to the cyclical turnover of materials within the biosphere. The recalcitrant substances, however, represent a linear flow of material; we have represented this situation in Figure 3.4.

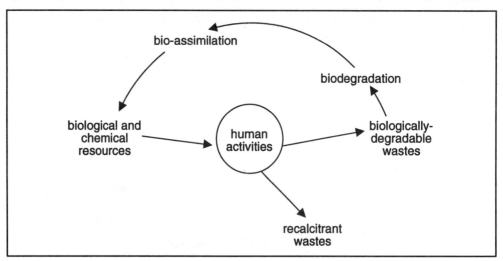

Figure 3.4 The generation and fate of chemical wastes generated by human activities.

∏ Are all recalcitrant wastes prevented from ever returning to biological recycling?

The answer is no. We can for example visualise a situation in which an organic chemical which cannot be metabolised could be incinerated and thus be converted to products such as CO_2, H_2O and NO_2 which could be utilised by biological systems.

∏ Does the biodegradation of chemical wastes always lead to the production of new materials which are non-toxic to biological systems?

Again, the answer is no. In fact the reverse might be true. Fairly innoxious materials may, through the activities of biological systems, be converted to more toxic compounds. The production of highly toxic and mobile methyl mercury from relatively innoxious mercury compounds through the action of micro-organisms is a good example. Thus, although in general terms we might conclude that the production of biologically-degradable waste is potentially less damaging than the generation of recalcitrant waste, the issue is in many instances less than clear cut.

xenobiotic chemicals or xenobiotics

As a consequence of industrialisation and the use of chemicals in modern agriculture, huge amounts of synthetic compounds, like polymers, pesticides and detergents, are dispersed throughout the biosphere. Many of these man-made compounds have chemical structures that are not encountered in nature. Therefore, these pollutants are often called xenobiotics which means that these compounds are considered to be unnatural. A much broader description of this term, which is in common use now, defines xenobiotics as man-made compounds which, although they may occur naturally, are present in the environment at unnaturally high concentrations.

∏ Give an example of naturally occurring compounds which should be considered a xenobiotic according to the currently used definition.

Metals and certain products of the oil industry, such as phenols and (polycyclic) aromatic hydrocarbons, are examples of naturally occurring compounds which should be considered as xenobiotics. These compounds cause environmental problems if their concentration in the environment is far above the natural one.

It is the potential environmental hazards which arise through the production of exotic synthetic compounds that arouse great concern. Each year, many hundreds of new chemicals are developed. For many of these, their toxicity is established during the research and development stages of production. However, these studies are done mainly to establish the occupational hazards which may arise through producing and using these chemicals. It is only relatively recently that the more subtle effects of these chemicals within ecosystems have become apparent.

occupational
hazards

DDT   We will illustrate this with just one, well documented, example. DDT (1,1-bis-(4-chlorophenyl)-2,2,2-trichloroethane) was once used extensively as an insecticide. It is a chemically stable compound whose structure may be represented as:

$$Cl \!-\!\!\! \left\langle \bigcirc \right\rangle \!-\!\! \underset{\underset{CCl_3}{|}}{\overset{\overset{H}{|}}{C}} \!-\!\! \left\langle \bigcirc \right\rangle \!-\! Cl$$

It is recalcitrant and is not readily biodegradable. About 25% of the applied DDT remains in the environment for 4-15 years. It is readily absorbed by biological systems. It was applied at low concentrations and, although absorbed by biological systems, was deemed to offer little threat to organisms apart from the target organisms (insects). You will recall from Chapter 2 that all organisms must be regarded as components of complex food webs. Thus when DDT is absorbed by, for example, plants and micro-organisms, it remains in the tissues and is not metabolised. If these organisms are subsequently eaten by other organisms, then their biological components are catabolised but the DDT remains intact. In their turn, these organisms are devoured by others. Again the biological components of these food organisms are catabolised, but the DDT remains unchanged. The consequence of this is that the concentration of DDT as a proportion of the biomass increases. This process, often referred to as biomagnification, may lead to such high concentrations of DDT in some organisms that the DDT begins to have toxic effects. In the case of DDT, the concentrations used to kill insects was not, originally, thought to be toxic to birds. However, it was subsequently shown that, through biomagnification, high levels of DDT accumulated in the livers of certain sea-birds (shags) with the result that they produced thin shelled (fragile) eggs.

biomagnification

The process of biomagnification is an important one. Many agrochemicals are consumed with food, volatile materials from industry and vehicle exhausts are inhaled and many chemicals in industrial effluents contaminate water supplies. Although humans and other biological systems may be exposed to only low concentrations of these compounds, the recalcitrant nature of the chemicals means that small increments of each of these chemicals over a period of time could lead to damaging consequences.

Even without the subtilities of biomagnification, it is well recognised that many chemicals we release into the environment, or products which are generated from them by natural processes, have harmful effects. Attempt the following SAQ which will test your general knowledge about the toxicity of chemicals. If you are unfamiliar with some of the named chemicals, you should find our response helpful.

SAQ 3.2

Assign each of the compounds listed below to one of the following categories:

Categories

carcinogens (mutagens)
asphyxiants
systemic poisons
pulmonary irritants
fibrosis producers

Compounds

sulphur oxides   chlorine
nitrogen oxides   hydrogen sulphide
lead   titanium
mercury   ethylenemine
cadmium   boron
asbestos   carbon monoxide
N-nitrosodimethylanine   ammonia
ozone

---

allergenic and other consequences of chemical pollution

The SAQ you have just attempted gives some examples of the potential consequences of releasing toxic chemicals into the environment. There are many more. Some, for example, produce allergenic reactions such as contact dermatoses. Others have consequences only in limited circumstances, for example high concentrations of nitrates may produce methemoglobinemia ('blue babies') in the very young. You should anticipate that these chemicals may also have effects, albeit different, on other life forms and ecosystems.

priority and 'black list' pollutants

A variety of parameters, such as the aquatic and mammalian toxicity, carcinogenicity, levels of production, reactivity, bioaccumulation and persistence, make a contribution to what is called the 'pollution potential' of xenobiotics. This measure is used to distinguish priority pollutants, compounds which are thought to cause severe environmental damage. Every country decides on their list of priority pollutants. In the Netherlands diverse compounds such as ammonia, nitrate, cadmium, polycyclic aromatic hydrocarbons, polychlorinated biphenyls, asbestos, carbon monoxide, ozone and radon gas are on this list. Pollutants which impose an immediate threat to humans and ecosystems, are placed on a 'black list'. Black list pollutants as established by the European Community are shown in Table 3.8.

I	heavy metals, metalloides and related compounds
II	halogenated organic compounds
III	organic phosphorus compounds
IV	organic stannous compounds
V	persistent mineral oils and hydrocarbons from crude oil
VI	other organic compounds
VII	cyanides
VIII	asbestos

Table 3.8 Categories of 'black list' compounds as established by the Europian Community.

In many instances, especially with non-biodegradable wastes such as coal and coke-dusts, it is difficult to visualise how biotechnology may be used to remove the hazard. However, if biotechnology was to lead to a reduction in the use of coal-based products by providing an alternative route to generating the products currently produced from coal, then biotechnology offers the potential to reduce this hazard.

We can visualise therefore that biotechnology offers two fundamentally different strategies to reduce the hazards arising from chemical wastes generated by current practices. These are:

- providing an alternative route to achieve desired objectives so that the hazardous waste is not generated in the first place;

- providing processes which transform hazardous wastes into a non-hazardous form (see Section 3.5).

Both of these strategies are being pursued. For example, the replacement of recalcitrant chemical pesticides by biological control agents removes (or reduces) the ecotoxicological problems which may arise from traditional practices (see Chapter 8). Similarly, the use of bioreactors to treat effluents from industrial processes to render them less hazardous before their release into the environment is becoming more widespread.

Π   Which one of the following chemicals in industrial effluents would be most amenable to biological treatment before release of the effluent into the environment.

Nitrobiphenyl
Dichlorophenylamine
Chloroform
Formaldehyde
Sucrose
N-Nitrosodimethylamine

You should have argued that sucrose is the most amenable to biological treatment because it is readily metabolised by many organisms. The others are less easily metabolised and are, in many instances, toxic to biological systems. N-nitrosodimethylamine is, for example, a mutagen.

Π   In addition to the chemical nature of the waste, what else will govern whether-or-not it is suitable for bio-treatment?

physical forms
of wastes and
bio-treatment

The factor we hope you identified is the physical form of the waste. Consider, for example, an industrial process in which a liquid effluent containing the waste is produced. In principle, the pipe carrying the effluent could be coupled up to a bio-process of some kind. The actual design of this process will, of course, depend on such factors as the volume of the effluent being produced, the concentration of the waste chemical and the ease by which it may be metabolised. In contrast, consider the use of pesticides to treat crops. Once the pesticide has done its job, the pesticide residues left in the soil and on the crop may be regarded as waste. Because this waste is dispersed and, in any one location, at very low concentration it is difficult to visualise how a bioprocess may be used to remove the waste especially if it shows some degree of recalcitrance to biodegradation. The fact that it is present in low concentration does not mean that it does not present an environmental hazard.

3.4.3 Hazards arising from physical entities

dusts,
humidity,
radiation,
mechanical
stresses,
collisions, fires
and electrical
storms

The physical hazards to life arising in the environment include such factors as dusts, humidity, radiation, mechanical stresses, collisions, fires and electrical storms. You may feel that these hazards fall outside of the scope of biotechnology. Although this is largely true, you should also recognise that the adoption of biotechnological processes may modify the extent of some of these hazards. For example, the replacement of nuclear energy programmes by biotechnologically-based energy generation processes could reduce the risks from radiation. However, we must also bear in mind that the biotechnologically-based processes will also create a new set of circumstances with important environmental consequences. If, for example, biomass is used as a source of energy and this energy is released by incineration, this may release a wide variety of oxides into the atmosphere.

3.5 The treatment of wastes

In the previous section, we described, in outline, some of the environmental hazards which arise through human activities. Although we predominantly focused on the consequences of the production of wastes and pollutants on human health, many of these effects will be paralleled in other organisms. Some systems will be more sensitive than others to these harmful effects. The environment was never free of all hazards. What human activities have done is potentially adding to and diversifying them. The realistic aim of environmental biotechnology is not to remove all hazards. It should be, however, to aim to encourage the development of processes and products which minimise environmental damage at the same time as enabling mankind to maintain a high quality of life.

In Chapter 2 we described this potential contribution of biotechnology as having two major and overlapping components:

- reducing the environmental damage through the treatment of hazardous materials to render them safe;

- replacing existing practices by more environmentally sustainable practices.

The difference between these two contributions to a given production process is depicted in Figure 3.5.

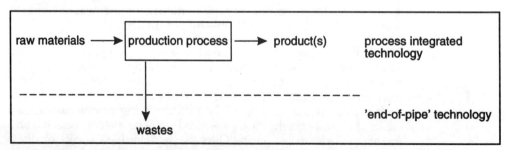

Figure 3.5 Schematic representation of the potential contribution of biotechnology to environmental management.

'end-of-pipe'
and process
integrated
technology

In this text, we mainly focus on the 'end-of-pipe' technology as this technology is inherent to the treatment of wastes. Although the term 'end-of-pipe' has a somewhat negative sound compared to process integrated technology, it should be realised that examples of integrated process technology are very limited. Before the implementation of a 'end-of-pipe' technology, the production process has to be evaluated and the waste streams minimised. Finally, it should be realised that zero waste technology will never exist.

In this section we will begin our discussion of the treatment of wastes by firstly describing some general principles followed by an outline of the treatment of wastes using landfill strategies.

As we have seen wastes can be generated as:

- gases;

- liquids;

- solids.

As it can be envisaged, the wastes may be:

- concentrated and localised;

- dilute and dispersed.

Furthermore, they may be:

- biodegradable;

- recalcitrant;

- a mixture of biodegradable and recalcitrant materials.

At present, we will not discuss the treatment of volatile and highly dispersed materials nor will we focus on the treatment of specific wastes from particular processes. We will, however, consider the treatment of wastes generated by a typical urban community in the developed world. Before doing this, it has to be made clear that the treatment of gases, liquids and solids are often overlapping. For example, during the cleaning of liquid wastes we usually produce solids as byproduct. Furthermore, removal of waste from water cannot actually be separated into aerobic or anaerobic processes: some treatments combine both techniques. For example, nitrification (conversion of ammonium in nitrite or nitrate) is traditionally carried out aerobically and denitrification (conversion of nitrite or nitrate to nitrogen) anaerobically.

In this text we will, only briefly discuss the treatment of gases which arise from industrial activities. Biological air treatment uses peat or some other support material suitable for growth of micro-organisms. The polluted air is blown through the support layer and the available micro-organisms breaks down the undesired components. Problems with these so-called biofilters, however, are that the biomass is difficult to control and that not all pollutions are converted by the micro-organisms. Consequently, the yield is variable and unpredictable. A new development is based on a two phase (gas/liquid) system separated by a membrane on which selected micro-organisms are immobilised on the liquid side. In this type of biofilters, the polluted components of the gas phase enter the liquid phase and are consumed by the micro-organisms.

water
reclamation

sewage
processing

The liquid waste, including domestic and industrial wastes, is usually treated in centralised units sometimes referred to as water reclamation or sewage processing units. These processes use the activities of micro-organisms to reduce the amount of organic material present in the water and to destroy pathogenic organisms that may be present.

∏ What materials present in the liquid reaching a water reclamation would be
   difficult to remove by microbial action?

settling and
sieving
processes
removes
particulate
materials

The main problems are, of course, the non-biologically degradable (recalcitrant) materials that may be carried along in the fluid flow. Particulate materials (for example grit and fibres) may be removed by sieving or by settling processes. Soluble non-biologically-degradable materials are likely to remain in the liquid passing through the process. In this text we have divided the discussion of the treatment of aqueous effluents into three aspects. In the first two, we examine the processes by which the readily biodegradable material can be removed from water. Chapter 4 deals with aerobic processes with emphasis on the technical aspects of the treatment and in Chapter 5 the anaerobic processes are described with emphasis on the micro-organisms used. In the third aspect (Chapter 6), we examine the use of micro-organisms to deal with the problems arising from recalcitrant chemicals.

The processes used to treat water-borne wastes usually produce solids (in the form of sludges) as byproducts. Although, in some instances these solids may be used as agricultural fertilisers, in many cases contamination of these sludges by other products of human activities (for example heavy metals and exotic xenobiotics) make their long term use as agricultural fertilisers unsatisfactory. Nevertheless, of the 10×10^6 tonnes of sludges produced from water treatment each year in Western Europe, about 30% is disposed of to agricultural land.

∏ Apart form the presence of toxic elements in the sludges, what else makes the
   practice of dispersing of sludge on agricultural land problematical?

Although water treatment reduces the numbers and varieties of disease organisms in water, some do persist in the sludges. The organisms of concern are various *Salmonella spp.*, *Bracella spp.* and viruses. The other concern arising form disposing of sludge to agricultural land is the offence given by strong odours. Because of these concerns, the disposal of sludges to agricultural land is strictly controlled.

∏ Instead of disposing of the sludges to agricultural land, what other options are
   available especially for an island state like the UK?

disposal of
sludge to the
sea

The sea has been used as a receptacle for sludge disposal especially by the UK and by other regions bordering the sea. This practice has been recently subjected to close scrutiny. Although the UK favours this form of sludge disposal (it is cheap to operate), other European states are less supportive and argue that the indiscriminate dumping of toxic materials is environmentally unsound and likely to have an impact on the marine environment. Although there remains disagreement about the environmental effects of sludge dumping, a number of international agreements (the London

Convention and the Oslo Convention) have been reached which control the quantity and quality of the material that may be dumped.

sludge disposal on land

Increasingly, steps are being taken to dispose of the sludge on land involving the sanitation of sludge by pasteurisation, thermophilic digestion, irradiation or incineration. In some instance, the solid residues produced during the treatment of water-borne wastes, are disposed of alongside the other solid wastes (refuse) produced by human activities. The most common method of disposal in contemporary sophisticated societies is by landfill.

However, biotechnology offers also possibilities to decontaminate polluted soils. In this respect two *ex-situ* techniques which are currently used in the Netherlands for the removal of more specific wastes which are attached or diffused in solids, such as oil and polycyclic aromatic carbohydrates (PACs) are worth mentioning: landfarming and bioreactors. For both techniques, the soil has to be removed from its original site as the natural process is rather slow and, therefore, the conversion conditions have to be optimised. In landfarming the polluted soil is brought to depots and can be treated there by ploughing the soil from time to time to assure availability of oxygen and/or rainfall upon the soil to enhance the transport of added nutrients and pollution removal. More or less the same is done in bioreactors, although this type of treatment has a more continuous character: the soil is placed in the reactor together with water containing nutrients where it is stirred and aerated. The most important drawbacks of biological decontamination of soil are the relatively long treatment times and the fact that as a result of the poor bioavailability, decontamination to very low values is not always possible. Nevertheless, biotechnology offers the possibility to reuse the decontaminated soil. This is not the case with landfill.

3.5.1 Landfill

In most instances, the ultimate product of waste treatment processes are solids. Even where waste treatment is by incineration, solid residues are produced. The traditional method used for disposing of these solids is by landfill. In principle, the process is simple. A site (usually either a natural area of low ground or hollow, or a man-made pit or area of derelict land) is chosen and the waste is simply dumped at the site. In practice, the process is much more complex. The tremendous increases in population, the increased use of disposable products, the production of hazardous chemical wastes arising from modern industrial processes together with the new environmentalism of the 1980's means that the practices of landfill are under close public scrutiny and under regulatory control. Even so, a significant portion of total wastes may still be disposed of by uncontrolled tipping. For example in Greece almost a fifth of the total wastes disposed of during the 1980's was done so in an uncontrolled way. In Eire the figure was as high as 33% whilst in France about 10% of total wastes disposal was uncontrolled. Despite the increased regulatory control, the trend of using landfill sites at greater distances from the sites of waste generation tends to encourage uncontrolled tipping.

3.5.2 Collecting materials for landfill

Although the details of the collection of materials that will ultimately be used for landfill differ in detail from region to region, the underpinning scheme is similar in most. A generalised scheme is shown in Figure 3.6.

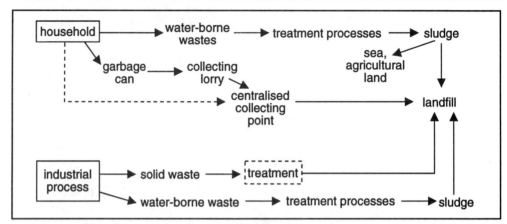

Figure 3.6 Generalised flow of solid and water-borne wastes in modern societies.

costs of refuse
collection

Most of us will be familiar with the practice of refuse collection and its transfer to a landfill site. The details of this differ from state to state. In the UK, the collecting service is a public-run utility, paid for out of taxes and managed by local authoring. Increasingly, in the UK, individuals are encouraged to take their own refuse to the centralised collecting point. Although this makes economic sense to the local authority, the use of many cars taking small bundles of refuse hardly makes environmental sense. With larger volume producers, such as major industries, there wastes are often transported directly to the landfill site. It is, however, beyond the scope of this text to discuss in detail the economics of refuse collection. Good guidance to economics forecasting in this area is given by P W Purdom, Environmental Health, Academic Press, London, 1990, if you wish to follow up this aspect.

There may be some pre-treatment of the waste before it is tipped. This may include mechanical pulverisation, incineration or some form of sorting. For example, in some regions, refuse is picked over for usable items. In some cases, this is driven by the shear poverty of individuals. In more affluent communities, this is simply a straightforward commercial operation. In some cases, the pre-treatment processes may be to render the waste less noxious (some examples are given in Chapter 6).

SAQ 3.3

Although the composition of the material deposited in landfill sites differs between countries, cultures and time, you should, however, be able to given a generalised description of the material deposited at a landfill site and to identify the hazards which this material potentially offers. Attempt to do this before reading on.

In the SAQ you have just done, you will have identified a range of hazards that might arise from the materials in a landfill site. These hazards will, naturally vary from site-to-site. Although the risk of fires is not to be ignored, the major general concern arising from landfills is the leaching of toxic and corrosive materials into surface and ground waters.

Π How may the toxic materials that may be present or generated in landfill sites be contained?

The obvious solution is to place the waste within an impermeable barrier. The more difficult question is 'what materials should be used to achieve this?' A wide variety of materials have been considered for this, including:

- clays;

- soil/cement mixtures;

- concrete;

- polymeric materials;

- asphalts.

Considerable controversy centres on the success, or otherwise, of the materials in containing the toxic materials. Of real concern is long-term containment. Thus, although a clay-lining of a landfill site may contain wastes for several years, the recalcitrant nature of some of these hazardous chemicals, means that they may outlive the efficacy of the protective barrier. Indeed there is a growing opinion that the tipping of some toxic waste without pre-treatment may pose a greater threat to public health then the burial of many radioactive wastes. For these reasons, pre-treatment of particular wastes is becoming recognised as being of vital importance.

3.5.3 Landfill practices

Landfill practices varies from country to country. One of the most common approaches is the use of the cell emplacement strategy. In this, the collected refuse is covered on all sides by soil at the end of each working day. There is, therefore, some degree of stratification albeit a rather irregular one (Figure 3.7).

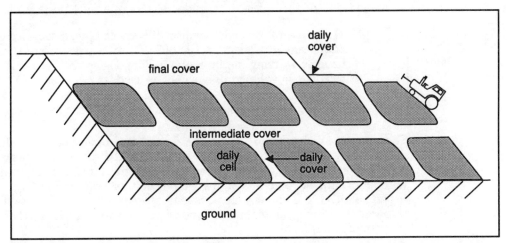

Figure 3.7 Cell emplacement landfill site. Shaded areas represent refuse, clear areas represent soil cover (not to scale).

cells covered by daily, intermediate and final covers

The size of the 'cells' depend on the daily volume that is tipped. Each cell is compressed and roughly levelled using mechanical bulldozers. Generally, the depth of each cell is limited to about 2-5 meters. The depth of the soil used to cover each cell at the end of each day is usually about 20 cm. If a multilayered site is used, the layers are usually separated by a thicker layer (30-40 cm) of soil referred to as an intermediate cover. As the site is completed it is covered by a 60-90 cm layer called the final cover.

∏ See if you can list two or three reasons why the refuse is covered each day.

There are several reasons but the main ones are to prevent (reduce) the risk and spread of fires, reduce infestations of disease vectors such as flies and to reduce the emission of odours.

∏ What factors govern the size of a landfill site that is required to service a community?

The main factor, of course, is the anticipated volume of compacted waste that will be generated by the community it serves. Population and the amount of refuse each person generates are therefore, contributory factors. Typically a population of 10000 in Western Europe generates 20-40 x 10^3 m^3 of compacted refuse y^{-1} but this figure is quite variable. Also the availability of the soils needed to cover the refuse is important since about 20% of the volume of a cell emplacement site is occupied by covering soils.

An alternative method to the cell emplacement strategy is the trench method also called the 'cut and fill' method.

In this, a trench is excavated below the ground surface and the refuse is placed in the trench and covered. The exact form of this method depends on the topography of the region. Commonly, two trenches are used simultaneously. The refuse is used to fill one trench while the soil excavated from the other is used to cover the material. This reduces the haulage costs incurred in soil collection but it does mean that excess soil has to be removed from the site unless it is decided to raise the level of the land.

SAQ 3.4

1) The refuse from a community of 10^5 people is to be disposed of using the cell emplacement strategy. A site of 8 x10^5 m^2 has been found. It is estimated to have as working depth of about 2.5m. Using the figures given in the text, calculate how long this site may be used.

2) Comment on the major assumption that has been made.

There are many issues governing the design and operation of landfill sites. These include the selection of sites and the regulations and codes of practice governing modes of operations. These largely fall outside of the scope of this text but we have included a number of suitable references in the 'Suggestion for Further Reading' section given at the end of this text if you wish to pursue this topic further. Here we will concentrate on some of the biological aspects of landfill and on the possibilities of exploiting the practice of landfill.

3.5.4 Turnover of landfill constituents

Some of the constituents of landfill rapidly become water soluble. This is especially true of cationic and anionic materials and law molecular weight constituents. The leachate from landfill will inevitably change over time and be dependent upon such factors as refuse composition, rainfall and soil type. We report some typical figures for a variety of parameters in the ground water in Table 3.8. Three sets of values are reported:

- background ground water;

- water from the landfill;

- water taken about 50 meters downstream of the landfill.

Parameter	Background (mg l^{-1})	Landfill (mg l^{-1})	Outflow (mg l^{-1})
total dissolved solids	600-700	6000-7000	1000-2000
COD (chemical oxygen demand)	20	1000-2000	70-80
sodium	30	800	300
pH	7.2	6.5-6.8	7.3

Table 3.8 Typical parameter values of groundwater of a landfill system. Date based on the EPA Landfill Guidelines (Brunner Dr and Keller D 'Sanitary Landfill Design and Operation' US Environ Prof Agency Washington DC 1972). Values have been approximated. See text for further details.

The COD (chemical oxygen demand) is a measure of the amount of oxidisable material present in the water.

∏ Will be materials present in the landfill remain largely unchanged?

The answer is no. The biodegradable components will be subjected to microbial degradation. Some of this will occur rapidly especially with such items as food scraps. Other articles (for example pieces of wood) may take much longer to be degraded.

There is a considerable potential for the degradation of materials within landfills.

In outline, this degradation can be divided into a number of phases. First complex macromolecular structures may be hydrolysed to release small soluble molecules. Some of these may be leached from the landfill, but in most instances they will be absorbed by micro-organisms and be catabolised. Initially this catabolism will be mainly aerobic leading ultimately to the release of CO_2. However, the activities of the aerobic organisms responsible for this catabolism will lower the availability of oxygen and may generate anoxic conditions. In these circumstances, the metabolic products of catabolism are simple organic acids and alcohols. These in turn may be used by other microbes to produce hydrogen which is subsequently used by methanogenic organisms to produce methane.

We will not elaborate on these metabolic processes further here because we will deal with them in more detail in Chapter 5. Nevertheless, what essentially is happening is that the carbon, hydrogen and oxygen atoms present in the degradable refuse are re-arranged to form carbon dioxide and methane. We can write this in symbolic form, thus:

biodegradable refuse $\rightarrow C_6H_{12}O_6 \rightarrow 3CH_4 + 3CO_2$

The actual ratio of methane and carbon dioxide released will be dependent on the average composition of the biodegradable refuse. The rate at which methane is released is highly variable and may extend over a considerable period. Historically, little attempt has been made to make use of the methane produced but more recently efforts have been directed towards methane collection and utilisation (see Chapter 7). Efforts have also been directed towards increasing in-place degradation of landfill aimed at maximising methane production. A theoretical maximum yield of methane has been calculated to be $0.266 \, m^3 \, kg^{-1}$ dry refuse although many believe a more realistic practical value to be about $0.05 \, m^3 \, kg^{-1}$.

Π What factors would you anticipate would influence the rate of methanogenesis?

temperature, pH, moisture and chemicals will influence the rate of methanogenesis

Apart from the composition of the refuse, all those factors which influence microbial processes (for example temperature, pH, moisture) would be expected to have an impact on methanogenesis. Many of these factors are not controllable in a landfill site. Furthermore, micro-organisms in general and methanogenesis in particular, are susceptible to inhibition by a wide range of organic and inorganic compounds. With the ill-defined materials present in landfills, coupled with the complexity of the processes involved in methogenesis it is extremely difficult to predict and control this process. Most experiments on these systems have attempted to manipulate moisture content and refuse density. However, the conditions needed for optimum gas production are still not known. Data derived from laboratory studies cannot often be applied with success to the field situation. Thus, if we see methane production from landfill as biotechnology on a large scale we must also see it as a rather unscientific and relatively inefficient process.

Apart from the problems of optimising methane generation, there are also many operational problems that need to be solved. These include the problem of how the gas is to be collected for this must be done both efficiently and without interfering with daily operations. Despite these difficulties some progress has been made and a number of processes are in operation in Europe and the Americas.

SAQ 3.5

The methane generated by landfills may be regarded as a usable energy resource yet it is hardly used for this purpose because the collection process is uneconomic. Why, despite these economics, might it be advisable to collect the methane?

Π Are CO_2, CH_4 and simple organic molecules the only products of microbial action on the biodegradable components of refuse?

The answer is no. The biodegradable constituents of landfill are largely biological in origin and, therefore, contain nitrogen (as amine and amino groups), phosphate, sulphur (as S-containing amino acids and sulphates) and a wide variety of minerals. As these materials are broken down, many of the elements are released in mineral form. This process is often referred to as mineralisation. Thus the nitrogen is released as NO_3^- and NH_4^+, the sulphur as H_2S and SO_4^{2-} and so on. Many of these are water soluble and are leached to the ground water. H_2S is volatile and may perculate upwards through the covering soil and escape to the atmosphere.

SAQ 3.6 In some landfill sites, the covering soil layer is often planted to provide an environmentally more attractive cover. Is the planting of trees or herbaceous plants more likely to be successful with:

1) a young landfill site;

2) an old landfill site?

(Read our response carefully).

3.5.5 Landfill exploitation

For the most part, we have considered landfill as a device for disposing of waste and reducing the hazards that may arise from creating wastes. However, we have implied that landfill may be exploited in two ways:

• for the generation of a usable energy source (methane);

• for the restoration of derelict land (see SAQ 3.6).

problems
associated with
the restoration
of derelict land
using landfill

We will be considering the biological processes and the engineering practices in methane generation in more detail later in this text so we will not consider it further here. The use of landfill to restore derelict land is of great importance. The principle is quite simple. Derelict land (for example, a disused industrial site or gravel pit) is, by its very nature, of little value and is undesirable. By filling such a site with refuse and covering it over with soil, the biodegradation of the refuse releases plant nutrients which will support plant growth. We saw, however, in our response to SAQ 3.6 that the ability of plants to grow on these sites depends to some extent on the nature of the material deposited in the landfill and on the age of the site. Despite these uncertainties, in many instances plant communities can be established. There are, of course, serious doubts whether such sites should be used by agriculture. The presence of toxins (heavy metals, organometalics, pesticides) co-deposited with domestic refuse which may be taken up by the plants raises problems concerning the use of these plants as food. Thus often these sites are used for other purposes (for example as wild life preserves). The success of using landfill to restore land is, however, problematic.

We will illustrate this using a single example. A landfill site at Gaston near Liverpool, UK was selected to be the site of the first UK National Garden Exhibition. The surface of the landfill was called upon to support a wide variety of plants. Initial plantings of many varieties were unsuccessful because of inhibitory (toxic) materials seeping from the deposited and degraded refuse. As a consequence, the upper surface of the site had to be sealed and covered by an additional layer of top soil.

Π See if you can think of one of the consequences of this process in terms of compound(s) produced within the landfill.

The site continued to generate methane. Whereas previously this diffused upwards through the upper layers and was vented to the atmosphere, sealing the upper surface reduced this process thus producing the possibility of a build up of methane within the landfill site thereby posing the potential risk of fire or explosion. The solution was to pipe the methane away and to burn it at a site adjacent to the exhibition site.

SAQ 3.7

In selecting a site for landfill, which of the following features give a site high value for waste disposal and which give a site a low value for waste disposal. Report each as either 'high' or 'low'.

1) Site able to take any kind of waste.

2) Site close to the production of the waste.

3) Site suitable for building development.

4) Site within an urban area.

3.6 Concluding remarks

In this chapter we have given an overview of the origins, composition and problems associated with the production of wastes by sophisticated modern communities. We have also included a brief discussion of the use of landfill to dispose of large volumes of solid wastes. In the next set of chapters, we will examine, in greater depth, the use of biological processes to treat water-borne wastes.

These processes are, as you will see, much better understood and controlled and there have been many advances made in the development of the technology associated with water treatment. The technology need to use of biological processes to treat gaseous (volatile) wastes is much less developed. The reduction of volatile emissions such as CO_2 release from coal-burning power stations is currently seen predominantly as requiring chemical or physical solutions, not biological processes. Nevertheless many of these solutions ultimately require biological treatment. For example, adsorption of undesirable volatile materials onto solid adsorbants often leads to the products that are disposed via landfill. Similarly, dissolving the waste volatile into water (or an organic solvent) leads to products which may be treated in a manner similar to that of other water-borne waste. We will be considering the use of biological processes to treat xenobiotic organic compounds, including some volatile organics, in Chapter 7.

bioprobes as environment monitoring devices

Biotechnology is not only making important contributions in the treatment of wastes, it also provides analytical tools that enable us to monitor the presence of a wide variety of chemicals within the environment in general and in specific effluents and emissions. This technology largely depends upon the use of enzymes and/or antibodies that interact with specific molecules. The principles and application of these so called bioprobes are described in depth in the BIOTOL text 'Technological Applications of Biocatalysts' so we will not examine this aspect in depth here. We will, however, draw your attention to the potential of bioprobe technology within the text (Chapter 7). In this context we should, however, remind you that the quality of an environment is reflected in the flora and fauna present in that environment. Thus the presence or absence of particular species are often indicative of environmental pollution. Again the biotechnological development of rapid identification procedures especially of micro-organisms can also be regarded as a positive contribution to the monitoring and management of the environment. Thus, despite the amounts and diversity of the wastes and pollutants generated by human activities, biotechnology has the potential to aid us in measuring and treating many, but not all, of these materials. However, biotechnology by itself cannot provide solutions to all of the environmental problems which arise from the products and processes of contemporary society. There is, as we pointed out in Chapter 2, a need for a fundamental re-think about the nature of the products and the processes used to produce them if we are to achieve sustainability.

Summary and objectives

In this chapter we have provided an overview of the origins, composition and problems of wastes produced by sophisticated modern communities. We began by examining the interactions between Man the environment and indicated how the production of wastes which arise as by products of human activities may significantly affect these interactions. We then considered the composition of domestic and industrial wastes and examined the hazards that these generate. We also indicated that the chemical constituents of waste may be broadly divided into biodegradable components and recalcitrant components. The hazards arising from waste production can be divided into biological, chemical and physical. Psychological and sociological consequences of waste production are also a great importance.

In the final part of the chapter, we considered general strategies for the disposal/treatment of waste with particular emphasis on the use of landfill. We described the general approaches to landfill and indicated some of the potential problems that may arise from landfill practice. We also indicated how landfill may be exploited to produce a usable energy source (methane) or to restore derelict land. This chapter provides a framework for the more detailed discussion of the applications of the biotechnology to waste treatment given in subsequent chapters.

Now that you have completed this chapter you should be able to:

- describe, in outline, how the production of wastes arising from human activities may influence the ability of the environment to provide life support systems;

- describe the nature and origins of waste produced by human activities;

- relate particular wastes to particular activities;

- give examples of the biological, chemical and physical hazards that may arise through waste production;

- distinguish between biologically-degradable and biologically-non-degradable waste;

- explain the importance of biomagnification;

- describe the cell emplacement and trench strategies of landfill, calculate life expectancy of landfill sites from supplied data;

- explain the hazards that may arise from landfill practices especially relating to flammable components and toxic leachates;

- describe, in outline, the processes which lead to methane production in landfill sites;

- explain the potential and the limitations of exploiting landfill as to restore derelict land.

Aerobic waste water treatment

Aerobic waste water treatment

4.1 Introduction

eutrophication

Water is used in considerable quantities by all developed countries. For example, in England and Wales to some $16 \times 10^6 \, m^3$ are used per day. This represents a rate of use of about 150 litres per inhabitant per day. Also some $20 \times 10^6 \, m^3$ are taken and returned directly to rivers by industry. Where water is contaminated, failure to treat it before discharge to rivers would result in extremely deleterious changes in the river ecology. Loading of the watercourse with organic and inorganic material (eutrophication) will lead to an upsurge in microbiological and plant activity with a subsequent demand on the oxygen present, when the organisms use this freshly available material. The resulting depletion of oxygen can lead to the death of fish and river fauna. Eutrophication can lead to the rivers being physically chocked by the proliferation of algal and plant species, and it is probable that the most general outcome will be a gross change of the river ecology both in terms of the species diversity and numbers. A more extreme example of an ecological change is the case of release of toxic discharge where, in worst case situations, the complete eradication of aquatic life may result. In addition to chemical interactions, physical effects can also change river ecology, an example being the presence of settleable inert solids. Water discharges of elevated temperature may also radically change the fauna and flora of a river.

trickle filters, biological contactors and dispersed growth reactors

In this chapter we will describe the aerobic treatment of waste water. We will begin by defining pollution in the terms commonly used in waste water treatment. We will then go on to give an overview of the components of a waste water treatment plant giving brief details of the unit operations. With this overview, we can then consider the aerobic treatment of waste water in greater detail. We will consider the design and operation of fixed film reactors especially trickle filters and rotating biological contactors. We will then examine the operation of dispersed growth reactors with special emphasis being placed on activated sludge processes.

Subsequently we will examine the microbiological processes which occur in the aerobic treatment process. In the final part of the chapter we will discuss the control and monitoring of the aerobic treatment process.

4.2 Key measures in defining pollution

There are many different measures used to define pollution in waste water. Most of these relate to the amount of organic matter suspended or dissolved in the water and to the amount of key inorganic nutrients (especially nitrogen and phosphorus). In particular circumstances, particular chemicals may also be monitored. Here we will concentrate on the main methods used to monitor the level of pollution in waste waters. These are the measurement of biochemical oxygen demand (BOD), the chemical oxygen demand (COD), the suspended solids, the ammonical nitrogen content and the level of phosphates.

Before we consider these, however, it should be realised that sewage consists of approximately 99.9% water and thus the amount of material suspended or dissolved in waste water represents only a small fraction of the total amount of material that needs to be handled.

4.2.1 Biochemical oxygen demand (BOD)

The biochemical oxygen demand is the amount of dissolved oxygen required by micro-organisms for the aerobic degradation of organic matter present in the waste water.

The oxidation of organic material by bacteria in a waste water sample results in a reduction in the concentration of dissolved oxygen. The amount of oxygen used after an incubation period of (usually) 120 ± 1 h at 20°C in the dark in a fixed volume of a filled aerated sample is a measure of the amount of biologically oxidisable material present. To increase precision, a series of dilutions are used to ensure that oxygen is not completely depleted (which might be the case if the concentration of organic material in the initial sample was too high).

nitrification

The assay, however, is not just simply a matter of inoculating and incubating dilutions of the waste water and determining the change in oxygen concentration. Other materials, especially ammonium, present in the water may also be oxidised by micro-organisms. If the concentration of NH_4^+ is high this can lead to appreciable consumption of oxygen and thus produce significant errors in the determination of the organic loading in water. The process of oxidation of NH_4^+ is called nitrification and is mediated by two distinct groups of chemoautotrophic bacteria. *Nitrosomonas* oxidises ammonia to nitrite and nitrite is oxidised by *Nitrobacter*. These two act syntrophically; in other words the product of one serves as a growth substrate for the other. We can, therefore, represent nitrification as:

$$NH_4^+ \xrightarrow[\textit{Nitrosomonas}]{O_2} NO_2^- \xrightarrow[\textit{Nitrobacter}]{O_2} NO_3^-$$

use of
nitrification
inhibitors

In conducting a BOD determination it is, therefore, usual to use inhibitors of nitrification. Allyl thiourea (ATU) is frequently used especially when incubations are conducted over an extended period (8-10 days).

The BOD test often takes 5 (sometimes more) days to obtain the results. More rapid assays have been developed but these do not replace the fundamental usefulness of this assay.

4.2.2 Chemical oxygen demand (COD)

The chemical oxygen demand (COD) is a measure of the total amount of chemically oxidisable material present in the water.

use of
potassium
dichromate

In this assay a known excess of acidic potassium dichromate is refluxed for 2 h with an aqueous sample containing organic matter. The remaining unreacted dichromate is measured by titration with ferrous ammonium sulphate to determine the amount of oxidisable matter originally present. Oxidation of most of the organics is achieved irrespective to whether the organic matter can be degraded biologically. The test is widely used and, for a given effluent, it is possible to develop statistical relationships between BOD and COD values.

| SAQ 4.1 | Why is the COD value for a given effluent always higher than the BOD value. |

4.2.3 Suspended solid matter

suspended solids

Effluents containing significant amounts of suspended solid matter can exert a long term oxygen demand on a river. They may form a blanket on the river bed close to their point of discharge, changing the distribution and exchange rates of both gas and soluble molecules under the surface. Measuring the concentrations of suspended solids is, therefore, important.

This is normally done simply by allowing the particulate matter in a sample of the water to settle out and to measure the volume of the settled material. The alternative is to filter the solids from a sample of the water.

4.2.4 Ammonical nitrogen

ammonical nitrogen

Ammonical nitrogen is toxic to fish in its unionised form, (NH_3) and the distribution of ammonia in this form is affected by the temperature and pH of the water. Ammonia is removed to a limited extent by ammonia and nitrite oxidising bacteria (the nitrifiers) naturally present in water courses. If a discharge contains appreciable amounts of ammonia, then the oxygenation demand due to nitrification can exceed that of the BOD. Ammonia can also interfere with the disinfection of water by chlorination.

Ammonical nitrogen may be assayed by conventional chemical assays. Thus by making a sample of the water alkaline the ammonium ions present are converted to ammonia. Thus:

$$NH_4^+ + OH^- \rightarrow NH_3 + H_2O$$

The ammonia is volatile and may be distilled from the water by gently warming and trapped in a receiving vessel. The trapped ammonia is either then determined by titration or by using a chromogenic reagent (for example Nessler reagent). The colour produced is measured spectrophotometrically and compared to calibration curves. In some instances, the use of ion specific (ammonium specific) electrodes may be used in much the same way as pH electrodes are used to determine pH. Although, in principle. these are simple to use, they are subject to errors especially arising from interference by other materials in the water.

∏ Do you think that ammonium electrodes should be used directly with the waste water or with ammonia distillates from the waste water? (Give reasons for your choice).

It is most sensible to use the electrodes to determine the ammonia in distillates for two main reasons. Firstly there is less contamination by other materials which may interfere with the sensitivity of the electrode. Secondly, the concentration of ammonia (ammonium) in the distillate can be adjusted to fall within the ideal range of the electrodes sensitivity. You could however argue that if the ammonium concentration of the waste water was in the appropriate range, it would be simpler to use the electrode simply with the waste water which would remove the need to carry out the distillation step. The actual choice of method depends on a variety of local factors including the actual composition of the waste water and the availability of equipment and trained staff.

Another measure is $N_{Kjeldahl}$ which measures both ammonical nitrogen and organic nitrogen (typically proteins).

4.2.5 Phosphates

The presence of phosphates in waste water is also important in the monitoring and treatment of waste water. In general, the concentrations of phosphates in waste water have increased over the past few decades primarily as a result of increased use of biodegradable detergents containing polyphosphates. If these phosphates are allowed to be discharged to receiving waters, they generally encourage the growth of algae and plants. For this reason, water treatment processes often incorporate a process specifically designed to remove phosphate. The measurement of phosphates in waste water and in treated water is usually conducted using conventional chemical assays involving the formation of coloured complexes that may be measured spectrophotometrically.

4.2.6 Assessing the effects of discharge

biological
indicators

There are many other ways of assessing the effect of a discharge on a receiving water course, for instance biological indicators of pollution can be used. This is basically a method involving sampling of the river fauna. The relative abundance of certain organisms in relation to other specified organisms in the community and the nature of their habitat can be an excellent indicator of pollution. This aspect is beyond the scope of this text and the reader is referred to Suggestions for Further Reading at the end of this text for further details.

20/30 standard

However, our primary concern is to understand the parameters defining pollution in order to appreciate the aerobic treatment of waste waters. A common standard for permissable effluent composition is known as the 20/30 standard, representing a maximum 20 mg/l BOD and 30 mg/l suspended solid concentration.

4.3 Outline of effluent treatment plant

The general principles are that an efficient effluent treatment plant should reduce organic matter, suspended matter and potential inorganic nutrients thus reducing the pollution effects described in Section 4.2.

We can identify five key stages in the treatment of waste water. These are:

- preliminary treatment; mainly to remove grit, heavy solids and floating debris;

- primary treatment; to remove a substantial portion of the suspended matter;

- secondary treatment; an aerobic or anaerobic (see Chapter 5) treatment step in which oxidisable organic material is removed by micro-organisms (bio-oxidation);

- tertiary treatment; to remove specific materials (for example, ammonia, phosphates);

- sludge treatment; designed to render safe and dispose of organic materials and organisms sedimented in other stages of the treatment processes.

We will discuss each of the first four stages in this scheme in this chapter. The treatment of sludge mainly involves anaerobic digestion and we will examine this in more depth in Chapter 5.

Figure 4.1 contains an outline of the processes involved in a typical waste water treatment works, and Figure 4.2 and 4.3 contain aerial photographs of the two main types of secondary (aerobic) treatment processes, namely that using trickling filter beds and that using activated sludge tanks. An appreciation of the scale of these processes can be obtained by noting the cooling tower seen in Figure 4.2 which is part of a power station near to the waste water treatment plant.

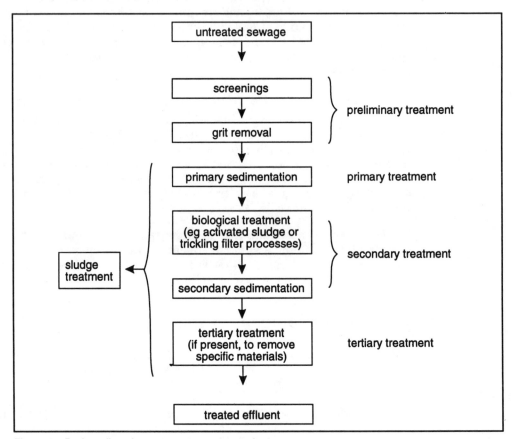

Figure 4.1 Basic outline of a waste water treatment plant.

trickling filter beds

associated humus tanks

Figure 4.2 Aerial photograph of trickling filter beds.

associated sludge tanks

associated humus tanks

trickling filter beds

Figure 4.3 Aerial photograph of activated sludge vessels (courtesy of Yorkshire Water).

Identify each of the following statements as being either true or false. If a statement is false briefly explain why.

1) The temperature of a waste water discharge may have an effect on the biological composition of the river it feeds.

2) A single distinct group of bacteria convert ammonia dissolved in waste water to nitrate.

3) The addition of allyl thiourea to waste water is likely to reduce the amount of oxygen consumed during a BOD determination.

4) The 20/30 standard for effluent composition relates to a maximum permissable BOD value and a maximum permissable suspended solids concentration.

4.3.1 Preliminary treatment

removal of grit, heavy solids and floating material

Preliminary treatment consists of processes used to remove grit, heavy solids and floatable material from waste water. These processes utilise grit settling, course screening (bar racks), medium screening and comminution or grinding. This will protect pumps, valves and pipelines from damage or clogging in the subsequent processes.

4.3.2 Primary treatment

Primary settlement of sewage removes up to 70 per cent of suspended matter and 40 per cent of the BOD by flocculation, adsorption and sedimentation. There are basically three types of settlement tank: upward flow, horizontal flow and radial flow.

The principle used in these primary settling tanks is the same. The flow rate of the waste water is slowed down and the heavier solids settle to the bottom of the tank whilst buoyant materials float to the top. Let us consider one design in a little more detail. A primary circular clarifier (radial flow) is shown in Figure 4.4.

Figure 4.4 Photograph of settlement tanks (courtesy of Yorkshire Water). The tanks shown are of the radial flow type and are sometimes referred to as circular clarifiers.

In the type of sedimentation tank shown in Figure 4.4, the bottom is in the form of a shallow cone (typical slope of 2.5 cm/meter). The conical bottom is equipped with a rotating mechanical scraper that removes sludge to a central hopper. The influent enters through a feed system near the centre of the clarifier which distributes it radially near the top. A peripheral weir overflow system carries the effluent out of the tank.

Settleable solids fall out of suspension by gravity during 15 to 45 minutes. A chemical coagulant may be added to enhance settling. Floating scum is trapped inside a peripheral scum baffle and removed into a scum discharge box.

4.3.3 Secondary treatment

The main theme of this chapter is to develop an understanding of the processes involved in the aerobic biological treatment of waste water. This involves two broad types of processes.

- fixed film (biofilm) processes;

- homogeneous aerobic processes (dispersed growth processes).

Fixed film processes include trickling filter beds, rotary biological contactors, fluidised beds and submerged filters. Homogeneous processes are most notably the activated sludge process and the more intensive waste paper treatment system, the deep shaft process.

Some of these processes will be covered in detail in later sections.

4.3.4 Tertiary treatment

Tertiary treatment is a further treatment of effluent from the biological stage to remove suspended matter thereby reducing BOD further or to remove specific materials (for example phosphate). It is used when more stringent standards are required. Since these are not generally biological in nature we will not discuss them in detail here.

The methods used include:

- lagooning to give further settlement;

- grass plot irrigation by means of channels or spray guns. Plots are underdrained and slope 1 in 60 to 1 in 100 towards collection channels;

- microstraining or microscreening. These devices work as large filtration devices. Particulate matters trapped as the water is passed through the strainer or screen. The trapped materials can be removed by back-washing. Mesh apertures may be 23-35 μm and so are only used for effluents of higher quality. Continuous backwashing with 2-5 per cent of the filtrate is necessary;

- slow sand filters. These are expensive in capital and operating costs;

- rapid downward flow filter. These can be loaded at 250 m^3 m^{-2} d^{-1} (cubic meters per square meter per day) at maximum flow. These are provided with backwashing facilities usually and the washwater containing grit must be returned to the inlet upstream of grit extraction;

- upflow sand filter. Loadings may be up to 400-900 m^3 m^{-2} d^{-1} depending on operational parameters;

- upward flow clarifiers. Pea gravel or wedge wire may be used at loading rates up to 42 m^3 m^{-2} d^{-1} for biological filter effluent or 30 m^3 m^{-2} d^{-1} for activated sludge effluent;

- phosphate removal. Usually achieved by precipitating the phosphate using lime or alum. The insoluble phosphates are removed in settling tanks (clarifiers).

Now that we have given you an overview of waste water treatment, we will move on to consider, in more detail, the aerobic treatment process.

4.4 Aerobic biological treatment of waste water

There are a number of different biological operations available to reduce the BOD of an effluent. They utilise different mechanisms to ensure that there is always a residual oxygen concentration. The microbial population may be classified as either fixed film or dispersed growth in these processes.

4.4.1 Fixed film processes

These represent the oldest form of waste water treatment system and include trickling filters and rotating biological contactors, which are considered individually in the following two sections.

4.4.2 Trickling filters

Reactor plan and aeration

Trickling filters have been in use for over 70 years. The reactor may be of rectangular or circular plan and is filled with permeable material. A rectangular trickling filter bed is shown in Figure 4.5.

traditionally rock and slag used as supports

Filter media must provide a large surface area for growth of a slime layer (the biofilm) and also a large void for oxygen transfer and liquid flow. Air distribution is by circulation through void spaces in the media. This is encouraged by temperature differences between air and waste water causing an up-flow of air through draught tubes located in the sides of the filter. Crushed rock or blast furnace slag (25-100 mm diameter) are traditionally used although this limits filter depth to 3.0 m. A more recent development is the use of lightweight plastic media in the treatment of high strength industrial wastes. Here filter depths may be up to 12 m.

Waste water is distributed mechanically over the media and a microbial film which develops over the surface of the media is responsible for the removal of organic material as the sewage passes down through the bed. To ensure efficient operation there must be an even distribution of settled sewage over the filter surface. In circular beds influent enters at the centre of the bed and passes into radial distributor arms above the bed surface. Discharge of sewage through sparge holes on the arms usually provides the necessary force to drive the arms around the central column. On rectangular filters the distributor is driven forward and backwards, the liquid being siphoned from a channel running along the length of the bed, as can be seen in Figure 4.5.

distributor arms with sparge
holes releasing waste water

Figure 4.5 Rectangular trickling filter beds (courtesy Yorkshire Water).

channel delivering waste water
to distributor arms

Ponding and sloughing off

Periodic dosing, wetting followed by a rest period, is necessary for successful filter operation. The condition to be avoided is an accumulation of solids at the filter surface such that voids become blocked and the filter 'ponds' (pool of effluent apparent at the surface).

Ponding often arises during winter months when low temperatures suppress activity of grazing organisms and there is an increase in film growth. Ponding reduces airflow in the bed and effluent quality. Growth may be limited at the top surface of the filter by reducing the dosing frequency from 0.25-2 minutes to once every 5-12 minutes. This not only produces a better distribution of nutrients deeper into the bed but also increases the time during which diffusion of nutrients from liquid to biofilm can take place. This effect can be such that growth of the film can in this way be limited by lack of nutrients. It is important that the periodicity of dosing can be varied to compensate for the increased microbial activity during warmer summer months.

Ideally a continual cycle of film growth followed by death and detachment from the media occurs thus preventing excessive build up of film and ponding. This 'sloughing off' produces a sludge which is normally carried away in the filter effluent. There is, therefore, a requirement for a sedimentation tank (humus tank) to settle out and remove solids from filter effluent (humus sludge). Sloughing is most pronounced at the change from winter to warmer seasons and may result in a short loss of efficiency of the filter until the new biofilm layer is established (which usually takes place rapidly).

<table>
<tr><td>SAQ 4.3</td><td>
1) Name the two main types (ie classifications) of aerobic processes involved in the secondary treatment of waste water.

2) Is grass plot irrigation a preliminary, primary, secondary or tertiary treatment process.

3) Is ponding on a trickling filter bed desirable or undesirable. Briefly explain your answer.
</td></tr>
</table>

Design of filters

A common approach for designing filters is to use the organic loading rate based on settled sewage.

organic loading rates

Settled sewage is, of course, sewage that has passed through the primary treatment process. The organic loading rate is a feature of the filter design. The filter may be used as a single passage device or part of the water being treated may be recirculated through the filter. Other designs include what are called alternating double filtration devices (ADFs) or two stage (primary and secondary) filters. The rates at which these will successfully remove organic materials is variable. The organic loading rates are a measure of the organic loading that can be treated by these filters. Organic loading rates are the amount of organic material (usually expressed as kg BOD) that can be treated per m^3 of filter volume per day. Thus the organic loading rates are usually expressed in terms of kg BOD m^{-3} d^{-1}.

∏ Do you think a single passage will have a higher or a lower organic loading rate than a recirculation type filter?

You should have concluded that the recirculation type filter will probably have a higher organic loading rate as the water being treated is passed more than once through the filter. In practice, however, the organic loading rate of a recirculation filter is only slightly higher than a single passage filter. Some examples of organic loading rates are given in Table 4.1.

Process	Organic loading rates (kg BOD m^{-3} d^{-1})
Single	0.06-0.12
Re-circulation	0.09-0.15
Alternating double filtration (ADF)	0.15-0.26
2 Stage filtration primary	1.56-2.29
secondary	0.04-0.12

Table 4.1 Examples of organic loading rates of various processes using trickle filter.

∏ In Table 4.1 we show a range of organic loading rates for each type of filter. What factors will influence the organic loading rates of each filter type?

The organic loading rate of a particular process will be influenced by such factors as the size and nature of the physical support material. These will influence such factors as oxygen diffusion rates and the attachment of biomass both of which are critical to the amount of organic material that will be oxidised. Temperature will also have an effect.

We would anticipate for example that the organic loading rates would be lower at low temperatures.

If we know the organic loading rate for a particular filter type, we can work out the volume (V) of the filtration medium that would be required to treat a particular volume of settled sewage. This can be done in the following way. The load applied to the filter (usually referred to as the applied load) can be calculated from the daily flow (ϕ) and the BOD (Li) of the flow. Thus the applied load = ϕ Li. (Note Li = input load).

The volume of filtration medium required is thus:

$V = \phi \, Li/R$ where R is the organic loading rate of the filter.

Π What are the applied units?

The units are usually kg BOD d^{-1}. (Note daily flow (ϕ) is in $m^3 \, d^{-1}$ and the BOD in kg m^{-3}).

Let us try a sample calculation.

Π The effluents produced by two factories are treated separately by trickle filters. Factory A produces 84 $m^3 \, d^{-1}$ effluent with a BOD value of 1.2 kg m^{-3}. Factory B produces 29 $m^3 \, d^{-1}$ effluent with a BOD value of 3.9 kg m^{-3}. Which of the factories will require the largest trickle filter and what will the volumes of these filters need to be. (Assume that the factories will install simple, single passage filters).

The load produced by factory A is:

84 x 1.2 kg d^{-1} = 100.8 kg d^{-1} (from applied load - ϕ Li)

The load produced by factory B is:

29 x 3.9 = 113.1 kg d^{-1}

Thus factory B would in theory require a slightly larger filter. Since the organic loading rate that can be used by single passage trickle filters is:

0.06-0.12 kg $m^{-3} \, d^{-1}$ (see Table 4.1), then the size of the trickle filters needed by the factories are:

Factory A

$$\frac{100.8}{0.12} \rightarrow \frac{100.8}{0.06} m^3 = 840\text{-}1680 \; m^3$$

Factory B

$$\frac{1.13}{0.12} \rightarrow \frac{113}{0.06} m^3 = 942\text{-}1883 \; m^3$$

Thus in practice, there is considerable overlap between the filter requirements of the two factories.

Other techniques are used as well to calculate the size of filter units required.

We will briefly describe a few of these but it is beyond the scope of this text to derive them all. Most of these have either been derived empirically from experimental data. If you are interested in examining this aspect of waste water treatment in more detail you will find some useful references in the section 'Suggestions for Further Reading' at the end of this text. A more theoretical approach to fixed film reactor design is given in the BIOTOL texts 'Operational Modes of Bioreactors' and 'Bioreactor Design and Product Yield'.

We will begin our brief description of the relationships used to calculate the size of filter units required using the National Research Council formula (1946) still in use in the UK.

This formula is:

$$\frac{Li - Le}{Li} = \frac{1}{1 + 0.44\,(\omega/(fV))^{0.5}}$$

where Li = BOD of influent, Le = BOD of final effluent, ω = BOD load (kg d^{-1}), V = volume of filter media (m^3), f is a re-circulation factor. For straight through system $f = 1$. If recirculation is used it has been shown that:

$$f = \frac{1 + \alpha}{(1 + 0.1\alpha)^2}$$

in which α is the recirculated flow: settled sewage flow ratio.

Yet another equation is the first basic order equation : $Le = Li\,\exp\,(-KS/Q)$. In this equation S = specific surface of medium (m^2 m^{-3}). Q = hydraulic loading rate (m^3 m^{-3} d^{-1}), and K = first order rate coefficient. This equation has been expanded by Bruce and Merkens, (1973) to include effects of temperature:

$Le = Li\,\exp\,(-K\theta^{(T-15)}S/Q)$ in which θ = the temperature coefficient and T is temperature.

Pike has made more modifications using multiple regression analysis of experimental data:

$Le = Li\,\exp\,(-K\theta^{(T-15)}S^m/Q^n)$

θ = temperature coefficient and the constants m and n relate the physical nature of the surface and its wetability. It must be noted that this method is limited to straight through filtration since values of K, m and n vary with the degree of sewage treatment.

Let us try out some of these relationships.

SAQ 4.4

1) Using the National Research Council Formula given in the text, what happens to the BOD of the filter effluent, if the volume of the filter bed is increased?

2) Using the National Research Council Formula, what will be the BOD of the final effluent if the BOD load is 50 kg d^{-1}, the volume of the filter bed is 200 m^3 and the BOD of the influent is 0.5 kg m^{-3} if a single passage filter is used?

3) If half of the effluent from the process described in 2) is recirculated through the filter, what is the BOD of the final effluent from the process?

4) Does recirculation lead to a substantional reduction of the BOD of the final effluent in the processes described in 2) and 3)?

High-rate filtration

synthetic supports developed to treat high strength industrial wastes

The problem of ponding has been avoided by using plastic media in a form that would give a larger voidage than is possible using rock packings. As has been previously mentioned these supports overcome the limit in filter bed depth imposed by rock filters and the tendency of the latter to 'pond' when conditions of overload promote a heavy growth of biofilm. Synthetic supports were originally developed for the treatment of high strength industrial wastes. The specifications which were sought were that they should remove more BOD at high loading rates and have a sufficiently open structure to facilitate oxygen transfer and reduce the possibility of filter blockages in comparison with their traditional counterparts. They were also designed to support the wet weight of the biofilm even when stacked up to 12 m depth, to be resistant to biological and chemical degradation and to be light enough to facilitate the construction of filter beds when these were initially installed. Their application is often found in treatment plants for a specific industrial waste. The reason for the continued prevalence of traditional supports in many waste water treatment plant lies simply in the age of the equipment. In a plant serving a large city it is possible that the rock media in the trickling filter beds is still the same support that was in use a generation ago.

Examples of high-rate filtration media currently in operation in Europe include modular sheet packing and random packing. It is probable that where plastic media are installed, the actual design of the structure of the plastic supports will be chosen to optimise the removal of BOD by using prior information about the treatability of the waste source in question and the characteristics of the plastic which will be employed.

4.4.3 Rotating biological contactors (RBC)

The aerobic secondary treatment process using rotating biological contactors works on a similar basis as the trickling filter bed. The microbiological film is adhered to the surface of a set of discs 2-3 m in diameter (see Figure 4.6). The discs are rotated by a motor at speeds of typically between one or two revolutions per minute, at right angles to the flow of settled sewage. The film of microbes on the disc will usually grow to a few millimetres thickness during operation. This biofilm continually cycles between contacting the nutrients dissolved in the waste water and the atmosphere where aeration is effectively achieved. The discs can be made from wood, metal or plastic and located separately or in groups where they form a mesh-like support for the biofilm. The process has two main advantages, a low land requirement (due to its compact size) and very low maintenance demands. It additionally has an excellent stability to changes in hydraulic and organic loading rates.

∏ Bearing these features in mind, where would you anticipate finding RBCs?

These features contribute to its use in isolated regions serving small communities where on-site maintenance is difficult to provide. Also where the waste water demands of a large city are increased by the creation of new domestic or industrial sites a RBC may be installed near to these new buildings to partially treat their effluent prior to discharge into the existing city treatment plant, reducing the increase on loading of the latter.

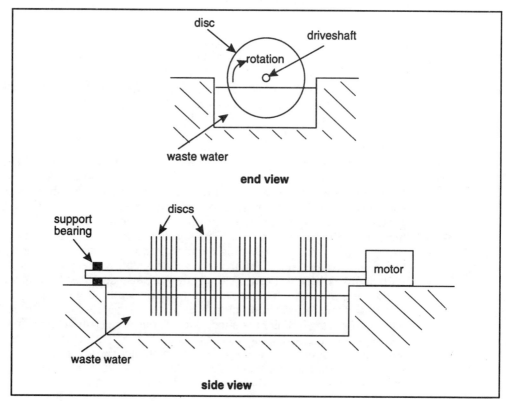

Figure 4.6 Schematic diagram of a rotating biological contactor.

4.4.4 Dispersed growth processes

The activated sludge process

This activated sludge system is a very popular method for carrying out biological oxidations of liquid wastes on a large scale. It comprises a large vessel with provision for aeration. This is constantly provided with organic matter and a mass of micro-organisms which grow in flocs. The flocs are constantly being washed out of the reactor to a sedimentation tank, where they settle under gravity. A fraction of this settled sludge is recycled back into the aeration tank to provide sufficient biomass to maintain the microbial population. A schematic diagram of the process is given in Figure 4.7.

There exists a wide number of variations on this basic process which can be modified to suite the nature of the incoming liquid waste. These include the addition of extra processing tanks where the effective removal of particular contaminants such as nitrogen or phosphorous is sought. These examples are specific, however, and more generally the main differences vary only in their method of oxygen supply.

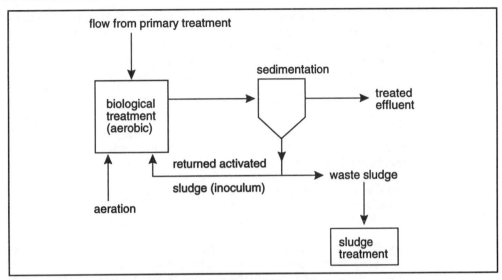

Figure 4.7 Schematic diagram of the basic activated sludge process.

The mass transfer processes and process engineering aspects of getting oxygen into solution in large 'reactors' are examined in detail in the BIOTOL text 'Operational Modes of Bioreactors' so we will not consider these aspects in detail here. However, we point out that mechanical aeration is the most common method and these systems transfer oxygen from the atmosphere by agitation of the liquid surface. Figure 4.8 contains a photograph of a simplex cone (which is driven by a large motor above) which aerates and aids mixing of the vessel contents. The immersion depth of the cone within the liquid must be maintained accurately to effect efficient aeration. This system can transfer over 2 kg of oxygen into the liquid per kilowatt hour provided to the motor. This efficiency being typical of other methods employing mechanical aeration.

mechanical
aeration

simplex cone

motor

Figure 4.8 Mechanical aeration of an activated sludge tank.

immersed cone

Also diffused air systems can be used to transfer oxygen into the liquor. There are two general types: fine and coarse bubble systems. Fine bubble systems produce a fine stream of bubbles by forcing the air through porous ceramic rocks. This form of system is generally more efficient in energy by up to a factor of two in aeration in comparison to a mechanical system. There is a disadvantage in that the pores in the bricks can become blocked, and the equipment often requires regular maintenance. This problem is overcome in the second type of diffused air system where the air is forced through small holes in pipes transversing the vessel floor; the coarse bubble system. In this latter case there is less efficient oxygen transfer at the interface of these larger air bubble surfaces and the liquid, their energy requirements being comparable with that of mechanical aerators in terms of efficiency.

Finally, pure oxygen can be introduced directly into the vessel as a fine stream of bubbles produced by an expansion jet and forced under pressure into the vessel. A well established example is known as the VITOX process.

To design an effective activated sludge treatment plant several parameters must be considered. Predictably the main consideration will be the organic loading rate. If this is not adequately calculated the effluent quality will be poor as a result of inappropriate sludge settling rates and the plant performance will be far from optimal. The organic loading rate is defined as the amount of BOD applied per unit volume of the vessel.

organic load rate (kg m^{-3} d^{-1}) =

$$\frac{\text{incoming flow (m}^3\text{ d}^{-1}) \times \text{BOD of incoming liquid (kg m}^{-3})}{\text{vessel volume (m}^3)}$$

This relationship is entirely analogous to the one we used when discussing trickle filters.

During a 24 hour period the organic loading rate which the plant will have to process will vary considerably. This will arise as a result of changes in the levels of discharge from industrial sources and also from the much reduced domestic discharge overnight. Therefore, the BOD value used in this equation will be a figure weighted to ensure adequate vessel function. This parameter will be calculated from many BOD values obtained throughout a typical day. Alternatively 24 samples of equal volume taken at 1 hour intervals may be pooled and the BOD value for this 'pooled' sample determined.

The time during which the effluent is aerated in an activated sludge tank also needs to be monitored. This residence time is simply calculated by dividing the reactor volume by the total incoming flow. Thus:

$$\text{Residence time (d)} = \frac{\text{reactor volume (m}^3)}{\text{total daily flow(m}^3\text{ d}^{-1})}$$

If this parameter is too low then the liquor will be poorly oxygenated and little BOD ultimately removed.

Perhaps the most useful loading parameter is the sludge loading rate. This is a measure of the ratio of the microbiological input (derived from returned settled flocs) to the nutrients, which is represented by the flow of influent of a particular BOD into the vessel. It is expressed as:

$$\text{sludge loading rate (ie nutrient/microbe ratio)} = \frac{\text{BOD of incoming liquor (kg m}^{-3}) \times \text{influent flow (m}^3\text{ d}^{-1})}{\text{reactor volume (m}^3) \times \text{reactor solids (kg m}^{-3})}$$

Note that the 'reactor solids' is a measure of the amount of microbes present in the reactor.

This parameter is particularly valuable as it can be readily controlled by altering the quantity of returned flocs. It is usually manipulated to effect the largest achievable reduction in BOD within the vessel, but if alternatively an effective degree of nitrification is sought the desired result may also be achieved by control of the sludge loading ratio.

∏ Why might it be desirable to achieve a high degree of nitrification?

Remember that waste waters may contain NH_4^+ and/or NH_3. Ammonium is quite toxic. Nitrification, carried out by *Nitrosomonas* and *Nitrobacter* (see Section 4.2), converts these into NO_3^- which is much less toxic. This process is aerobic and will, if sufficient oxygen is available, take place in activated sludge processes.

If elevated nitrification is sought the sludge age will have to be high as the bacteria which achieve nitrification have low growth rates (we will discuss this more fully in a later section). The sludge age is given by the equation:

$$\frac{\text{sludge age (d)}}{} = \frac{1}{(\text{sludge yield} \times \text{organic loading rate}) - k_d}$$

where:

the sludge yield is: kg sludge produced/kg BOD used; and k_d: the decay coefficient.

It is apparent from this equation that an increase in loading rate will decrease the sludge age. This is because the organic material enables the further growth of biomass.

SAQ 4.5

1) How may the sludge loading rate be maintained at a constant value if the influent flow rate into an activated sludge process is increased?

2) Nitrification is shown to only take place to a limited extent in an activated sludge process. It does not appear that the supply of oxygen is limited in this process. What steps might be taken to improve the amount of nitrification taking place in the unit?

Removal of the microbial flocs produced in the aeration tank

After treatment in the aeration tank the waste water is allowed to stand so that the flocs settle to the bottom. In reality this process does not occur in totally still water but rather in ponds of a large surface area with very low flow rates. Figure 4.9 shows a photograph of such a process and in this vessel adequate settling is aided by the surface water having to pass over comb-like protrusions which can be clearly seen. The solid wastes (ie sludge) can be further treated by anaerobic digestion in fermentation tanks or other disposal methods which are not primarily involved with the aerobic processes contained in this chapter (see Chapter 5). Some of the sludge is fed back into the aeration tank.

Figure 4.9 Humus tank.

4.5 The microbiology of aerobic waste water treatment

In biological waste water treatment the capability of micro-organisms to mineralise a great number of different (in)organic compounds is exploited. It should be realised however that not all components in waste water (ie some detergents) can be completely degraded. Microbiological processes in waste water treatment are fundamentally similar to naturally occurring processes: micro-organisms use specific pollutants as source of energy and for the formation of cell mass. The main products of aerobic treatment processes are therefore CO_2, H_2O, biomass and nitrogen compounds such as ammonium and nitrate.

∏ One of the disadvantages of aerobic waste water treatment is that it produces much more sludge than the anaerobic process does. Can you explain why?

In aerobic oxidation processes degradable organic matter is totally dissimilated to CO_2 and H_2O, while in anaerobic processes the end products are CO_2 and CH_4 (Chapter 5). Much more energy is generated if the energy source is completely oxidised. Therefore, aerobic metabolism of substrates has a higher yield. Typically aerobic processes may yield 0.5-1.5 kg biomass for every kg BOD, whereas anaerobic processes yield only 0.1-0.2 kg biomass per kg BOD.

Π   Can you distinguish one important difference between microbial processes in
    natural environments and in waste water treatment systems?

In waste water the concentration of (polluting) (in)organic matter is much higher than
in natural aquatic habitats. To fulfil the aims of biological water treatment, ie to
transform pollutants into compounds innocuous for the environment within an
acceptable period of time, one should optimise process conditions in order to intensify
microbial activities.

Π   Give some examples of process optimalisation in aerobic water treatment
    systems.

In trickling filters, optimilisation is reached by the use of filter media which provide a
large surface area for biofilm formation and also a large void for oxygen transfer and
liquid flow. In the activated sludge process microbial decomposition of organic matter
is intensified by the partial return of settled sludge and the application of different
aeration systems to provide sufficient oxygen.

4.5.1 Microbial processes and important micro-organisms

The heterogeneous chemical nature of domestic waste water provides an excellent
growth medium for a multitude of micro-organisms.

Π   Which nutritional type of micro-organisms do you expect to be predominant in
    sewage water treatment sites? If you are not familiar with these types, you should
    refer to the BIOTOL text 'In Vitro Cultivation of Micro-organisms'.

Probably all nutritional types have their representatives among the microbial flora
encountered in filter beds and activated sludge. Most abundant however are
heterotrophic forms, which use organic matter as energy and carbon source. In the
complete mineralisation of complex organic compounds different heterotrophic species
act in succession. Other important groups are:

*   lithotrophic organisms. These organisms, to which only bacterial species belong, use
    inorganic compounds (eg ammonium and hydrogen sulphide) as their source of
    energy. Most species grow autotrophically (CO_2 as a carbon source), but
    heterotrophic species are also present.

*   facultative autotrophic and mixotrophic organisms. An example of the first type are
    some algae which grow photoautotrophically in daylight, but grow
    chemoheterotrophically in the dark. Mixotrophic bacteria are
    chemolithoautotrophic organisms which can adapt to growing on an organic source
    of carbon. An example of the latter type is *Beggiatoa*, a genus of sulphur oxidising
    bacteria.

*   phototrophic organisms. To this category belong algae and cyanobacteria. However,
    this group is less important since light is absent inside the filter bed and the activated
    sludge tank. They may be present on the surface of the filter bed and fix inorganic
    nutrients into biomass by their autotrophic growth.

complex
microflora
develop in
treatment
processes

The diversity of participating micro-organisms ensures a complex ecosystem will exist in secondary and also tertiary treatment processes. Subtly different ecosystems can be found to exist within one processing step, an obvious example being found in a trickling filter bed where the species found near the surface which intimately reacts with both the air above and sunlight are very different from those deep in the bed where dissolved oxygen can be highly depleted in local environments. The diversity of micro-organisms which must be considered encompass all the main groups of micro-organisms; namely bacteria, protozoa, fungi and viral species. As mentioned earlier, phototrophic organisms (algae and cyanobacteria) are less important and will not be discussed here.

Bacteria represent the most abundant form of micro-organisms in waste water treatment processes, their presence often being far in excess of 10^{12} (one million million) cells per litre.

Within primary treatment a substantial factor of the total 30-40% BOD removed at this stage is contributed to by heterotrophic bacteria. It should be noted that a large proportion of this will ultimately be degraded by anaerobic digestion of the waste settled during these processes, or by the residues being composted (see Chapter 5).

During secondary (trickle filters and activated sludge) treatment the contribution of specific bacterial species is evident, and at the end of these processes typically 85-90% of the BOD will be removed by these largely aerobic stages. An important bacterium is *Zoogloea ramigera* which secretes a polysaccharide with a mucus-like consistency. This secretion brings about both the attachment of the bacteria to the surface of the bed matrix in trickling filter beds and the adhesion of other bacteria, fungi, algae, insect larvae and nematodes (worm species). The complex community resulting converts organic molecules such as proteins, carbohydrates and amino acids to carbon dioxide and mineral salts of nitrate, sulphate and phosphate. *Zoogloea* also has an essential function in the activated sludge process whereby its slime-like secretion holds aggregates of other micro-organisms together to form flocs. The degradation of organic nutrients by the species in these flocs increases their numbers. As these flocs provide an effective inoculum for further waste water treatment, settled flocs are usually recycled (as described in Section 4.3).

The settling characteristics of the flocs are critical for their efficient removal. Poor settling characteristics of the sewage sludge (a condition called 'bulking') is a serious problem in waste water treatment caused by proliferation of specific filamentous bacteria such as *Sphaerotilus*, *Beggiatoa* and *Thiothrix*. Also filamentous fungi (*Geotrichum*, *Cephalosporium*) and protozoa may play a role.

Some species of bacteria are capable of resisting dramatically severe changes in their environmental conditions by forming spores, a form of structure which allows the organism to exist in an essentially dormant state. Bacteria contributing to waste water treatment which can adapt by this alteration in their morphology are of two main types, *Bacillus* and *Clostridium* species. Accordingly these bacteria can resist large depletions in the presence of nitrogen, carbon and phosphorus (the three elements which are essential in media to support microbiological growth).

Other bacteria which appear consistently to contribute to waste water treatment are those of the genera *Sarcina*, *Staphylococcus*, *Streptococcus*, *Pseudomonas*, *Escherichia*, *Salmonella*, *Shigella*, *Aerobacter*, *Vibrio* and *Desulphovibrio*. Their combined action results in the removal of carbohydrate, protein and fat from the surrounding media.

conversion of
ammonia to
nitrate by
nitrifying
organisms

Bacteria of two chemolithotrophic genera, *Nitrosomonas* and *Nitrobacter*, (as their names suggest) play a central role in the fate of nitrogen in waste water. Sewage effluent is a major source of nitrogen pollution. Ammonia is a toxic species and is undesirable in a watercourse its presence being unwanted if found in substantial concentrations in drinking water. The conversion of ammonia to nitrate by these nitrifying organisms has already been briefly described. Although less toxic than ammonia, nitrate is not desirable in water courses; it leads to eutrophication and high concentrations are undesirable in drinking water. A diagram indicating the pathways where nitrogen can be interconverted is shown in Figure 4.10. Throughout the course of any biological waste water treatment process up to 30% of the total nitrogen will be removed by the microbiological synthesis of cellular structures, including protein and DNA, primarily from the uptake of ammonium.

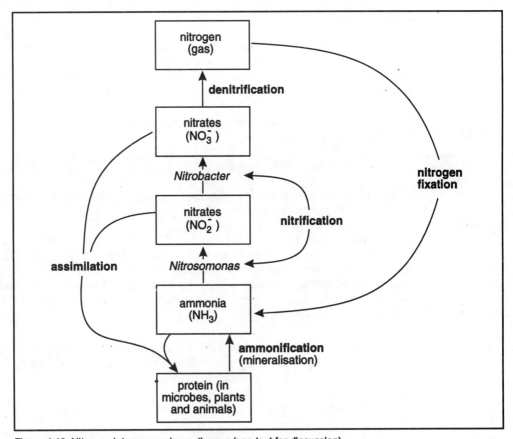

Figure 4.10 Nitrogen interconversion pathways (see text for discussion).

nitrogen and
eutrophication

The discharge of substantial nitrogen levels in effluent is associated with the formation of algal blooms and severe eutrophication of the receiving watercourse. In such cases additional removal of nitrogen must be introduced into the sewage treatment works. As we have seen members of the genus of *Nitrosomonas* catalyse the oxidation of ammonia to nitrite whereas *Nitrobacter* further oxidises nitrite to nitrate. However, both these groups of bacteria are exceptionally slow to grow and multiply. This information is of importance in the design of waste water treatment regimes. If a high degree of ammonia removal is required care must be taken to ensure adequate populations of these bacteria exist. In fixed film processes such as trickling filters this involves the limitation of the organic flow rate to less than 0.2 kg m^{-3}. At this loading up to 80% of the ammonia will be removed whereas at twice the loading rate very little nitrification occurs at all. Fixed film processes must also have little variation in flow rate and ensure the media is continuously wetted if *Nitrobacter* and *Nitrosomonas* populations are to be maintained. In activated sludge processes their slow growth rate necessitates the use of return sludge of an increased age where populations of *Nitrobacter* and *Nitrosomonas* have had time to become significant or to use an additional aeration process.

denitrifcation

The final step in this pathway resulting in the removal of nitrogen from the watercourse is denitrification, namely the conversion of dissolved nitrate to gaseous nitrogen. The bacteria which undertake this process include various *Alcaligenes*, *Achromobacter*, *Micrococcus* and *Pseudomonas* species. Their growth, when it occurs, is under conditions where oxygen is absent, such as in environments deep within processing tanks or filter beds which are anaerobic, because denitrification is an anaerobic process. Sometimes denitrification is achieved by recurrent periods of aeration (alternating aerobic and anaerobic conditions). Although denitrification is an anaerobic process it is worthwhile mentioning that not all of these genera are capable of complete conversion of nitrate to nitrogen and other gaseous products can be produced including various oxides of nitrogen.

SAQ 4.6

Make a list of the possible fates of nitrogenous materials which enter a water treatment process.

∏ Are trickle filters likely to produce greater proportion of N$_2$ to other nitrogenous products then are activated sludge processes?

The answer is yes. The conditions in a trickle filter are inhomogeneous and localised areas of anaerobiosis are likely to be generated within the biofilm. This encourages denitrification. Activated sludge processes are usually well mixed and the aim is to maintain reasonable dissolved oxygen concentrations throughout the reactor. This will disfavour denitrification. Nevertheless, denitrification may still occur in less well mixed, poorly aerated parts.

SAQ 4.7

Classify each of the following interconversions of molecules containing nitogen which are produced by the activity of bacteria as denitrification, nitrification, nitrogen fixation or ammonification.

1) $NH_3 \rightarrow NO_2^-$.

2) Protein → free NH_3.

3) $NO_2^- \rightarrow NO_3^-$.

4) N_2 (nitrogen gas) → NH_3.

5) $NO_3^- \rightarrow N_2$.

fungi

Fungal cells are eukaryotic cells and although found in waste water treatment plants their numbers can be low with no single characteristic species routinely contributing to the biological treatment of the water. Where present they are often found as an external biofilm of fungi (sometimes associated with algae) on the surface of flocs (in the activated sludge process) or on the surface of the microbiological film on filter beds. Their contribution may be in the removal of the nutrients nitrogen and phosphorus from the surrounding media.

protozoa   Protozoa are eukaryotic organisms. Some species exist as multi-cellular organisms, however, most types exist as unicellular entities. Protozoa are typically of a relatively large size, their cellular composition being readily seen under a light microscope. They all lack a cell wall and their cytoplasm contains prominent nuclei, mitochondria and vacuoles, all the features characteristic of eukaryotic micro-organisms. Three classes (phyla) of protozoa are found in waste water treatment.

flagellates   *Flagellates* (subphylum mastigophora) are single-celled protozoa that move by means of flagella. The protozoa within this classification are the most abundant type found in aerobic waste water treatment in both the activated and trickling filter processes.

ciliates   The *ciliates* (subphylum ciliophora) form the largest group of protozoa representing more than 7000 currently recognised species. They also represent the most diverse group found in waste water treatment. However, they are present in lower population numbers than the mastigophora. One genus within this classification called *Vorticella* is often found to adhere to flocs in the activated sludge system.

amoebal forms   The last of the three main types of protozoa found in waste water are the *amoebas* (subphylum sarcodina). In essence, an amoeba is a single-celled microscopic organism that floats or creeps through an aquatic environment. An amoeba is motile by means of pseudopodia (literally 'false feet') which serve not only as a means of locomotion but also as a means of gathering food. The envelopment of nutrients by engulfing is known as phagocytosis and is diagrammatically depicted in Figure 4.11. Protozoa also can feed by absorbing nutrients through their plasma membrane. A final nutritional mode which exists in protozoa is predation. This is mainly found in cilliate species which can feed directly on bacteria.

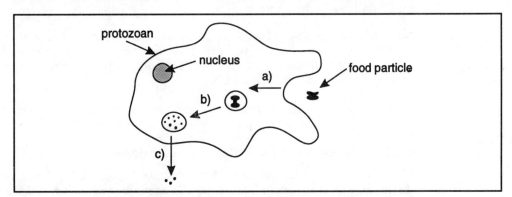

Figure 4.11 Digestion of nutrients by phagocytosis. a) Particle engulfed by pseudopodia and vacuole formed. b) Enzymic breakdown of nutrients within vacuole. c) Undigested remains expelled.

The protozoa are important in water treatment because, by engulfing bacteria by phagocytosis, the biomass is transformed from small particles (bacteria) to larger particles (protozoa).

These, being bulkier, are easier to sediment at the settling stage. The protozoa can also engulf particulate dead organic residues and, by a process called pinocytosis, take up soluble nutrients.

We can therefore imagine a succession of events in the aerobic process.

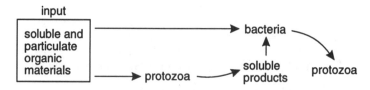

Thus we can visualise a mixed population of protozoa and bacteria taking up and oxidising the organic debris. Then a second population of predatory protozoans engulfing the bacteria.

Some protozoans are sessile and produce stalks (for example *Vorticella* and *Stentor*).

∏   Are these types desirable in activated sludge process?

The answer is no. The production of stalks means that when we come to separate the protozoan (and other micro-organisms) in the settling tank, they will not pack down tightly and produce loose flocs. This is called bulking and it makes it difficult to separate the treated water from the biomass produced in the activated sludge process.

viruses

Viruses are forms of microbes which are clearly differentiated by their complete dependence on host organisms to replicate. For example, a virus called bacteriophage lambda infects the bacteria *Escherichia coli* and can only be grown on a 'medium' made up of a culture of this bacteria.

bacterial numbers may reduce by bacteriophages

Naturally occurring bacteriophages (viruses of bacteria) exist within waste water treatment processes but do not contribute to the treatment process. Where a new virulent bacteriophage enters the waste water supply it may severely reduce the numbers of bacteria which it infects.

Viruses in waste water treatment are however, important in another context. Some are the causative agents of important diseases. It is this aspect we will consider next.

4.5.2 Water-borne microbiological pathogens

Many diseases can be transmitted by microbes which are present in waste water discharges. Most originate from a faecal origin although urine is the source in some bacterial examples.

The importance of waste water treatment processes in controlling the transmission of infectious agents cannot be over emphasised. When water treatment is inadequate, we often hear of outbreaks of a variety of water-borne diseases. The aim of water treatment has historically been aimed at reducing such outbreaks. It is beyond the scope of this text to review this aspect of epidemiology but we will briefly remind you of some of the major types. We have provided some Suggestions for Further Reading at the end of this text if you wish to follow this aspect up in further detail.

Most of the organisms in the human alimentary tract are harmless. For example, most strains of *Escherichia coli* are non-pathogenic and strains of this species are present in large numbers in all humans. Some organisms of the alimentary tract are, however, pathogenic.

Specific bacteria are the cause of typhoid fever, shigellosis, salmonellosis, cholera and many other bacterial species cause varying levels of indisposition by gastro-enteritis. Protozoan derived water-borne disease includes amoebic dysentry, giardiasis and blantidiasis.

Viral diseases include hepatitis, poliomyelitis and gastroenteric disease.

Although a virus will not die in the absence of a nutrient supply (by virtue of its absence of metabolism), species which are human pathogens are relatively fragile and will be degraded by enzymatic and chemical processes rapidly in a treatment plant. The vast majority of microbes present in the aerobic stages of a waste water treatment plant are non-pathogenic, these stages, therefore, ensure the final plant effluent should provide no health hazard in terms of biological discharge.

SAQ 4.8	You are proposing to check whether the water leaving a secondary (aerobic) water processing unit is likely still to be contaminated by human pathogens. Explain how you would attempt to do this.

4.6 Plant control, monitoring and associated technology

Although the treatment of waste water can be achieved by relatively little equipment, such as a primary screening stage and RBC (rotating biological contactor) serving a small rural community, most treatment in Europe is accommodated by much larger and well integrated sites. The photographs in this chapter are of a large plant: the Knostrop treatment plant of Yorkshire Water, in Leeds, UK. These sites are by necessity highly automated. The three main components of this control system are the input devices (transducers) the computer which receives input and ultimately produces outputs which feed to control devices (servo-systems).

∏ What sort of information will an operator require to control the processing of water using the processes we have described earlier.

The information which an operator will require will include information on vessel volumes, flow rates throughout the site, power used in stages requiring a significant consumption of electricity and the dissolved oxygen concentration of liquor in certain aerobic vessels. We will provide an outline of this process control technology here. Other aspects of process control are covered in the BIOTOL text 'Bioreactor Design and Product Yield'. The transducers involved in these operations are represented in Figure 4.12.

Volume is measured by registrating the time needed for sound waves emitted by an ultrasonic transmitter and reflected by the surface of the vessel contents to reach an associated receiver. The lower the level of liquid in the container the longer it will take before the reflected wave reaches the detector. This delay is converted electronically to a signal which is fed to the computer which inputs this data as a numerical value representing the accurate vessel volume. The reason behind the choice of this ultrasonic variety of transducer over other possible optical or mechanical probes is that it is relatively maintenance free, while the latter techniques would be prone to breakdown as a result of the viscosity and general lack of consistency in the liquid to be measured. Flow meters also use this ultrasonic technology. They rely on the increase in depth of a liquid passing over a constriction when the flow rate rises.

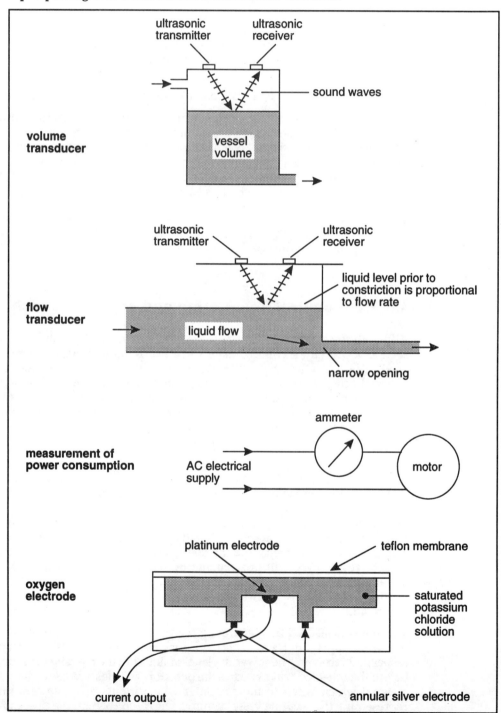

Figure 4.12 Transducers used in waste water management.

The measurement of power consumption by plant stages is monitored by measuring the current consumed. The voltage feeding equipment is constant and the power consumed can be calculated by the equation:

Power [watts] = current (amps) x voltage (volts)
         [measured]   [constant]

The period time a particular power level is consumed, multiplied by the power, provides the energy consumption which is usually measured in kilowatt hours (kWh).

The final type of transducer considered is the oxygen electrode, its construction is also shown in Figure 4.12. When a small voltage is applied across the electrodes the current which flows is related to the dissolved oxygen concentration in the liquid in contact with the transducer membrane. Industrial probes are encased in stainless steel sheaths because of the relatively fragile composition of the electrode components.

Upon obtaining plant information from various transducers, the operator receiving the data in the control room needs to be able to alter plant operations, the servo-systems which are available include:

Valves: regulating flow rates and also the routes of flows between possible stages (eg discharge, or further treatment vessels). These are mainly controlled by small motors.

and Motors: such as those driving pumps or mechanical aeration devices.

The final component of the control system is represented by the computers which these services address and which are contained in the control room. An outline of the entire plant status may be obtained on the computer screen as indicated in Figure 4.14. This overview indicates the main plant operating conditions. The display contains all the immediate details the operator needs. Here we need to consider only the most important aspects, which are as follows.

We will use the display shown in Figure 4.13 and the process carried out at the Knostrop treatment plant in Leeds, UK as the basis for our description.

The incoming sewage is pumped by four screw-type pumps through primary treatment stages. These pumps (and indeed most pumps on the site) can be placed under computer control (eg auto-run) or switched to the control of operators near the vessel.

More detailed information about any of the site operations can be obtained by selecting areas of the screen with a cursor which accesses further screen pages and diagrams. These details include vessel volumes, flow rates, dissolved oxygen concentrations and valve settings.

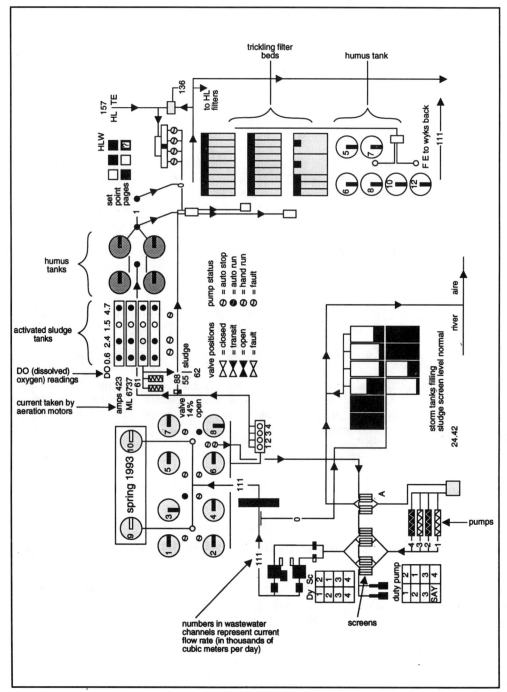

Figure 4.13 A stylised display of a water treatment control panel. (Based on the control panel at the Knostrop treatment plant in Leeds).

After primary treatment the flow is normally passed onto secondary treatment stages. However, when the volume of incoming waste water is excessive, as will occur after a period of heavy rain, this loading would have a detrimental effect if passed entirely to secondary treatment vessels. The flow can be diverted by closing a sluice gate limiting this flow to these vessels and diverting most of the flow to storm tanks or for discharge as untreated effluent. Figure 4.14 indicates two channels involved in a related process, and shows the motorised gates which can be operated at distance via a servo-control, to regulate the passage of waste water. During 'storm' conditions the BOD of the incoming waste water is usually very low as is the suspended solids level, consequently some of this water can be discharged directly to the plant effluent.

Figure 4.14 Servo-control of flow of effluent.

The activated sludge tanks displayed on the computerised display of the entire plant operating conditions have an associated display of dissolved oxygen levels (see Figure 4.13), this information providing an instantaneous output which relates to both aeration efficiency and BOD removal. These aeration tanks subsequently feed their associated settling tanks.

The wiring running through the plant resulting from the automation of its operations is usually electrical, but more recently fibre optical cable is being chosen. This is as these cables can carry more information (by virtue of their higher bandwidth) and can supply signals from video cameras installed at specific site locations to the control room in addition to signals involved in site maintenance.

A consideration of the operation of such a highly automated plant reveals that the real practical day to day operations of such a site are about as far removed from the public perception of waste water treatment being both a relatively messy, odoriferous and physically demanding procedure, as could be imagined.

4.7 Concluding remarks

In this chapter we have given an overview of the treatment of waste water with particular emphasis on the aerobic processes used to reduce BOD. The primary and secondary treatment processes are often coupled to tertiary treatment. There is an increasing tendering to demand a greater removal of contaminants from water before it is discharged from a treatment process. Thus phosphate may be removed by precipitation with lime or alum. Nitrogen (as ammonia) can be removed by distillation. Alternatively if the nitrogen is in the form of NO_3^-, it may be passed to an anaerobic bioreactor and be converted to N_2 by denitrifying organisms using a cheap carbon substrate. We should emphasise that the combined use of aerobic and anaerobic systems to remove particular components (such as nitrogen, phosphate and specific xenobiotics) is becoming of increasing importance in water treatment process.

Inevitably each additional stage in the treatment of waste water increases the cost of treatment. This we have shown in Figure 4.15. You will notice that the major part of the BOD is removed by the primary and secondary treatment processes.

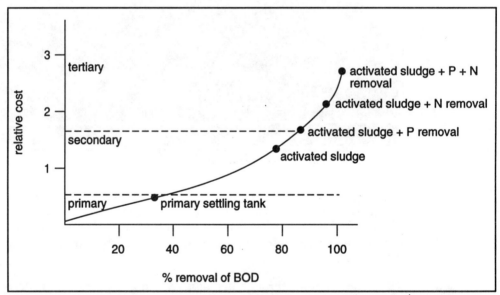

Figure 4.15 Relative cost of waste water treatment against the removal of BOD and other contaminants.

The major product of water treatment is, of course, clean (or cleaner!) water. Also produced is sludge from the various settling tanks. This may be disposed of via landfill, dumped at sea, used as an agricultural fertiliser (composed) or digested anaerobically. It is this latter treatment that we will consider in our discussions in the next chapter.

SAQ 4.9

1) An activated sludge process smells strongly of hydrogen sulphide. Would you expect that the effluent from this process would contain a high or a low ratio of NH_4^+ to NO_3^-? Give reasons for your choice.

2) Which of the following might you use to grow denitrifying organisms under anaerobic conditions to remove NO_3^- from an effluent by denitrification?

   Methanol, oxalic acid, glucose, phenol.

Summary and objectives

Aerobic waste water treatment is essentially a biologically based process which is long established. There are two main types of treatment; fixed film processes and homogeneous growth processes. The most widely used fixed film process is the trickling filter bed, and for homogeneous growth methods, the activated sludge process. Other specialised treatment regimes exist, and are often applied to the treatment of specific industrial wastes.

The choice of a treatment process is aided by the ability to define the nature of the waste water to be processed by the use of established parameters. These include BOD, COD, ammonical nitrogen and suspended solid matter.

Knowledge of fundamental equations and processes ultimately enable the design of vessels to handle known quantities of waste water of a known composition, and the optimisation of these stages for the removal of BOD or other desired effects when they have been built and are being operated. This understanding also enables the process to be run cost-effectively, for example in the choice of oxygenation methods in dispersed systems.

The skills needed within a waste water treatment plant are multi-disciplinary and include civil engineering, chemical engineering, electrical engineering and obviously, biology. A biotechnologist should appreciate the integration of these disciplines and have some understanding of their individual contributions.

Now that you have completed this chapter you should be able to:

* understand why the treatment of waste water is necessary and describe the parameters used to assess the effects of pollutants on a river (BOD, COD, ammonical nitrogen and suspended solid matter);

* recognise that a number of different stages are involved in the treatment of waste water including the aerobic (secondary) treatment process;

* describe reactor types employed in fixed film aerobic water treatment processes including trickling filter beds and rotary biological contactors; describe homogeneous aerobic reactors, and appreciate different methods of oxygen transfer in the activated sludge process;

* be familiar with the important microbiological species present in treatment processes and their impact on operational aspects in these processes;

* appreciate aspects of plant control in large treatment works.

Anaerobic waste water treatment

Anaerobic waste water treatment

5.1 Introduction

products of
aerobic waste
water treatment

In the previous chapter we provided an overview of the treatment of waste water. We described how particulate materials in waste water may be removed by using settling tanks and filtration devices. The main focus of the chapter was, however, on the use of micro-organisms in aerobic systems to remove the remaining organic (soluble and fine particulate) materials from water. In these processes, micro-organisms use the aerobic oxidation of some of the organic materials present in the water to provide the energy needed to assimilate the remainder of the organic material into cell biomass. In such processes, therefore, the organic contaminants in the water are converted either to CO_2 or into cellular biomass. The cellular biomass produced in this way is relatively easy to separate from the water by settling or filtration devices. There are, therefore, two main products of such processes, CO_2 and cellular biomass. We can represent aerobic waste water treatment in a rather simplified way as:

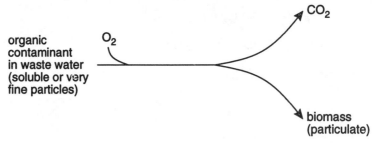

We also indicated that reduced nitrogenous materials (ammonia) may be oxidised by some micro-organisms (the nitrifying organisms) to nitrate.

The key to the successful operation of such processes relies on how effectively oxygen is supplied to the system and the nature of the micro-organisms that are ultimately produced. Larger cells (eg protozoans) are easier to separate from the liquid than smaller cell types (eg bacteria). Motile organisms and organisms which will not pack tightly (eg stalked protozoans) are not desirable since they also pose difficulties in separating them from the treated water.

In this chapter, we will examine anaerobic processes used to treat waste water. In these processes, the micro-organisms do not have access to oxygen or other inorganic electron acceptors such as nitrate which may be used to oxidise organic materials. Instead they derive the energy they need from the rearrangement of the organic material in which some of the organic carbon is oxidised whilst the remainder is reduced. If we assume that the organic material is in an oxidation state equivalent to carbohydrates (that is xCH_2O), then in theory it is possible to convert this into a mixture of CO_2 and CH_4. Thus:

$$2CH_2O \rightarrow CH_4 + CO_2$$

In principle, part of the carbon is oxidised to CO_2 whilst the remainder of the carbon is reduced (CH_4). Both of these products are, of course, volatile and will escape from the water. This is the essential basis of the anaerobic treatment of waste water.

We will begin the chapter by providing a brief history of the use of anaerobic waste water treatment and drawing your attention to some of the main differences between aerobic and anaerobic waste water treatment processes. We will then give a rather detailed account of the micro-organisms involved in anaerobic waste water treatment including some details of their metabolic capabilities. The major part of the chapter is devoted to a discussion of the operation of the bioreactors (digesters) used in anaerobic waste water treatment.

5.2 Historical background

septic tanks

The use of anaerobic water treatment is not new. In Europe, the use of septic tanks was introduced about 100 years ago. In septic tanks, solids present in the incoming water are allowed to settle as a sludge. There is no oxygenation of the sludge and the sludge is digested by anaerobic micro-organisms. CO_2 and CH_4 are released from the sludge and the sludge volume is reduced. At the same time, the numbers of pathogens in the sludge are greatly reduced.

The original septic tanks were single tanks in which both the settling of the sludge and sludge digestion took place (see Figure 5.1a). Such tanks needed to be large to accommodate the volume of water being treated and to provide a reasonably long residence time for the sludge. Inevitably the volume of sludge builds up over a period of time and sludge had to be removed from time to time.

discharged to adsorption fields

These simple devices are still used especially in isolated rural areas. The 'cleaned' water output from the septic tanks is usually discharged through pipes, to an adsorption field (see Figure 5.1b).

designs of Travis and Imhoff

separation of sedimentation and digestion

Subsequent developments included the incorporation of baffles in the septic tank to facilitate sedimentation. Various designs were adopted especially those of Travis and Imhoff. Ultimately, however, devices were introduced which separated sedimentation from the digestion process (Figure 5.2). This enabled a reduction in the size of the digestion unit needed to treat particular volumes of waste water. It also enabled the development of digesters which operated at temperatures slightly above ambient. Typically these digesters operate at 35°C and use long hydraulic retention times. Retention times were usually longer than 20 days. Little attempt was made to retain biomass (sludge) in these vessels and thus the retention times for biomass (and sludge) was similar to that of the incoming material.

∏ Examine Figure 5.2 carefully and see if you can predict what will be present in the water leaving the settling tank.

You should have realised that the water output from the settling tank will contain soluble and colloidal material which failed to separate out. In larger waste water treatment plants, this output from settling tanks may be passed to either an aerobic treatment unit (for example a trickling filter or activated sludge process) or to an anaerobic stirred tank reactor.

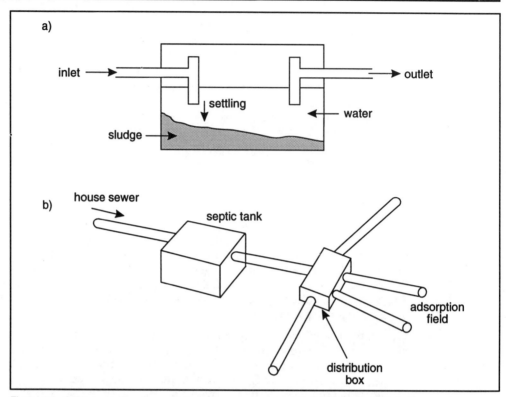

Figure 5.1 a) Cross-section of a septic tank. In some cases, septic tanks may be fitted with a vent.
b) Arrangement of a septic with a distribution system in which treated water is passed into an adsorption
field.

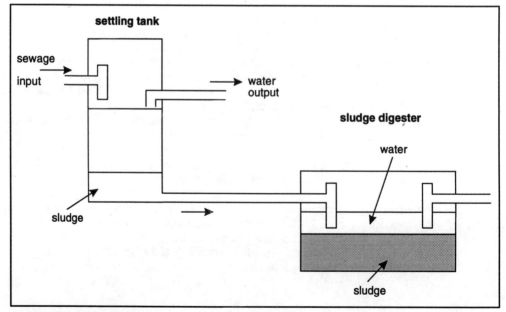

Figure 5.2 Stylised anaerobic water treatment process in which sludge separation and digestion are carried
out in separate processes.

high rate
anaerobic
processes

The subsequent development of anaerobic treatment processes involved optimising the process conditions (for example, temperature, pH) in order to reduce the size of the reactors needed to treat particular volumes of waste water. The development of such high rate anaerobic processes enabled the use of anaerobic processes for larger volumes of waste water.

Π Apart from optimising the conditions for anaerobic digestion, what else should be achieved to improve the efficiency of an anaerobic digester in terms of the volume of waste water that can be treated per unit volume of digester?

The key to improving the efficiency of the digester is to find a way of maintaining a high concentration of the active biomass in the digester.

Π See if you can think of at least one way in which the active biomass could be maintained at a high concentration in the digester.

Two common approaches to this problem have been developed. In the first, digesters were fitted with solid supports on which the anaerobic biomass would grow and be retained in the vessel. There have been many developments based on this process. Here we will illustrate this with specific examples. Some anaerobic digesters have included sand as the physical support medium. If the flow through this sand is upwards but slow, the sand remains packed as a bed. We might describe such a system as an anaerobic filter. If, however, the flow is fast, the sand will cease to remain packed as a bed and the system will become a fluidised bed process.

anaerobic
filters and
fluidised bed
process

recycling of
biomass

An alternative approach to using solid supports to retain the active biomass is to recycle the active biomass or to design the sludge digester such that biomass does not wash out at the same rate as the liquid. A process involving the recycling of active biomass is illustrated in Figure 5.3.

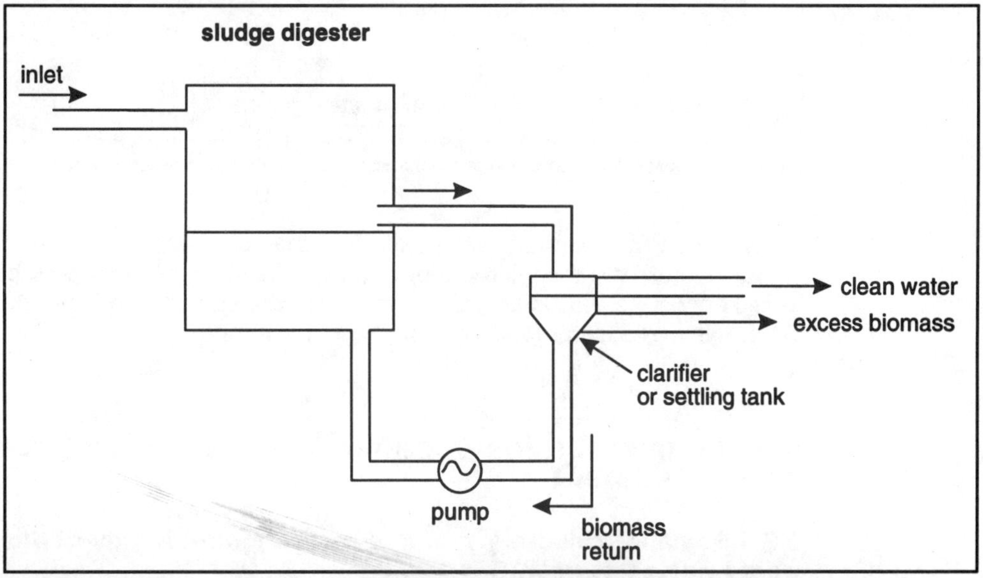

Figure 5.3 Stylised representation of an anaerobic sludge digester with biomass recycling device. In practice different devices may return a varying proportion of the biomass to the digester.

Thus the development of anaerobic treatment processes can be divided into a number of phases. We have summarised these in Figure 5.4. You should, however, realise that these phases are not mutually exclusive and devices representing all of the phases represented in Figure 5.4 are still in operation.

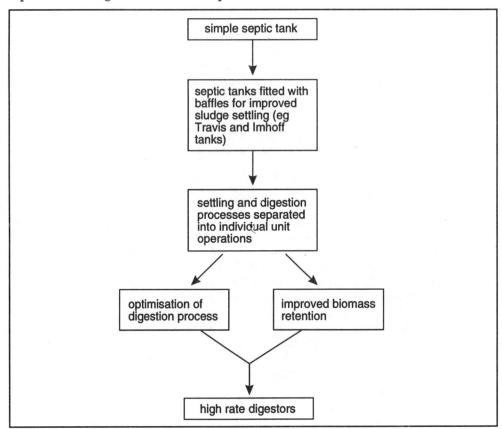

Figure 5.4 Phases in the development of the anaerobic treatment of waste water.

Before we further examine the operation of anaerobic waste water treatment processes, we need to explain the microbiology and biochemistry of the processes in more detail. This knowledge will enable us to consider the design and operation of aspects of anaerobic waste water treatment in a more meaningful way.

5.3 The microbiology and biochemistry of anaerobic waste water treatment

5.3.1 Some introductory remarks on the microbiology of anaerobic waste water treatment

Advances in the microbiology of anaerobic waste water treatment have, historically, depended upon studies on pure cultures of micro-organisms. Such studies have enabled us to elucidate metabolic pathways and to establish the influence of physical parameters such as pH, temperature and oxidation- and reduction potential on the growth of particular organisms. Such studies are valuable but are strictly limited in terms of their application to understanding and improving anaerobic treatment processes.

Anaerobic treatment processes invariably involve a wide variety of organisms which show a great complexity of interactions. These interactions have yet to be fully elucidated but sufficient is now known to enable us to explain the activities of the major groups of micro-organisms involved in anaerobic digestion.

The overall process of anaerobic digestion can be represented as the conversion of organic waste, consisting mainly of proteins, lipids and carbohydrates, into methane and carbon dioxide. In general terms, three groups of micro-organisms can be identified in terms of the nutrients they use. These are:

- hydrolytic organisms which break down the complex molecules in the waste water and produce small molecules, especially short-chained organic acids, as end products of their catabolism;

- hydrogen producing organisms which mainly utilise the products of catabolism of the hydrolytic organisms. They also produce acetic acid and are frequently referred to as heteroacetogens;

- methanogens, organisms which produce methane. These may be divided into hydrogen utilisers and acetic acid users.

We have illustrated the relationships between these groups of organisms in Figure 5.5.

∏ Examine Figure 5.5 and decide which taxonomic group (eg bacteria, protozoa, fungi etc) dominates the microflora of the anaerobic digestion process.

From your knowledge of microbiology you should have come to the conclusion that the microflora of the anaerobic digestion process is dominated by bacteria.

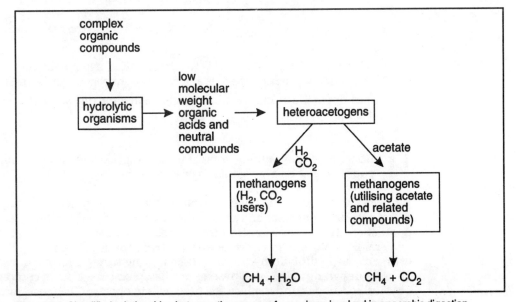

Figure 5.5 Simplified relationships between the groups of organisms involved in anaerobic digestion.

∏ Figure 5.5 is a great simplification of the overall relationships between the bacterial groups involved in anaerobic digestion. If the input into the digester contains SO_4^{2-} and NO_3^- residues what other groups of organisms might you anticipate would be present and what would be their metabolic products?

Desulphovibrio

Some micro-organisms are able to oxidise organic nutrients and to use SO_4^{2-} as the electron acceptor. These, are the so called sulphate-reducers (for example *Desulphovibrio spp.*) and they produce S^{2-} as a metabolic product. Thus:

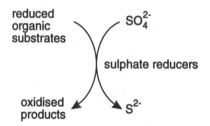

Thus, if SO_4^{2-} is present in the input, we might anticipate that sulphate-reducers would be present in the digester and S^{2-} to be amongst the products of the digester.

denitrifiers

Similarly other organisms (the denitrifiers) may use nitrate as an electron acceptor. Their catabolic activities may be represented by:

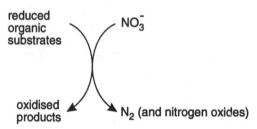

The level to which NO_3^- is reduced is to a large extent dependent upon pH. At neutral pH, N_2 is the major product whereas at lower pHs, nitrogen oxides are formed. Thus, if NO_3^- is present in the input, we must anticipate that the digester will contain a number of denitrifiers.

∏ Apart from the nutrient input into the digester, what else will influence the population of micro-organisms present in the digester?

We hoped that you would have suggested that physical parameters such as the pH of the input and the temperature of the operation would influence the microbial population. These certainly effect the species which are present. But, however, over a wide range of physical parameters we still obtain species representing the three groups of organisms (hydrolytic, hydrogen-producing, methanogenic) illustrated in Figure 5.5. Thus, in designing a digestion process, we have some freedom of choice regarding the physical conditions under which the process is to operate. The most commonly altered variable is temperature. Anaerobic digesters may be operated at psychrophilically (<20°C) temperatures, mesophilically (20-45°C) or thermophilically (50-70°C).

☐ See if you can list some advantages and disadvantages of using thermophilic temperatures for anaerobic digestion.

thermophilic processes are less stable

The main advantage is that the rate of digestion is higher at these elevated temperatures. However, experience has shown that thermophilic processes are less stable and small fluctuations in temperature leads to great fluctuations in performance. Thus recovery of performance of a thermophilic digester after a small temperature shock is slow compared with the recovery of a mesophilic process. It also costs money to heat the digester and the increased rate of digestion observed at higher temperatures does not usually offset this additional cost. Furthermore, the diversity of micro-organisms capable of growing at elevated temperatures is much less than that at mesophilic temperatures. Thus, if the composition of the input into the digester is variable (which is usually the case), the thermophilic system has a lower ability to respond to changing substrate input.

most process are mesophilic

For these reasons, most anaerobic digestion processes operate at mesophilic temperatures. Frequently temperatures in the range of 35-40°C are used. This provides a reasonably high rate of digestion, an ability to respond to a wide range of substrates by allowing for broad species diversity and the heat input required is sufficiently low to remain cost-effective. It is for this reason that we will mainly consider mesophilic processes.

SAQ 5.1

Identify each of the following as either true or false.

1) Inclusion of high levels of SO_4^{2-} in the input to an anaerobic digester is desirable.

2) High levels of N_2 in the gas venting from an anaerobic digester is indicative of high levels of protein in the input to the digester.

3) The same biochemical processes are carried out by different species of bacteria in mesophilic and thermophilic digesters.

4) All of the organisms found in an anaerobic digester are heterotrophic organisms.

5) We should anticipate that the gas vented from an anaerobic digester will contain only CO_2, CH_4 and N_2.

5.3.2 The hydrolytic organisms of anaerobic digestion (the primary fermenters)

We might anticipate that there will be a great diversity of hydrolytic organisms present in the digestion milieu and that the size and composition of the microbial population will depend upon the concentration and composition of the input. This is indeed what is observed. Table 5.1 lists some of the common genera encountered in anaerobic digesters.

Acetobacterium	Lactobacillus
Bacteriodes	Peptococcus
Bifidobacteria	Peptostreptococcus
Clostridum	Spreptococcus
Eubacterium	

Table 5.1 Genera often encountered in anaerobic digesters.

∏ If you examine Table 5.1 and your microbiological knowledge is sufficient you should be able to identify Gram-positive rods and cocci, spore formers and non-spore formers amongst the genera listed.

Typical Gram-positive rods are represented by *Clostridium spp.*, *Eubacterium spp.* and *Lactobacillus spp.* whilst the streptococci are typically Gram-positive cocci. *Clostridium spp.* are good examples of spore formers. Even if you could not identify each of these, the point we are trying to establish is that the hydrolytic organisms present in digesters are very diverse.

∏ Given that much of the input into a digester is macromolecular (eg proteins and polysaccharides), what might you anticipate with respect to the enzymes used by these organisms to hydrolyse the incoming material?

hydrolytic enzymes include proteases, amylases, cellulases, pectinases and lipases

Many of the hydrolytic enzymes produced by these organisms function as exo-enzymes. We might expect that these exo-enzymes secreted by the microflora will include a range of proteases, amylases, cellulases, pectinases and lipases required to hydrolyse the available nutrient. The only common input into anaerobic digesters which is rather recalcitrant to hydrolysis is lignin. The breakdown of lignin is relatively slow and, in many anaerobic processes most of the lignin remains intact during anaerobic digestion. Long residence times are required to achieve significant lignin degradation.

Here we do not propose to examine all of the metabolic pathways which lead from the hydrolysis of macromolecules to the production of the low molecular weight products of anaerobic catabolism. These pathways are described elsewhere in the BIOTOL series ('Principles of Cell Energetics' and 'Energy Sources for Cells') and in other good biochemical and microbiological texts. The main products of such metabolism are low molecular weight products such as acetic acid, propionic acid, butyric acid, lactic acid and ethanol.

There are, however, two main points that we need to make about this group of organisms. Firstly, there is only a relatively small production of cells from this phase of organic breakdown.

∏ See if you can explain, in thermodynamic terms, why there is a small production of cells (Hint, think about the yield of yeast grown anaerobically on glucose compared with the yield of yeast grown aerobically on the same amount of glucose).

usable energy derived from anaerobic and aerobic catabolism

The answer is quite straigtforward. Under anaerobic conditions, the organic substrate can only be rearranged, it cannot be completely oxidised. Thus, in the case of yeast using glucose, the catabolism can be represented as:

$$\text{glucose} \longrightarrow \text{2 ethanol} + 2CO_2$$
$$(C_6H_{12}O_6) \qquad (2C_2H_5OH)$$

The energy yield is quite modest ($\Delta G^{o'}$ = -244 kJ mol^{-1}). Under aerobic conditions, glucose may be completely oxidised to CO_2 and H_2O. Thus:

$$\text{glucose} + O_2 \longrightarrow 6CO_2 + 6H_2O$$
$$(C_6H_{12}O_6)$$

The energy yield is much higher ($\Delta G^{o'}$ = -2870 kJ mol^{-1}).

Therefore, under anaerobic conditions only a small amount of usable energy is generated from the catabolism of substrates and thus little cell synthesis occurs. In other words, we can anticipate a high turnover of substrates by a small amount of biomass.

biomass yields are low in anaerobic processes

This is a major difference between aerobic and anaerobic treatment processes. Anaerobic processes generate much less biomass than aerobic processes. Aerobic processes generally lead to yields of 0.5-1.5 kg of biomass for every kg of BOD removed. Anaerobic process tend to produce only 0.1-0.2 kg of biomass for each kg of BOD removed; typically a mesophilic sludge may contain 10^5-10^9 hydrolytic organisms ml^{-1}. This is, of course, greatly influenced by the concentration of the input nutrients.

The second point we need to make is that primary fermenting organisms may adjust their metabolism in response to the conditions in the digester. Of particular importance are the pH and hydrogen concentration in the vessel. Amongst the fermenting organisms are some which can generate hydrogen gas. For simplicity we can represent the metabolism of these organisms by the equation:

$$\text{glucose} \longrightarrow \begin{array}{l}\text{oxidised} + H_2 \\ \text{organic} \\ \text{products}\end{array}$$

For example, the dissimilation of glucose to acetic acid and hydrogen may be written as:

$$\text{glucose} + 2H_2O \longrightarrow \text{acetic acid} + 4H_2 + 2CO_2$$
$$C_6H_{12}O_6 \qquad\qquad 2CH_3COOH$$

methanogenic bacteria use hydrogen

We will see later that the hydrogen produced by such organisms may be utilised by methanogenic bacteria. However, let us assume that there has been a sudden surge of glucose (or a source of glucose) in the input to the vessel. The hydrogen-producers described above increase rapidly with a concomitant rise in the concentration of hydrogen. There would be a time lag before the hydrogen utilisers would be able to respond. Thus for a time there would be an excessive concentration of H_2 and acetic acid in the vessel. These would depress both the oxidation-reduction potential (E_h) and the pH in the vessel. If this continued, the fall in pH would inhibit the growth and metabolism of the micro-organisms present and the digestion process would come to a halt. Fortunately many of the primary fermenting organisms present in the sludge may utilise alternative catabolic pathways which are switched on at lower pH and E_h values. Under E_h and pH stress some of the organisms switch on pathways which consume H_2. For example, propionic acid may be produced from glucose in a H_2 utilising pathway. Thus:

$$C_6H_{12}O_6 + 2H_2 \rightarrow 2C_2H_5COOH + 2H_2O$$

The point we are making is that, although fluctuations in the input to a digester should be avoided, the diversity of the organisms present together with their ability to switch to alternative pathways means that the process is to some extent self-adjusting. We will return to this point later.

5.3.3 The heteroacetogens

many different species are heteroacetogens

The heteroacetogens are organisms whose primary products of catabolism are acetic acid and hydrogen. They utilise the products of the hydrolytic organisms described in Section 5.3.2. There are many species of organisms that may be included in this group but as yet there is no general agreement as to the identity of all of the heteroacetogens encountered in anaerobic digesters. Amongst the genera representing this group are *Acetobacterium*, *Syntrophobacter* and *Syntrophomonas*.

Below are two examples of the stoichrometry of the process catalysed by these types of organisms.

propionic acid CH_3CH_2COOH + $2H_2O$ $\longrightarrow$ acetic acid CH_3COOH + CO_2 + $3H_2$

butyric acid $CH_3CH_2CH_2COOH$ + $2H_2O$ $\longrightarrow$ acetic acid $2CH_3COOH$ + $2H_2$

∏ From our previous discussion and from your knowledge of thermodynamics, is there likely to be a high energy yield from these reactions?

The answer is no. In fact, of course, the energy yield is greatly dependent upon the relative concentrations of the substrates and products.

(Remember that $\Delta G = \Delta G^{o'} + 2.303\ RT \log \frac{[products]}{[reactants]}$)

the energy yield is influenced by the concentrations of reactants and products

The $\Delta G^{o'}$ value for the metabolism of propionic acid described above has been estimated to be in the range of 48-72 kJ mol^{-1}. In other words, using equimolar concentrations of propionic acid, acetic acid, CO_2 and H_2, the formation of propionic acid is favoured. It is only at low concentrations of H_2 or high concentration of propionic acid that propionic acid dissimilation is favoured. Providing the hydrogen concentration remains low (for example by its removal by methanogens) then the dissimilation of propionate is favoured. These thermodynamic constraints imply that there must be a close relationship between the heteroacetogens and the methanogens.

∏ What would happen to the heteroacetogens if the methanogens in the digester were killed?

The death of the methanogens would mean that there would be a build up of H_2. This would make the metabolism of the heteroacetogens thermodynamically unfavourable and thus their metabolism would also cease.

Π Bearing in mind what we have said about the thermodynamics of heteracetogenic metabolism, should we anticipate a low or high population density of heteracetogens?

The answer is that the population density will be quite low. The thermodynamic yield is low, therefore, only limited growth will be achieved. Again, the actual population density will depend upon the input and residence time in the vessel but, typically, the heteroacetogen population falls in the range of $1-5 \times 10^6$ cells ml^{-1} sludge.

| SAQ 5.2 | Bearing in mind that the metabolism of methanogenic bacteria can be represented by the equations $CO_2 + 4H_2 \rightarrow CH_4 + 2H_2O$, and $CH_3 COOH \rightarrow CH_4 + CO_2$, explain why heteroacetogens and methanogens should be regarded as symbionts. |

5.3.4 Methanogens

methanogens use a limited range of substrates

The range of compounds that serve as energy sources for the methanogens is very limited. H_2 and CO_2 or acetic acid are the most common substrates used, but formate, methanol, methylamines and ethylamines are also utilised by some. The energy-yielding reactions used by methanogens are shown in Table 5.2. Here we will mainly focus on the first two reactions shown in this table. Note that we have used HCO_3^- instead of CO_2 in the consideration of the energetics of methanogens in the first reaction; this is after all, the form in which the CO_2 is utilised by the cells.

Reaction	ΔG° (kJ mol^{-1} CH$_4$)
$4H_2 + HCO_3^- + H^+ \rightarrow CH_4 + 3H_2O$	-138
$CH_3 COO^- + H^+ \rightarrow CH_4 + CO_2$	-38
$4H COO^- + H^+ \rightarrow CH_4 + 3CO_2 + 2H_2O$	-146
$4CH_3 OH \rightarrow 3CH_4 + CO_2 + 2H_2O$	-318
$4CH_3 NH_3^+ + 3H_2O \rightarrow 3CH_4 + 4NH_4 + HCO_3^- + H^+$	-314

Table 5.2 Energy-yielding reactions performed by methanogens.

Π The relatively low energy yield of the reactions shown in Table 5.2 should suggest to you something about the population of methanogens in the digester. What is it?

cell concentrations are low

It should have suggested that the methanogens are probably present in low numbers. The low energy yield, together with the low concentration of H_2 will result in only a low cell number and relatively slow growth rates. Typically methanogen concentrations in mesophilic sewage sludge is of the order of 10^6-10^8 cells ml^{-1} and doubling times may be as long as 1-10 days.

Taxonomically, the methanogens constitute a diverse group of organisms belonging to several genera (Table 5.3). They are distinguishable according to their reaction with Gram's staining, cell wall structure, lipid content and the base composition of their DNA. The taxonomy of these organisms is under active examination and the groups listed in Table 5.3 are subject to constant review. This taxonomic instability reflects the fact that these organisms have become of increasing interest and are being studied intensively.

Genera	Gram reaction	Cell wall structure	Characteristic lipid	G+C content (as %)	Examples of substrates used
Methanobacterium	+	Pseudomurein	C_{20} diethers and C_{40} tetraethers	32-50	H_2, some use formate acetate
Methanobrevibacter	+	Pseudomurein	C_{20} diethers and C_{40} tetraethers	27-32	H_2, some use formate acetate
Methanococcus	-	Proteinaceous	C_{20} diethers	30-32	H_2, formate acetate
Methanospirillum	-	Proteinaceous	C_{20} diethers and C_{40} tetraethers	45-47	H_2, formate acetate
Methanosarcina	+	Heteropoly-saccharide	C_{20} diethers	38-51	H_2, formate methanol, acetate

Table 5.3 Genera of methanogens. (Adapted from Stamer, Ingraham, Wheels and Painter 1989, General Microbiology, 5th Edition, MacMillan, Basingstoke). Note that the taxonomy of the methanogens is relatively unstable and is subject to updating.

key cofactors in methanogens; methanopterin; methanofuran and coenzyme M

The autotrophic metabolism of methanogens in which H_2 and CO_2 are the substrates has been most actively investigated. An outline of the pathway of methanogenous metabolism in which CO_2 (as HCO_3^-) is converted to CH_4 is shown in Figure 5.6. A feature of methanogens is that they contain several cofactors not found in other bacteria. Three of these - methanopterin (MP), methanofuran (MF) and coenzyme M (CoM) - are carriers of the one carbon unit during its reduction from CO_2 to CH_4. Factor$_{430}$ (F_{430}) is the prosthetic group of methyl-CoM reductase, the last enzyme in the pathway.

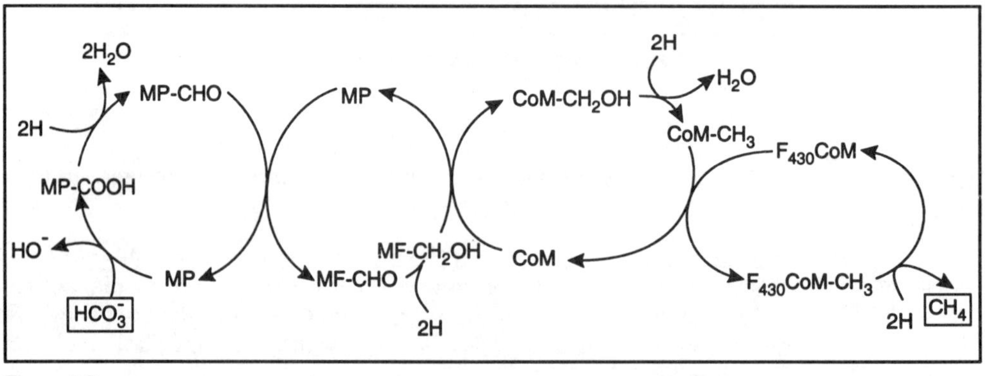

Figure 5.6 Pathway of methanogens showing the conversion of CO_2 (as HCO_3^-) to methane. Note that the reduction occurs in a stepwise manner (see text for further details).

mechanism of ATP synthesis in methanogens is not known

The mechanism by which methanogenesis is coupled to ATP synthesis is not known. The absence of cytochromes and quinones in most methanogens suggests that classical electron transport is not involved. Incubation of methanogens with ionophores leads to a rapid decline in ATP concentrations and thus it is generally assumed that a proton motive force is essential to drive ATP synthesis in these organisms.

Many favour a model for ATP production in these organisms in which hydrogen is oxidised on the outer surface of the plasma membrane resulting in the generation of protons outside of the cell. Inside of the cell, CO_2 reduction consumes protons (see Figure 5.7).

The net result is the generation of a proton gradient across the plasma membrane. This proton gradient may then be used to drive ATP synthesis using a plasma membrane bound ATPase in much the same way as is in other systems.

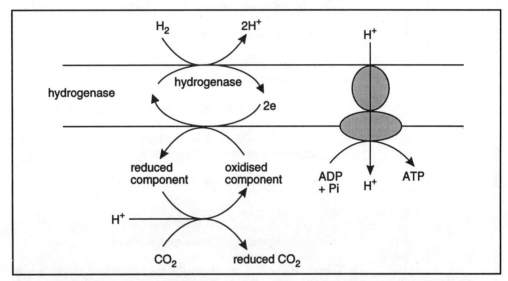

Figure 5.7 A possible model for the generation of ATP using a proton motive force generated by the oxidation of hydrogen during methanogensesis.

In addition to the reduction of CO_2 to CH_4, most methanogens can use CO_2 as the sole carbon source for the synthesis of cellular material. Using radioactively labelled precursors (eg ^{14}C-CO_2), it appears that the CO_2 assimilation pathways differ in different species. However, all seem able to reductively carboxylate acetyl CoA and many of the reactions common to the tricarboxylic acid cycle seem to be involved.

An example is the pathway of CO_2 assimilation in *Methanospirillum* shown in Figure 5.8.

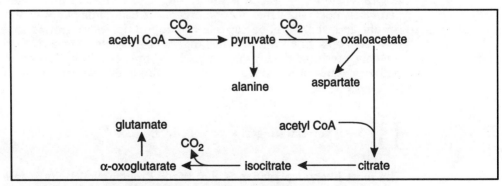

Figure 5.8 Outline of the CO_2 assimilation in *Methanospirillum*.

SAQ 5.3	1) Without looking at Figure 5.6, place the following in the order of their involvement in methanogenesis CoM-CH$_2$OH, MP-COOH, HCO$_3^-$, Fe$_{430}$-CoM-CH$_3$, CH$_4$, MF-CHO.

2) Which of the following statements are compatible with an organism being a methanogen.

   a) The organism lacks cytochromes.

   b) The organism assimilates CO$_2$.

   c) Growth of the organism is stimulated by increasing H$_2$ concentrations.

   d) The organism grows in highly reducing environments.

   e) Growth of the organism is stimulated by the presence of fermentative micro-organisms.

   f) The organism is insensitive to cyanide.

   g) The organism generates CH$_4$.

   h) The organism contains methanopterin.

3) Which of the statements in 2) prove that the organism is an methanogen.

Let us now return to the activities of methanogens in digesters.

∏ Given that the energy yield of the reactions $4H_2 + HCO_3^- + H^+ \rightarrow CH_4 + 3H_2O$ and $CH_3 COO^- + H^+ \rightarrow CH_4 + CO_2$ are ΔG_o = -138 kJ mol^{-1} CH$_4$ and $\Delta G°$ = -38 kJ mol^{-1} CH$_4$ respectively (see Table 5.2). Which pathway should, in theory, contribute most to CH$_4$ generation in a digester?

Logically the $4H_2 + HCO_3^-$ pathway ought to contribute most. Since this pathway generates most Gibbs free energy, organisms using this pathway should derive most benefit and, therefore, be capable of supporting faster growth rates. Experimental measurements, however, with real digesters indicates that nearly three quarters of the methane produced during the anaerobic treatment of sludge is produced by the acetate pathway, while only about a quarter to a third of the methane is derived from the H$_2$ + HCO$_3^-$ pathway.

∏ See if you can suggest a reason why this is so?

The most likely explanation is that the hydrogen availability is rate limiting. Thus, although the use of hydrogen is thermodynamically most favoured, its lack of availability reduces the rate at which this pathway operates.

5.4 Some theoretical and practical considerations in anaerobic fermentation

5.4.1 Predicting the performance of digesters based on elemental balances

For convenience we will refer to the gas generated from anaerobic digestion of sludge as biogas. Biogas contains mainly CH_4 and CO_2 with varying proportions of other gases (especially N_2 and H_2). In principle, we can determine the composition of biogas if we know:

- the elemental composition of the substrate (sludge);

- the proportion of the sludge which is biodegradable;

- the proportion of the sludge which is assimilated into biomass.

Let us examine why this is so. If we can identify the amount and elemental composition of the biodegradable portion of the sludge, then this will ultimately end up as biomass, CO_2 or CH_4 (for the moment we will ignore other elements). We can represent this reasoning diagrammatically in the following way:

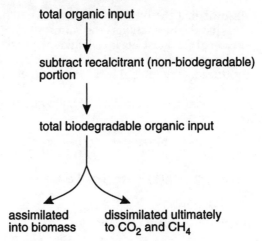

Since the elements cannot be destroyed, it is logical that:

- the carbon of the biodegradable input must end up as the assimilated biomass carbon or as the carbon of CO_2 and CH_4;

- the oxygen of the biodegradable input must end up in the biomass, in CO_2 or in H_2O;

- the hydrogen of the biodegradable input must end up in the biomass, in CH_4 or in H_2O.

We can, therefore, write relationships such as:

| total C in the biodegradable input | = | total C assimilated in the biomass | + | C in the CO_2 produced | + | C in the CH_4 produced |

Likewise for oxygen we could write a balance.

| total O in the biodegradable input | = | total O assimilated in the biomass | + | O in the CO_2 produced | + | O in the H_2O produced |

and for hydrogen, we could write a balance.

| total H in the biodegradable input | = | total H assimilated in the biomass | + | H in the CH_4 produced | + | H in the H_2O produced |

Thus by knowing how much of these three elements are present in the biodegradable input, the composition of the biomass and the proportion of carbon assimilated into biomass, it should be possible to calculate the ratio of CO_2:CH_4 produced.

Let us do a sample calculation.

∏ Assume that the biodegradable substrate has the elemental composition CH_3O and that the biomass has the composition (CH_2O). Also assume that 20% of the carbon of the substrate is assimilated into biomass. Assuming that the remainder of the carbon is dissimulated as CO_2 and CH_4, calculate the ratio of CO_2:CH_4 produced. (Try to do this before reading on).

We can represent the conversion of substrate to biomass, CO_2 and CH_4 by the word equation:

$$\text{substrate} \longrightarrow \text{biomass} + x\ CO_2 + y\ CH_4 + z\ H_2O$$
$$(CH_3O) \qquad\qquad 0.2\ (CH_2O)$$

Note the coefficient (0.2) in front of the biomass since 20% of the carbon is converted to biomass.

We can write an elemental balance for C as:

$1 = 0.2 + x + y$

For H, the balance will be:

$3 = (0.2 \times 2) + 0 + 4y + 2z$

For O, the balance will be:

$1 = 0.2 + 2x + z$

Thus we can solve for x, y and z.

Thus $x = 0.8 - y$.

and $4y = 2.6 - 2z$, thus $y = 0.65 - 0.5z$.

and $z = 0.8 - 2x$.

Substituting $y = 0.65 - 0.5z$ into $x = 0.8 - y$

$x = 0.8 - 0.65 + 0.5z$

and, since $z = 0.8 - 2x$

$x = 0.8 - 0.65 + 0.5 (0.8 - 2x)$

$x = 0.15 + 0.4 - x$

So $2x = 0.55$ and $x = 0.275$

Therefore $y = 0.8 - 0.275 = 1.525$

and $z = 0.8 - 2 (0.275)$

$= 0.25$.

Thus we can write our equation as:

$$\text{substrate} \longrightarrow \text{biomass} + 0.275\ CO_2 + 0.525\ CH_4 + 0.25\ H_2O$$
$$(CH_3O) \qquad\qquad 0.2\ (CH_2O)$$

Thus the ratio of $CO_2:CH_4$ production is, therefore, 0.275 moles CO_2:0.525 moles CH_4 (that is 1 mole: 1.91 mole).

This example is, of course, a great simplification of the determination of the overall stoichiometry of such a bioconversion of substrate to biomass, CO_2 and CH_4. In real life we have to take into account other elements (especially N and P) which are major components of biomass, and rarely are the sole products of anaerobic digestion biomass, CO_2 and CH_4. Frequently other products (NH_4^+, organic acids) are released from the digester. Let us consider another example, this time including N.

If the input into the digester has the empirical composition of $CH_3ON_{0.5}$ and, in the digester, this is converted to CO_2, CH_4 and NH_4^+ and to biomass of empirical composition $CH_2ON_{0.2}$, then we can also write an elemental balance for N.

The overall reaction can be written as:

$$CH_3ON_{0.5} \longrightarrow a(CH_2ON_{0.2}) + x\ CO_2 + y\ CH_4 + z\ H_2O + w\ NH_4^+$$
$$\text{substrate} \qquad\qquad \text{biomass}$$

Then the balance for C is:

$1 = a + x + y$

For O, the balance is:

$1 = a + 2x + z$

For H the balance is:

$3 = 2a + 4y + 2z + 4w$

For N, the balance is:

$0.5 = 0.2a + w$

Thus we have five unknown coefficients (a, w, x, y and z) and four relationships between them. If we measure one of these unknown coefficients, then we can mathematically determine the remainder.

SAQ 5.4

1) Using the example described in the text and the knowledge that 20% of the carbon in the substrate is assimilated into biomass, calculate the coefficients x, y, x and w and then write a stoichiometric equation for the overall reaction.

2) What is the ratio (mole:mole) of CO_2 to CH_4 produced in this system?

The type of modelling we have just carried out using elemental balances is particularly useful in situations where the substrates and products are well defined. It is therefore, especially applicable in circumstances in which a pure culture of an organisms is used using a pure substrate in which the end products are known. In circumstances such as in anaerobic digestion in which the input may be ill-defined and may vary in composition and in which it is not easy to define all of the end products, such modelling is less precise in enabling us to predict the outputs from such processes. Nevertheless if the major outputs (for example, biomass, CO_2, and CH_4) are known and if the elemental composition of the input does not flucculate enormously from an average composition, such modelling has some value in predicting the performance and outputs of an anaerobic digester.

We will leave our description of the use of elemental balances in predicting the outcomes of anaerobic digesters here. If, however, you would like to learn more about the use of elemental balances to model microbial fermentations, we recommend the BIOTOL text 'Bioprocess Technology: Modelling and Transport Phenomena'. This latter text also covers the kinetic aspects of modelling.

importance of using pilot scale operations

In the two simplified elemental balances we have described, the ratios of $CO_2:CH_4$ produced were 1:1.91 and 1:0.58 (see in text activity and SAQ 5.4). In practice, the values obtained vary between 1:1 and 1:3. In part this variation may be accounted for by differences in the elemental composition of the feedstock and its potential biodegradability. Not all of the variation can be accounted for in this way and in designing large scale processes, it is still desirable to build and test pilot scale digesters before investing in large scale equipment.

5.4.2 Predicting the performance of digesters based on yield coefficients

The yield coefficient (Y) is the ratio of biomass produced per unit of substrate consumed. Thus:

$$Y = \frac{\text{biomass produced}}{\text{substrate consumed}}$$

Typically units for Y are expressed as kg kg^{-1} or simply as a dimensionless coefficient.

determination of Y from theoretical considerations

Values for Y can be determined from theoretical considerations. If, for example, the amount of ATP produced per mole of substrate consumed can be calculated because the pathway of catabolism is known, and the amount of ATP consumed per unit of biomass is also known, then it becomes possible to calculate Y.

This type of calculation has been done for a wide range of cultures growing on an array of substrates. (Details of this type of calculation are given in the BIOTOL text 'Bioprocess Technology: Modelling and Transport Phenomena').

Similar calculations have been done for organisms involved in anaerobic digestion using a variety of substrates by Mosey (Mosey FE, (1983), Kinetic Descriptions of Anaerobic Digestion, Proceedings of 3rd International Symposium on Anaerobic Digestion). Some examples of these data are given in Table 5.4.

Substrate	Yield (kg kg^{-1})
Carbohydrates	0.23
Proteins	0.13
Fats	0.42

Table 5.4 Examples theoretical biomass yield coefficients of mixed cultures (abstracted from Mosey (1983); Proceedings of 3rd International Symposium on Anaerobic Digestion), see text for details.

Y values can be determined in terms of BOD or COD consumed

In most anaerobic digesters, we are not using single, well defined, substrates and it is perhaps more useful to describe yields, not in terms of mass of substrate used, but in terms of BOD or COD consumed. This is, of course, more applicable to the objective of anaerobic digesters to remove BOD or COD from a waste water.

If, for example, we had a sludge composed entirely of carbohydrates, proteins and fats, and if we know the proportions of these then, using the data presented in Table 5.4, it is possible to calculate a mean biomass yield coefficient. It should also be possible to calculate the BOD value for the same sludge. Thus it is possible to calculate, from theoretical consideration, a value of Y based on BOD.

Try the next SAQ which provides a useful example of the type of theoretical calculation that can be done.

SAQ 5.5

We have a sludge (waste water) containing 1g l^{-1} carbohydrate as the only oxidisable component. Under aerobic conditions this would be oxidised in the following way: $C_6H_{12}O_6 + 6O_2 \rightarrow 6CO_2 + 6H_2O$.

Use this information to determine:

1) Y in terms of biomass produced g^{-1} carbohydrates consumed in an anaerobic digester (use Table 5.4);

2) Y in terms of biomass produced per mg l^{-1} BOD.

The type of calculation we did in SAQ 5.5, shows that we can express biomass yield coefficients in terms of biomass produced per unit of BOD consumed.

By analogy, we could also express yield coefficients in terms of biomass produced per unit of COD. Thus if we know the BOD or COD value of readily biodegradable wastes and the yield coefficients for these wastes, we would be able to predict the yield of biomass from such wastes. For example if the BOD of a waste was determined as 1000 mg l^{-1} and the yield coefficient as 0.2 mg l mg^{-1} then we should anticipate that the yield of biomass would be 0.2 x 1000 mg = 200 mg. This, of course, depends upon all of the BOD being consumed and the yield coefficient remaining constant.

yield coefficients depend on the organisms and on physical and chemical parameters

Yield coefficients, of course, reflect the efficiency of the metabolism of the biomass and we could visualise the displacement of some organisms in the digestion by different organisms with different metabolic efficiencies. Also non-metabolisable factors such as salt concentration and elevated temperatures may influence yield coefficients.

∏ See if you can explain why salt concentrations and temperatures may influence yield coefficients.

importance of energy used for maintenance

All organisms consume energy to perform maintenance functions such as maintaining an osmotic balance or repair/replace denatured and inactivated cell constituents. This consumption of energy is not reflected in any increase in biomass. Thus organisms which have to expend considerable amounts of energy in maintenance will have lower yield coefficients than those which expend only small amounts of energy in maintenance. Factors, such as salt concentration and elevated temperatures, influence maintenance functions and thereby will influence yield coefficients.

5.4.3 Effects of chemical factors on digester performance

It should be self-evident that the performance of a digester will be largely governed by the nature (composition and biodegradability) of the organic input. There are, however, both inorganic and organic materials which exert an effect far greater than their proportions in the input.

∏ See if you can predict, what types of chemicals in the input may exert a major effect on digester performance.

From our discussion in Section 5.4.2, you probably thought of chemicals which may influence the maintenance function of cells. Although this in true, usually such factors only influence the performance of digesters in a rather minor way by slightly changing growth yield coefficients. Much more important is the presence or absence of essential trace elements and pre-formed biochemicals (for example vitamins) and the presence of toxins. We will deal with each in turn.

Essential nutrients

macro- and micro-elements

All organisms have a requirement for a broad spectrum of minerals. In addition to the major elements (carbon, hydrogen, oxygen, nitrogen, phosphorus, sulphur, magnesium, potassium and calcium), a wide variety of micro (trace)-elements are also required.

The methanogens particularly need iron, zinc, manganese, nickel, molybdenum and cobalt many of which act as cofactors for enzymes or are part of the prosthetic groups of enzymes.

If the incoming sludge is deficient in any one of the macro- or micro-elements or if these are present only in a form that cannot be utilised by cells, then the performance of the digester will be greatly impaired.

vitamins and
essential
amino acids

Similar, some micro-organisms are unable to make all of the biochemicals they require for cell synthesis. Typically amongst these compounds are the vitamins (which act as cofactors in enzyme-catalysed reactions) and amino acids. The presence or absence of these in the incoming sludge and the failure, or otherwise, of organisms to produce and release these within the sludge, will influence the types of organisms which will grow within the sludge. Thus, in turn, will influence the performance of the reactor.

∏ Which of the two waste waters listed below will most likely support active anaerobic digestion? 1) The outflow from a sweet factory containing mainly carbohydrates, but with some potassium phosphate. 2) The slurry from a pig farm consisting mainly of pig excreta.

The answer should be fairly obvious. The slurry from the pig farm will consist of a complex of organic materials and will contain the major elements in various forms. This will include compounds present in undigested food, the biochemical components of the gut organisms of the pigs and so on. It will therefore be biochemically rich and is likely (but not guaranteed) to contain a reasonable proportion of the major elements (such as nitrogen and phosphorus) and the trace elements. Thus it is likely to support a mixed and flourishing microflora within the digester.

In contrast the outflow from the sweet factory would appear to be deficient in many of the macro-elements such as nitrogen and sulphur and does not contain a diversity of biochemicals. It is, therefore, most unlikely to support a flourishing microflora.

The point we are making is that in setting up an anaerobic treatment process it is essential to think of the composition of the material to be treated in terms of the requirements of the desired microflora.

optimum
concentrations
of minerals
depends upon
the media and
organisms

It should be self evident that addition of essential elements to an otherwise deficient sludge should stimulate microbial activity. However, there is an optimal concentration for each element. Exceeding this concentration tends to be inhibitory. Although precise optimum concentrations depend, to some extent, on the operating conditions and the composition of the sludge, some general guidance can be given as to the optimum concentrations of various minerals in anaerobic digesters (see Table 5.5).

Metal	Optimum concentrations (mmol l^{-1})
Potassium	2-5
Calcium	2-5
Magnesium	3-8
Sodium	4-10
	Optimum concentrations (nmol l^{-1})
Molybdenum	40-70
Cobalt	40-70
Nickel	80-150

Table 5.5 Optimum concentrations for various metals in anaerobic digesters.

If the concentrations of these minerals greatly exceed their optimum concentrations, then they become inhibitory.

∏ Would you expect inhibitory concentrations of minerals to affect all the organisms present in the digester in the same way?

sensitivity to the toxic effects of minerals differs in different organisms

The answer should be no; some organisms have natural resistance to the toxic effects of minerals or do not carry out processes which are particularly sensitive to particular cations. For example, high levels of zinc inhibit electron transport chain activities. Organisms not using electron transport chains (for example fermentative organisms) are generally less sensitive to these ions.

adaption of toxic metals

Also some organisms have adaptive mechanisms which allow them to circumvent the toxic effects of metal ions. For example, some organisms begin to produce chelating agents in the presence of high levels of particular ions. These agents sequester the toxic ions.

Toxins

In the previous section, we indicated that high levels of, otherwise, essential ions may be toxic to the microflora of a digester. Other ions, especially those of the heavy metals, have no apparent biological function and are especially toxic. Mercury and lead are good examples of these.

lead and mercury

many xenobiotic may inhibit anaerobic digestion

The toxins which may enter an anaerobic digester are not, however, simply inorganic in nature. Other toxins include man-made chemicals especially halogenated organics and phenols produced in a variety of processes and used for example as pesticides and disinfectants and solvents such as chloroform and carbontetrachloride may have a major inhibitory effect on anaerobic digester performance. These types of compounds are often regarded as recalcitrant (non-biodegradable) or xenobiotic.

presence of toxins in municipal waste waters is highly variable

This area is extremely complex and, as yet, we only have a limited knowledge of the effects of xenobiotics on anaerobic digestion. If you think for a moment of the great variety of materials that may arrive at a municipal water treatment unit, you will realise that there is the potential for enormous variation in the concentrations and types of potential toxins that may arrive at the treatment unit. Thus, in areas where textiles are manufactured, we might anticipate, for example, the presence of dyes; in areas of mineral manufacture we would expect to find higher levels of metal ions in the waste water whilst in areas of petrochemical manufacture we might anticipate the presence of hydrocarbon-derived xenobiotics.

There is increasing pressure to remove such materials from waste water prior to its arrival at the central municipal water treatment plant. We will examine some of these processes in greater detail later (see Chapter 6).

the effects of toxins are difficult to predict

Before we leave the topic of the effects of toxins on digester performance, we need to make one further point. From our earlier discussion, you will realise that the presence of a toxin or toxic concentrations of the essential elements influences the metabolism of organisms in the digester and may cause the replacement of some organisms by different species or strains. This, in turn, may influence the sensitivity of digester to other factors in the sludge. Thus the presence of one ion may alter the effect other ions have on the process.

We can, therefore, anticipate that in some cases an increase in the concentration of one ion may increase or decrease the sensitivity of the digester to other ions. Likewise, the presence of one type of organic toxin may increase or decrease the sensitivity of the digester to other materials. We are, therefore, dealing with an extremely complex situation in which the response of the system to a particular input depends very much on what else is present.

As yet, this complexity has only been partially unravelled. For example, we know that high levels of sodium ions reduce the response of the system to potassium, magnesium and calcium whilst the presence of calcium increases the response of the system to magnesium. We can distinguish two types of interactions:

- antagonistic interactions: those interactions in which the presence of one component reduces the effect of another;

- synergistic interactions: those interactions in which the presence of one component increase the effect of another.

We have represented these situations diagrammatically in Figure 5.9.

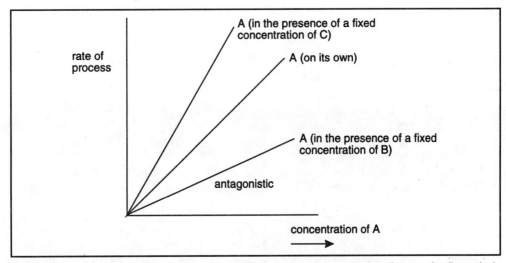

Figure 5.9 Diagrammatic representation of antagnostic and synergistic interactions (see text for discussion).

The representation of synergistic and antagonistic interactions shown in Figure 5.9 shows how the rate of the process is effected by increasing concentrations of A in the presence and absence of fixed concentrations of B and C.

∏ What would you anticipate to be the effect if we used a higher concentration B?

The most probable effect would be that the slope of the line would be lower (that is the higher the concentration of B the greater the antagonism).

∏ What would be the effect if we added both B and C?

This would be extremely difficult to predict. The effect of B might be completely over-ridden by the effect of C or alternatively, the effects of B and C may only partially cancel out each other. If you keep in mind the number of molecular constituents present in a typical digester and the range of concentrations of each that might be encountered, then you will realise that modelling the effects of adding or removing particular constituents is, at this time, very difficult. Although this area is under active investigation, it is usual to investigate such effects using pilot/bench top devices before changing to large scale processes.

5.4.4 The kinetics of anaerobic digestion

So far we have considered anaerobic digestion from a biochemical/microbiological perspective. It is, however, essential that we consider the kinetics of the process. We need to know, for example, the residence time of the sludge in the digester needed to achieve effective degradation. This, together with our anticipated volume of feed to the digester, will enable us to calculate the size of the vessel needed to carry out digestion and the rate of output from the vessel.

Traditionally, the Monod relationship has been used to describe the rate of substrate consumption. This may be written as:

$$\frac{d[S]}{dt} = -q[X]\left(\frac{[S]}{K_s + [S]}\right)$$

where:
[S] = substrate concentration;
[X] = biomass concentration;
q = maximum rate of substrate consumption per unit of biomass (maximum specific substrate utilisation rate);
K_s = Monod constant or substrate concentration at which the rate of substrate consumption is half its maximum rate.

(Note that the derivation of this equation is given in the BIOTOL texts 'Bioprocess Technology: Modelling and Transport' and in 'In vitro Cultivation of Micro-organisms'). You will note the similarity of this equation with the Michaelis-Menten equation relating the velocity of an enzyme catalysed reaction to substrate concentration

$$v = \frac{V^{max}[S]}{K_M + [S]}).$$

In principle, therefore, if the substrate concentration, K_s, q and biomass concentrations are known, we can calculate the time required to reduce the concentration of substrate to a pre-fixed low value. To do this we need to use an integrated form of the Monod equation. If we assume we are operating a batch process and that the biomass concentration remains constant, then:

$$q[X]t = 2.303\ K_s\ \log\frac{[S_o]}{[S_t]} + ([S_o]-[S_t])$$

where [S_t] = substrate concentration at time t and [S_o] = initial substrate concentration and [X] is the biomass concentration.

This equation also assumes that the activity of the biomass remains the same throughout. Thus, if we know q, [X] and [S_o], then we can calculate the time needed to achieve a particular reduction in the substrate concentration.

Try the following calculation.

A batch operated process is set up. It has a biomass concentration of $1g\ l^{-1}$ and $1g$ of biomass can consume substrate at a rate of $0.1g\ h^{-1}$. If the K_s value for this substrate is $10mg\ l^{-1}$ and substrate is supplied at a concentration of $10g\ l^{-1}$ how long will it take for 90% of the substrate to be consumed? (Assume that there is no increase in biomass)

The types of relationships between biomass, substrate uptake rates, incubation time and changes in substrate concentration described above are generally useful for bioreactor systems but are largely inappropriate for anaerobic digesters.

$\prod$ Try to explain why this is so?

the substrate used in anaerobic digesters in a chemically complex mixture

Firstly the substrate present in sludge consists of a wide variety of different compounds, used by a variety of different organisms. We should anticipate that q, K_s and [X] will be different for each group of input chemicals. For example, the lipid portion of the input may be metabolised by only a portion of the biomass whilst other organisms metabolise the proteins or the cellulose in the sludge. Some constituents of the sludge will be more readily degraded (ie high q values) than others (ie low q values).

residence time required depends on what is to be metabolised

Thus, we must anticipate that the residence time required to degrade readily metabolised intermediates would be much shorter than that required by more recalcitrant materials. This is, in fact, what is observed. Easy to metabolise materials such as low molecular weight organic acids usually require a residence time of the order of 5-7 days while hydrocarbons and long chain fatty acids may require residence times of the order of 1-2 months.

A decision has, therefore, to be reached about the required residence time. A short residence time (5-10 days) will bio-convert the lower molecular weight components but will leave the more recalcitrant material in the sludge. A longer residence time (40-60 days) will achieve better biodegradation but will require much larger vessels to treat the same volume of sludge. Inevitably, a compromise has to be reached based on the composition of sludge, the desired degree of degradation of the sludge and the capital and operational costs of the process.

It is now time for us to turn our attention to the design and operation of the digesters.

5.5 Digester design and operation

In this section we will examine the actual devices used as digesters. We indicated earlier (Section 5.2) that a variety of devices have been developed. They range from simple septic tanks through to more complex stirred tank and fixed film reactors. Here our main focus of attention will be on a general description of the devices rather than on the engineering aspects. These aspects and the principles of process technology applicable to a wide range of bioreactors are examined in depth elsewhere in the BIOTOL series. In particular we draw your attention to 'Bioprocess Technology: Modelling and Transport Phenomena', 'Operational Modes of Bioreactors' and 'Bioreactor Design and Product Yield'. Although not specifically focused onto anaerobic processes, these texts provide details of the problems associated with mass and heat transport and explain the effects of process parameters on performance.

We will first briefly examine septic tanks before moving on to larger scale municipal sludge digesters. We will then examine so called stirred tank reactors. We will also describe single chamber digesters, clarifier processes and upflow sludge bed (blanket) processes. All of these processes essentially involve the use of suspended biomass. Because the retention of biomass in the digester offers potential advantages in terms of the loading rate that may be used, fixed film reactors offer some attractions. We will complete this section by examining some fixed film formats.

5.5.1 Septic tanks

Septic tanks are unstirred vessels which are usually operated at ambient temperatures. The volumes vary enormously but for sewage treatment about 200 litres are required per person. Most commonly the tank is divided into two chambers. The inflow enters the larger chamber (see Figure 5.10). This is fitted with a sloping floor and retains the sludge. The smaller chamber is usually about ⅓ of the total volume of the tank.

Thus the flow enters the larger chamber, is digested in the sludge, the 'cleared' water spills over into the second chamber and is discharged.

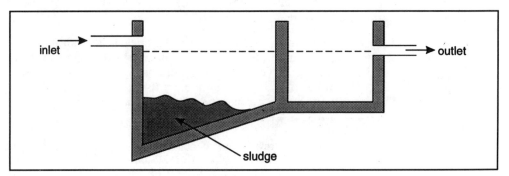

Figure 5.10 Diagrammatic representation of a typical septic tank.

The sludge is removed periodically (usually annually) although a portion is left to re-seed the reactor.

Typically about ⅕ of the sludge may be left. Thus the biomass remains in the tank for quite some time (on average, approximately ⅕ x 365 days = 73 days). With such a long residence time, the sludge retains little methanogenic activity and so little methane is generated.

| **SAQ 5.7** | In the UK, the design criteria used for septic tanks include the capacity of the tank should be calculated from the following equation: |

$$C = 180\,P + 2000$$

where C = capacity in litres, P is the population generating the 'feed' for the tank. (Data from British Standards Institution: British Standard Code of Practice for Design and Installation of Small Sewage Treatment Works and Cesspools BS6297 (1983)).

1) What would the capacity of a septic tank need to be to treat the waste water produced by 50 individuals?

2) Given that each individual uses about 120 litres of water per day, on average, how long will this water remain resident in the septic tank.

5.5.2 Large scale municipal sludge digesters

Larger scale digesters are used in urban areas and are often associated with other water treatment processes (see Chapter 4). Primary sludges and collected activated sludge from aerobic processes are used as the feed for these devices. After digestion, the sludge is disposed of either to landfill or as agricultural fertilisers. This latter use is strictly controlled since the persistant use of spent sludge as a 'fertiliser' may result in the build up of toxic ions in soils with concomitant losses in crop productivity.

The main aims of large scale municipal sludge digesters is to:

• reduce pathogens;

• reduce odours;

• stabilise the solids.

All of these aims facilitate the final disposal of the sludge.

A wide variety of designs has been generated but the most common one involves the use of two tanks. In the first tank, the sludge is often maintained at about 30-40°C (mesophilic) and it is here that the digestion takes place. The second tank acts as a kind of storage vessel in which the stabilised sludge is allowed to compact. This second vessels is unheated and unmixed. A diagrammatic representation of a municipal sludge digester is given in Figure 5.11.

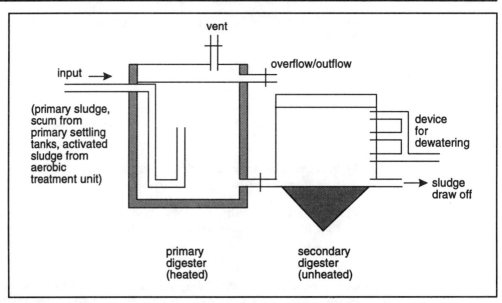

Figure 5.11 Simplified representation of a two tank municipal sludge digester.

∏ Examine Figure 5.11. It is a great simplification of the real device. For example we have not shown control valves, taps, pumps, mixing and heating devices. How do you think that heat is supplied to the primary digester?

It is usually delivered by circulating warm water through pipes within the digester.

The complex and variable nature of municipal waste water means that it is extremely difficult to apply a rigorous process engineering approach to the design of these processes. A rather empirical approach is used. Of primary value are 'rules of thumb' relating to volumetric loading rate, % solids, biodegradability and retention times. In Table 5.6, we list some of the difficulties which are encountered in predicting the probable performance of municipal digester.

Factor	Comment
Toxins	Toxins may reduce bioconversion rates - particularly important in highly industrialised areas.
Climate	Fluctuations in ambient temperature may disrupt the process.
Seasonal changes	Fluctuations in input (volume, content) may disrupt process (eg local annual holidays may markedly change input temporarily).
Inhomogeneity	Large vessels are usually inadequately mixed. This coupled to the presence of large particles leads to inhomogeneity and may lead for example to localised depression of pH.
Presence of recalcitrants	Poorly degraded substrates (eg the lipids in the scums from primary settling tanks) may influence reactor performance.

Table 5.6 Factors which make it difficult to accurately model municipal anaerobic digesters.

SAQ 5.8

The sludge in a municipal anaerobic digester is composed of the sludge from a primary settling (sedimentation) unit and sludge from an aerobic waste water treatment process. The composition of these sludges is given below.

Constituent	Primary sludge*	Activated sludge*
Nitrogen	5.0	8.2
Protein	30.0	50
Total P	1.5	3
Crude fibre	20	0
Polysaccharides (starch)	3	0
Lipid (ether extractable components)	28	3
Ash	14.0	15.0

* Figures are % dry solids

After anaerobic digestion, the sludge had the following composition.

Constituent	Digester product*
Nitrogen	4.4
Protein	21.2
Total P	2.8
Crude fibre	2.5
Polysaccharides	0.5
Lipid (ether extractable components)	6.0
Ash	27.0

* Figures are % dry solids

1) Explain why the crude fibre content is higher in primary sludge then in activated sludge and in the digester product.

2) Why is the ash as % dry solids greater in the digester product than in primary and activated sludge.

3) Explain why there is a high level of lipid material in primary sludge and in digester product, but a relatively low amount in activated sludge.

5.5.3 Specialised stirred tank reactors

The municipal anaerobic digesters described in Section 5.5.2 have to treat extremely complex and variable mixtures of materials. The waste waters from specific industries have particular compositions and especially those from food production and processing units often have high concentrations of biodegradable materials. Examples of these are from vegetable oil extraction units (which may produce waste waters rich in plant residues) sugar extraction units, brewing operations and intensive animal farming units.

Each of these effluents pose particular problems but, because their compositions are more or less defined, it is possible to optimise design and operational parameters. For example, the extraction of sugar from sugar beet produces pulps rich in cellulose. It is possible to improve the biodegradability of these pulps in anaerobic digesters by pretreating with cellulases or with cellulase-producing organisms such as various *Trichodeima spp*. Similarly, the high concentration of low molecular weight organic materials in the waste waters from breweries and soft drinks manufacture means that shorter residence times may be used.

In some industries, such as in palm oil extraction, the waste water leaves the production unit at an elevated temperature (50-70°C) thus a thermophilic digestion process is most appropriate in this case.

It is attractive to collect the methane produced and to use this as a fuel thereby reducing the net cost of water treatment. We will not extend the discussion of this aspect here as we will be examining this aspect of biotechnology in a later chapter. However, the production of methane by anaerobic digestion has received particular attention in the treatment of waste waters from pig and poultry farms. A variety of stirred tank reactors have been designed for this process. Typically the slurry from pig farms contains about 2-5% solids. To treat the slurry from 1000 pigs requires a reactor with a capacity of about 600m^3 and to use a retention time of about 10-20 days. The gas from such units contains 60-70% methane and may be use to provide heat and to generate electricity.

This important area is still poorly understood and often the performance of such operations fall short of the design expectations. Frequently, the feed rates and retention times have to be adjusted sometimes making the operations uneconomical. We should anticipate, however, that the combined effort of microbiologists and process technologists working together will lead to important improvements in this area of technology.

The general design of such processes is illustrated in Figure 5.12.

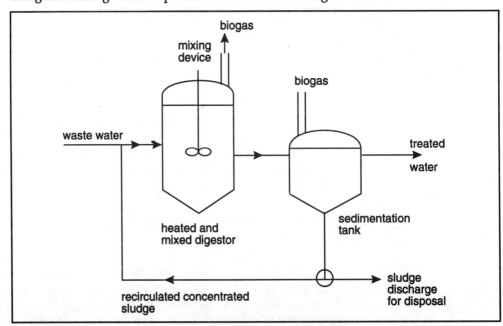

Figure 5.12 Diagrammatic representation of a two stage stirred tank digester.

Note that a portion of the sludge is returned to the digester in order to use short residence times. If sludge was not returned, the slow growth rates of the acetogens and methanogens would force the use of slow feed rates. In the following section we will describe some alternatives to sludge recirculation.

5.5.4 Single chamber digester/clarifiers

A variety of devices have been designed in which a single chamber is used for digestion and clarification. A stylised version of these is shown in Figure 5.13.

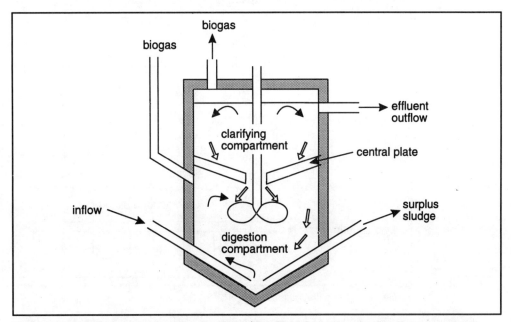

Figure 5.13 Stylised representation of a digester/clarifier. The inflow enters the digestion compartment where it is mixed with the sludge. Some of this passes through the centre plate and enters the clarifying compartment. Here the sludge settles back to the centre plate and slides back into the digestion compartment. Clarified water leaves the outflow at the top.

These vessels are usually divided into two compartments. The lower compartment acts as a digester. Sludge passes through the central plate which separates the two compartments. The liquid in the upper compartment is unmixed and sludge settles back to the central plate and is allowed to fall back into the digestion compartment. In this way the biomass is retained within the vessel. Usually the central plate is scraped to facilitate the return of settled sludge to the digester compartment.

These devices have been successfully used in treating distillary wastes, reducing the high BOD (20 000 mg l⁻¹) by 98% using a retention time of about·a week and an operational temperature of 33°C.

5.5.5 The upflow anaerobic sludge blanket process (UASB)

This is a further refinement of the digester/clarifier developed in the Netherlands in the 1970s. The main problem with the clarifier system is that the sludge only settles very slowly and is easily disrupted and resuspended. It was argued that if the biomass was produced as larger granules, perhaps of 1-5mm diameter, then they would settle more rapidly.

The ability to produce granules (flocs) depends upon the properties of the feedstock and the reactor conditions and granulation has not been achieved using raw sewage and slaughterhouse wastes. The formation of granules enables greater retention of biomass. However it is rather important to create regions in which gas is disengaged otherwise the gas bubbles will tend to keep the granules suspended. To achieve this, the reactor is fitted with baffles which collect the gas (see Figure 5.14). Careful design is necessary to either prevent or to physically remove, scum.

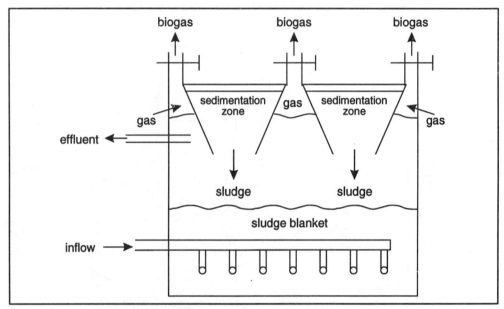

Figure 5.14 Stylised representation of a UASB reactor.

UASB reactors can generally cope with 12-15 kg BOD m^{-3} d^{-1} and, because biomass is retained, hydrolylic loadings can be high (typically up to 17m^3 m^{-2} d^{-1}).

5.5.6 Fixed biomass reactors

An alternative approach to retaining biomass in a vessel is to either allow it to grow attached to a solid support (carrier) or to entrap the biomass within a solid support.

Π Why is entrapped biomass with a solid support unlikely to be successful for the anaerobic treatment of many waste waters?

If the waste waters contain polymeric and particulate materials, then these will not come into contact with the entrapped biomass and will not, therefore, be degraded. Thus it is more common to fix the biomass within the reactor/digester by allowing it to grow as a film on the surface of supports producing what are sometimes referred to as biofilms. The simplest format is the anaerobic filter. A description of the formation of biofilms is given in the BIOTOL text 'Bioreactor Design and Product Yield'.

Anaerobic filters

In principle, these are similar to trickle filters except that the filter bed is enclosed and the feed is usually supplied by upward flow (Figure 5.15).

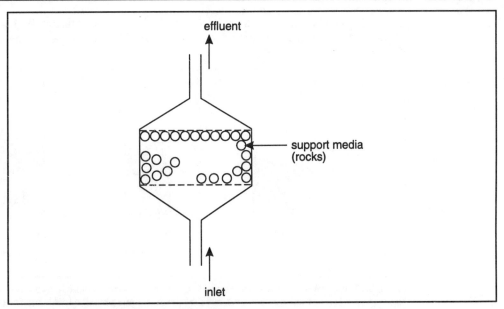

Figure 5.15 General design of an anaerobic filter.

The smaller the support granules, the greater the surface area for biomass retention but the greater the propensity to clog. Also flow rates have to be slow in order to prevent detachment of biomass.

∏ See if you can list three reasons why these devices have found only limited application.

There are several reasons including:

- the propensity to clog and the need to periodically backwash the filter bed;

- the slow flow rates;

- the high costs of support media with the desirable characteristics;

- they are not effective in removing particulate materials or in treating waters with a high organic content.

Another important fact is that other configurations show appreciable advantages.

The type of anaerobic filter illustrated in Figure 5.15 can also be fed by a downflow. The production of biogas in these systems facilitates mixing and distribution in the vessel.

Anaerobic fluidised bed reactors (AFBRs) and anaerobic expanded bed reactors (AEBRs)

This area has seen some important developments in process engineering. In essence, the biomass is grown attached to a suitable support (for example sand). The waste water is then pumped upwards through the bed of sand at such a rate as to cause the sand to become fluidised. Note the distinction between expanded and fluidised beds is not well

distribution
between
expanded and
fluidised beds

defined. Usually if the bed is expanded by 10-20% of its static height it is regarded as an expanded bed, if the expansion is greater than 30%, it is regarded as a fluidised bed. This is illustrated in Figure 5.16.

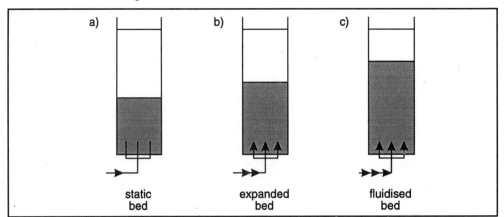

Figure 5.16 The distinction between expanded and fluidised beds. a) represents the bed with no media flow. In b) the flow of medium forces the bed particles to separate and the bed is expanded. In c) the high flow rate of the medium separates the bed particles even more and the bed becomes fluidised.

In operation, fluidised and expanded bed digesters are usually associated with a recycling device (see Figure 5.17) which also serves as a settling tank to return any sludge which is expelled from the digester. In some instances, these recycling tanks are fitted with ports for the addition of specific nutrients if these are required.

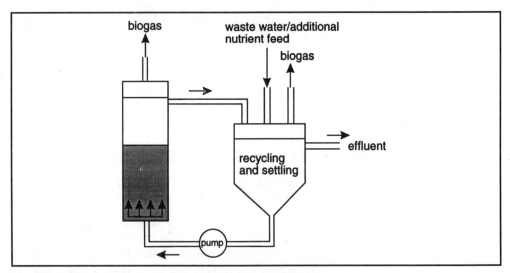

Figure 5.17 Diagrammatic representation of a fluidised bed digester.

Π Try to make a comparison between an anaerobic filter and a fluidised bed. Think in terms of the propensity to become clogged, the thickness of the biofilm on the particles and the power input required to operate the process.

In fluidised bed digesters, the particles are physically separated and are free to move, whilst in anaerobic filters the particles are static and tend to adhere to each other. Thus particulate matter tends to clog the spaces between the particles in anaerobic filters but is less likely to do so in fluidised and expanded bed digesters. In anaerobic filters the flow rate is low thus sheer forces are low as well and, as a consequence, the biofilm on the particles becomes thick. Under these conditions, the diffusion of substrates into the biofilm may become rate-limiting. In fluidised beds, the sheer forces are much greater, the biofilm remains thin and is thus less likely to be diffusion limited. We can illustrate this in the following way:

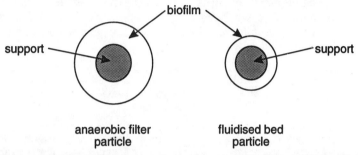

anaerobic filter
particle

fluidised bed
particle

effectiveness
factor

Thus, in an anaerobic filter a substantial portion of the biofilm may be inactive due to substrate limitation. In process technology terms, we can speak of the effectiveness factor (η) of the biofilm. The effectiveness factor is the ratio of the actual activity of the biomass: the maximum activity of the biomass. Thus:

$$\eta = \frac{\text{actual activity}}{\text{maximum activity}}$$

Clearly the greater the degree of diffusion-limitation, the lower the value of η. We point out that the BIOTOL text 'Operational Modes of Bioreactors' provides an indepth discussion of the formation and behaviour of biofilms and discusses the modelling of immobilised systems including the determination of the effectiveness factor. Additional discussion of biofilms is given in the BIOTOL text 'Bioreactor Design and Product Yield'.

It should be self evident that the power required for expanded and fluidised bed digesters is greater than that required for simple anaerobic filters.

The achievement of successful fluidised bed digestion requires substantive process engineering input. For instance, it is desirable to generate a uniform flow across the reactor and to achieve high flow rates without causing turbulence which will cause biomass detachment from the particles. These engineering aspects are largely outside the scope of the text but may be followed up using the BIOTOL texts mentioned above and the appropriate texts mentioned in the 'Suggestions for Further Reading' provided at the end of this text.

5.6 Choice of anaerobic digesters

In the previous sections, we gave brief descriptions of a range of anaerobic digesters ranging from the simple septic tank through to the highly engineered fluidised bed digesters. All of these have some advantages and some disadvantages and the choice is to some extent influenced by circumstances. For example, it would be a nonsense to use a highly engineered and expensive fluidised bed digester to treat the sludge derived from a small, remote rural community, a simple septic tank would be sufficient. In contrast, the long residence times needed for septic tank operations makes these impractical for the treatment of waste water from a large urban population.

Considerable data have been accumulated relating to the capabilities of the various digester designs to remove BOD and the residence times required. These features, of course, depend not only on the design of the digesters and on the operational parameters (for example temperature) but also on the composition of the material being treated. Here we have elected not to provide a lot of specific figures for specific processes, but to provide you with a generalised comparative summary. Examine Table 5.7 carefully.

Feature	Septic tanks	Municipal sludge digesters	Stirred tank reactors	Anaerobic filters	Upflow blanket digesters	Fluidised bed digesters
Material treated	Household sewage farm effluent	Municipal sewage	Most frequently specific industrial wastes	Waste waters with low organic loadings	Mainly specific waste waters from industry	Newly introduced, full capabilities yet to be exploited
Capital costs	low	low	medium	medium	relatively high	relatively high
Running costs	low	low	medium	medium	medium	high
Typical residence time	50-70d	14-40d	2-20d	4-10d	5-14d	variable, short
Ability to handle chemically and physically diverse inputs	high	high	variable, recalcitrants not degraded	tending to clog with particulate materials	variable, recalcitrants not degraded	variable, depends on design and operational parameters
Biomass retention in the digestevr	needs to be removed periodically	no	no	yes - but some leaching	yes - but some leaching	yes
CH_4 generation	extremely low	low	yes	yes	yes	yes
Typical loadings (BOD m^{-3} d^{-1})	low	low	< 20 kg	< 20kg	< 15 kg	< 40 kg

Table 5.7 Generalised comparison of various designs of anaerobic digesters.

The comparisons made in Table 5.7 are very generalised since the actual performance of these digesters depend to a large extent on the details of the each design and the operational parameters being employed. It also, of course, depends very much on the chemical and physical nature of the materials to be digested. Thus, for example the performance of a stirred tank reactor depends on whether the liquor being treated contains mainly soluble, easy to degrade sugars or cellulose. Pretreatments of the liquor may also greatly influence performance.

Bearing this in mind, attempt the final SAQs of this chapter.

| **SAQ 5.9** | Although most waste waters are treated in aerobic digesters, we would like to test your understanding of the features of the major types of anaerobic digesters. Select the most appropriate anaerobic digester for the treatment of waste water for the following circumstances. |

1) A factory producing confectionary products. Its waste water contains high levels of soluble sugars and starch.

2) A small village community (50 people) in an isolated region in the Alps.

3) A large pig farm in a marsh land region.

4) A small town (population 20 000) in a rural region of the Netherlands.

5) A major city in Europe.

| **SAQ 5.10** | Answer either 'anaerobic digestion' or 'aerobic digestion' to each of the following questions. |

1) Which type of digestion generates most sludge?

2) Which type of digestion general incurs most running costs?

3) Which type of digestion may generate a usable fuel?

4) Which type of digestion requires the shorter residence time?

Summary and objectives

In this chapter, we began by giving a brief account of the historical development of anaerobic digesters before moving on to a consideration of the microbiological and biochemical background of anaerobic waste water treatment. We explained how anaerobic digestion can be divided into three phases involving hydrolytic organisms, heteroacetogens and methanogens and we described some of the features and activities of these organisms. We then turned our attention to some theoretical considerations in anaerobic fermentation. In particular, we examined the use of elemental balances and yield coefficients to predict digester performance. We also described the effects of chemical factors on digester performance and distinguished between nutrients which enhanced activity and toxins which inhibited digestion. We also briefly considered the kinetics of digestion. We then briefly described a range of digester designs including aseptic tanks, municipal anaerobic digesters, single chamber digester/clarifiers, upflow anaerobic blanket digesters and fluidised bed digesters. We drew attention to the major features of each and concluded the chapter by making a generalised comparison of various designs.

Now that you have completed this chapter, you should be able:

- describe the three phases of anaerobic digestion which leads from the organic waste to the production of CH_4, CO_2 and biomass;

- predict the consequences of inclusion of reducible inorganic ions such as NO_3^- and SO_3^{2-} in anaerobic digesters;

- list advantages and disadvantages of using mesophilic digestion rather than thermophilic digestion;

- list examples of the micro-organisms that might be encountered in anaerobic digesters;

- explain why biomass yields are lower in anaerobic digesters than in aerobic digesters;

- describe some of the distinguishing features of methanogens;

- develop simple elemental balances from supplied data and to use these balances to predict outcomes from anaerobic incubations;

- use supplied data to calculate growth yields in terms of biomass produced per unit of BOD consumed;

- explain why nutrients may also become toxic and explain the terms antagonistic and synergistic in relation to the minerals present in an anaerobic digestion;

- describe, in outline, a wide range of anaerobic digesters and make a comparison of the uses and performances of these digesters;

- select appropriate digesters to suit particular circumstances.

The biodegradation of xenobiotic compounds

The biodegradation of xenobiotic compounds

6.1 Introduction

microbial infallibility

In 1965, Alexander put forward the principle of microbial infallibility which stated that no natural organic compound is totally resistant to biodegradation given the right environmental conditions. Indeed, we do not observe the accumulation of large quantities of natural organic materials in the biosphere because they are subject to microbial recycling. Fossil fuels, such as oil reserves, do accumulate but only because the conditions of their location are unfavourable to microbial metabolism.

The development of synthetic organic chemicals during the last 60 years has, however, resulted in the production of a wide variety of synthetic organic compounds that have inevitably found their way into the environment either deliberately or accidentally. Most of these synthetic organic compounds are susceptible to biodegradation due to their similarity to naturally-produced organic compounds; other are less biodegradable.

xenobiotic compounds

recalcitrant

Man-made compounds that are detected in the environment at unusually high concentrations are termed xenobiotic. This term is also applied to those compounds that do occur naturally but, due to Man's activities, are deposited in the environment at unnaturally high concentrations. Such xenobiotic compounds are not readily biodegradable since their molecular structures or bond sequences are not readily recognised by existing degradative enzymes. The net result is that these compounds are either resistant to biodegradation or are metabolised incompletely or slowly and consequently persist and accumulate in the environment. Compounds that persist (or are persistent) in the environment are termed recalcitrant.

In this chapter, we will begin by examining the range and sources of xenobiotic compounds and discuss the problems which arise from the release of xenobiotics into the environment. We will then go on to give an overview of the metabolic pathways involved in xenobiotic degradation. We will consider what factors influence the recalcitrance of these compounds and consider some of the practical issues relating to the utilisation of biodegradation to reduce the environmental problems which may arise from the production and release of xenobiotics.

This is quite a long chapter containing a lot of detailed information so do not attempt to study it all in one sitting.

6.2 Types of xenobiotic/recalcitrant compounds and their source

∏ Before continuing, list examples of the types of organic compounds that are resistant to biodegradation.

halocarbons
are stable

There are many examples you could have listed. Here we will include some of the main ones. As a general rule, compounds which are substituted, insoluble or of high molecular weight are often very resistant to enzyme attack. Many compounds are resistant to biodegradation due to the incorporation of halogens in the molecule. The carbon-halogen bond is a highly stable one and cleavage of this bond requires a substantial amount of energy to be put in (endothermic reaction). Halocarbons are thus chemically and biologically stable. Stability is of course a desirable property for their application, for example, as pesticides, but it is undesirable with respect to the 'environmental behaviour' of the compound as it results in persistence. The substitution of hydrogen on carbon skeletons with other chemical groups, such as nitro-, amino- and sulphonate groups also renders them resistant to biodegradation.

Cyclic structures, aromatic compounds, cycloalkanes, and heterocyclic compounds are also more resistant to biodegradation than linear ones due to the chemical stability of their ring structures. Often, the more complex the unnatural compound the more resistant it is to biodegradation. Other compounds are resistant to biodegradation due to the actual size and solubility of the compound, this is true, for example, for polymeric compounds such as plastics.

Figure 6.1 lists examples of the types of compounds which are recalcitrant (resistant to biodegradation) together with a description of their chemical structures.

Π Study the structures in Figure 6.1. List those compounds that are halogenated and therefore potentially toxic. Secondly, select those compounds that contain a cyclic or aromatic ring structure and are, therefore, potentially toxic and insoluble in water and finally select those compounds that are probably poorly soluble in water due to their large molecular size.

Halogenated, potentially toxic compounds are types 1a, 1b, 2, 3, 4, and 5. Those containing cyclic or aromatic ring structures are types 2, 3, 4, 5, 6c, 7 and 8. Finally, the large molecular weight compounds giving poor solubility are 6a, 6b and 6c.

Other reasons for recalcitrant behaviour, not apparent from structural information, include:

• failure of the compound to induce the synthesis of degrading enzymes, even though it may be susceptible to their action;

• failure of the compound to enter the microbial cell, due to its molecular size;

• lack of permeases required for its transport across the membrane;

• insolubility in aqueous transport media, or adsorption to external matrices such as soil particles;

• excessive toxicity of the compound, or its metabolic products, to microbial cells.

Let us now briefly discuss the sources of these various pollutants.

	Type of compound	Example	Structure
1)	Aliphatic halocarbons	a) Chloroform	$\begin{array}{c} Cl \\ | \\ H-C-Cl \\ | \\ Cl \end{array}$
		b) Dichchloropropionate ('Dalapon')	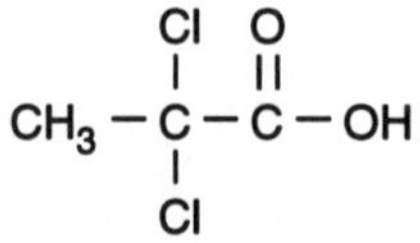 $CH_3-\underset{Cl}{\overset{Cl}{C}}-\overset{O}{\overset{\|}{C}}-OH$
2)	Cyclic halocarbons	Hexachlorocyclohexane ('Lindane')	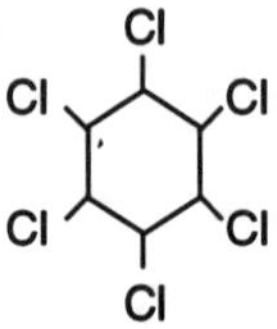
3)	Aromatic halocarbons	Pentachlorophenol	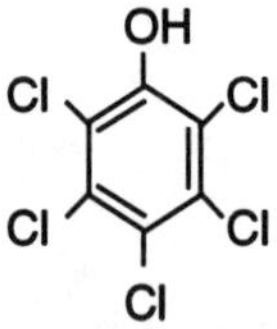
4)	Polychlorinated (X = H or Cl) biphenyls (PCB's), over 200 variations		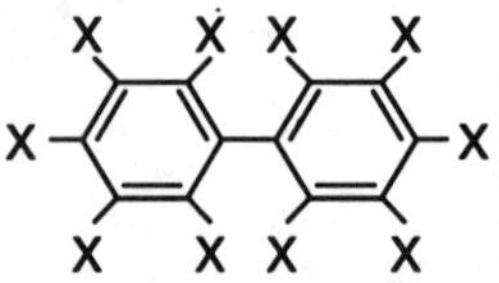
5)	Dioxins	Tetrachlorobenzo-dioxin (TCDD)	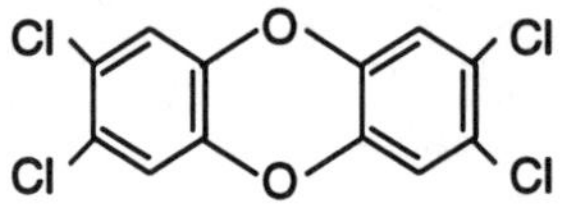
6)	Synthetic polymers	a) Polyethylene	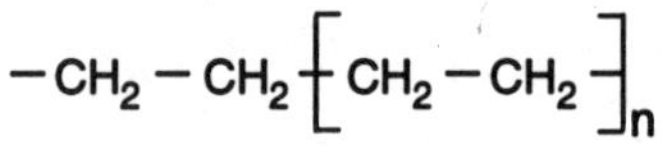 $-CH_2-CH_2\left[CH_2-CH_2\right]_n$
		b) Polyvinylchloride (PVC)	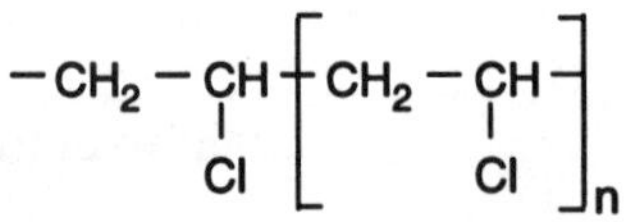 $-CH_2-\underset{Cl}{CH}\left[CH_2-\underset{Cl}{CH}\right]_n$
		c) Polystyrene	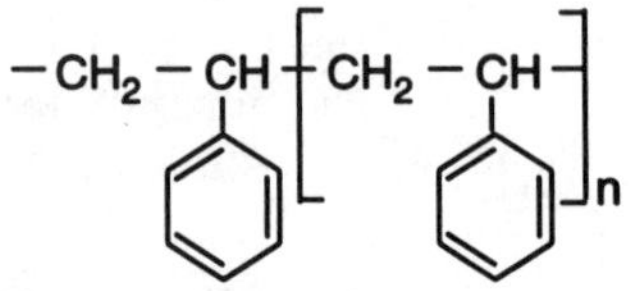
7)	Alkyl benzyl sulphonates		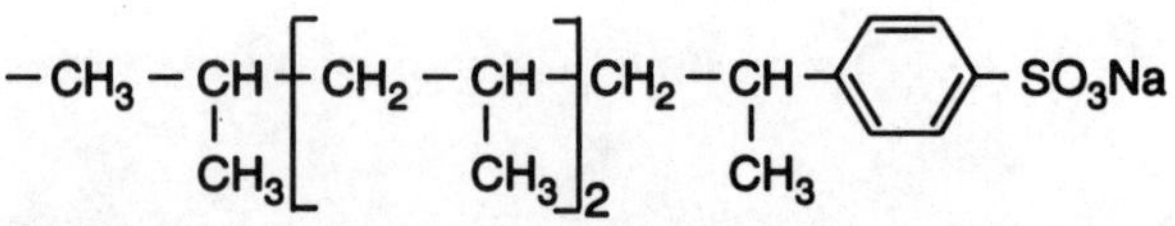 $-CH_3-\underset{CH_3}{CH}\left[CH_2-\underset{CH_3}{CH}\right]_2 CH_2-\underset{CH_3}{CH}\!\!-\!\!\bigcirc\!\!-SO_3Na$
8)	Oil	Mixtures of aliphatics, aromatics, cyclic hydrocarbons and polycyclic alkanes/alkenes	

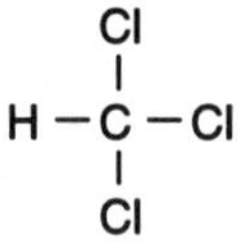

Figure 6.1 Examples of xenobiotic compounds.

Halocarbons

use as
solvents,
propellants
and in cooling
systems

Halocarbons are produced as solvents and aerosol propellants in spray cans for such things as cosmetics, paint, insecticides and in the condenser units of cooling systems such as refrigerators. Of most concern are the C_1-C_2 haloalkanes such as chloroform and the Freons (CCl_3F, CCl_2F_2, $CClF_3$, CF_4) which have been used as refrigerator coolants. They are no longer used for this purpose but are still present in existing systems. The biological inertness of these compounds in conjunction with their volatile nature means that they eventually find their way into the atmosphere where they cause destruction of the ozone layer surrounding our planet with the consequence of a decrease in the protection against harmful UV radiation.

many
pesticides are
halocarbons

Pesticides, such as herbicides or insecticides, are sprayed onto agricultural land and can leach from the land into water courses. These include halogenated aliphatics such as Dalapon; halogenated cyclic alkanes, for example Lindane; halogenated aromatics such as pentachlorophenol as well more complex molecules such as dichlorodiphenyltrichloroethane (DDT).

Polychlorinated biphenyls (PCB's)

PCB's used as
plasticisers,
heat and heat
exchange fluids

Polychlorinated biphenyls (PCB's) are produced industrially as plasticisers, insulator coolants in transformers and as heat exchange fluids in general. To a lesser extent, they are used in inks, paints, sealants, insulators and flame retardants. Their use is a consequence of their chemical and biological inertness which increases with the number of chlorine atoms present in the molecule.

Synthetic polymers

polymers used
as garments,
disposable
goods and as
wrapping
materials

Synthetic polymers are produced industrially as plastics (for example, polyethylene, polyvinychloride, polystyrene) and nylons which have widespread application as, for example, garments, disposable goods and wrapping materials. Their large molecular size and insolubility in water is mainly responsible for their recalcitrant behaviour.

Alkylbenzyl sulphonates

alkylbenzyl
sulphonate
used as
detergents

Alkylbenzyl sulphonates are produced industrially as surface-active anionic detergents. These molecules have polar (sulphonate) and non-polar (alkyl) ends. These properties make them superior to conventional soaps for the emulsification of fatty substances and hence in cleaning. The sulphonate group renders the molecule resistant to microbial attack from that end and if the alkyl end is highly branched, it resists biodegradation by β-oxidation. Nowadays, compounds with alkyl chains which are not branched are used because they are susceptible to β-oxidation and are, therefore, biodegradable.

Oil mixtures

Oil is a natural product composed of many components all of which are susceptible to biodegradation though at varying rates. Indeed it is accepted that constantly, low-level seepage of oil occurs but biodegradation avoids a build up. Problems arise due to the deposition of large quantities of oil into the environment, usually due to human error. Oil is considered as a xenobiotic due to its persistent nature caused mainly by its poor solubility in water and the toxic nature of some of its components.

Other xenobiotics

These include the multitude of pesticides other than halogenated aromatic pesticides. Such pesticides are based on aliphatics, cyclic ring structures that have substituents of nitro-, sulphonate, amino-, methoxy-, and carbamyl groups, often in addition to halogen substituents, which also render the carbon-skeleton resistant to biodegradation.

6.3 The potential problems of xenobiotics

pollution
potential

Some xenobiotic compounds can be a serious threat to the biosphere/environment due to their chemical nature or toxicity. Chemicals are thus often assessed according to their pollution potential which considers the following parameters:

- the aquatic and mammalian toxicity;

- carcinogenicity;

- level of production;

- reactivity;

- bioaccumulation (biomagnification);

- persistence or recalcitrance.

Such pollution potentials are used to assess those compounds that pose an immediate threat to the environment/biosphere.

Table 6.1 below summarises the major xenobiotics considered to be priority pollutants by the US Environmental Protection Agency (EPA). (We will deal with the regulatory aspects of this more fully in Chapter 7).

∏ Examine Table 6.1 carefully and see if you can decide why these compounds have been identified as priority pollutants?

Priority pollutants	No. of classified chemicals
Chloroaliphatics (eg chloroform)	31
Pesticides (eg polychlorobiphenyls)	26
Polycyclic aromatic hydrocarbons (eg anthracene, phenathracene)	17
Chloroaromatics (eg 2, 4, 6-trichlorophenol)	15
Simple aromatics (eg bis (2-ethylhexyl) phthalate)	13
Nitrogen-containing compounds (eg cyanide, 2-nitrophenol)	13
Metals (eg copper)	13

Table 6.1 The major xenobiotics considered to be priority pollutants by the US Environmental Protection Agency.

The answer should be fairly self-evident. They obviously score highly on the parameters used to assess their pollution potential (for example, toxicity, carcinogenicity, level of production, bioaccumulation, persistence etc).

Certain xenobiotic compounds, particularly the halogenated and aromatic hydrocarbons are inherently toxic and present a danger to the lower life forms such as bacteria, lower eukaryotes and shellfish as well as to higher animals, and humans. Such compounds at high concentrations can be fatal to mammals but more likely are the effects of low concentrations of these compounds which can result in various skin problems or loss of normal reproductive capabilities. Certain halogenated hydrocarbons have been implicated in the causation of cancer.

Many xenobiotic compounds tend to be dispersed by environmental forces over large areas such that the compound is only present in very low concentrations (less than a few parts per billion) compared to its initial deposition concentration. Such a dilution may render many chemicals no longer a toxicity threat. However, certain chemicals which include DDT and PCB's are subject to a process called biological magnification (or biomagnification). We described this process briefly in Chapter 3. It is, however, so important that it is worthwhile recapping on the main principles. Biomagnification occurs for those pollutants that are both persistent and lipophilic. Their lipophilic character means that they are readily absorbed from dilute aqueous systems into the lipids of both prokaryotic and eukaryotic cells. Thus by continuous absorption throughout the generation or life-span of the organism, the persistent nature of the compound means that its concentration will gradually increase within the organism. These organisms will, in turn, be ingested by higher life forms and so on until the organisms at the end of the food chain will contain concentrations of the compound which may substantially exceed the initial deposition concentration. Increases in concentration of 10^4-10^6 can be expected by biomagnification. DDT and other various toxic chlorinated hydrocarbons have been implicated in the causation of death of various animals and birds due to biomagnification.

biological magnification

biomagnification

Π We will examine the case of DDT in a little more detail. Examine the data in Table 6.2. Calculate the overall increase in concentration of DDT from that present in the aquatic environment for each of the biological localities and write your answer in the space provided in the table. The calculated value for plankton is given as an example and shows that plankton adsorbs DDT from the aqueous environment to a concentration 100 fold greater than that of the external environment.

DDT localisation	DDT concentration	Overall increase
Aquatic environment	0.3 ppb (parts per billion) (that is 1000 million)	0
Plankton	30 ppb	x 100
Small fish	0.3 ppm (parts per million)	x ___
Large fish	3 ppm	x ___
Ospreys	30 ppm	x ___

Table 6.2 Data relating to the biomagnification of DDT (see intext activity).

Small fish which feed on large amounts of plankton further concentrate the DDT from 100 to 1000 times that originally present in the aquatic environment. Large fish feeding on the small fish further concentrate it another 10 fold and ultimately ospreys which feed on the large fish result in a final DDT concentration in their bodies of 30 ppm which is 100 000 times higher than that originally present in the aquatic environment.

DDT and PCB's have been found in high but sub-lethal concentrations in humans, in those countries where they have been used, despite the people never having been in direct contact with the chemical.

SAQ 6.1

Decide whether the following statements are true or false and give reasons for your decision.

1) Halogenated aromatic pesticides are often compounds that are not readily degraded naturally and are therefore termed 'recalcitrant'.

2) Cyclic and aromatic compounds are always considered as xenobiotics.

3) Polyvinylchloride is considered as being xenobiotic mainly because of the toxic nature imposed by the halogen substituents on the molecule.

4) Dilution of potentially toxic chemicals is a common and safe method of disposal since compounds are not toxic to life forms unless they are above a critical concentration.

5) Plasticisers, inks, insulator coolants for transformers and paint sealants are considered as particularly dangerous pollutant sources because they contain high concentrations of unsubstituted aromatics.

6.4 The biochemistry of xenobiotic biodegradation

Because xenobiotics include a complex variety of compounds, we must anticipate that their degradation involves a wide variety of metabolic pathways. It is beyond the scope of this text to examine all of these pathways in detail. Here, we will first establish some general principles and then examine some specific pathways in more depth. If you are interested in learning more about this area of metabolism, we recommend the BIOTOL text 'Energy Sources for Cells' which provides an indepth discussion of the catabolism of hydrocarbons and man-made chemicals.

6.4.1 General principles of the catabolism of xenobiotics

In general, xenobiotics are either recalcitrant because they are chemically stable or their catabolism leads to the production of toxic compounds. With stable compounds (for example alkanes and aromatic hydrocarbons), it is usual for organisms to first introduce reactive groups (especially hydroxyl groups) using a mixed oxygenase. Frequently, this process involves cytochrome P_{450} or rubredoxin. We have illustrated this process using the oxidation of n-alkanes to primary alcohols in Figure 6.2.

cytochrome P_{450} and rubredoxin

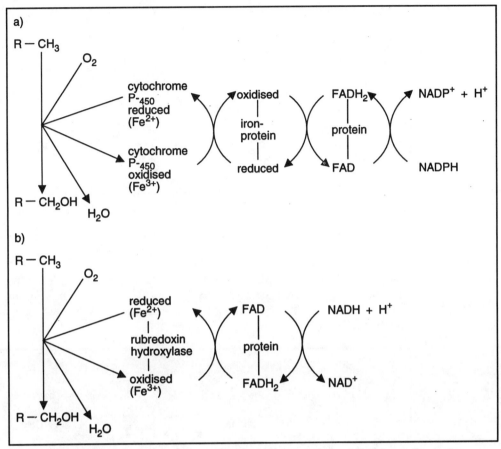

Figure 6.2 Mechanism for the oxidation of n-alkanes to primary alcohols. a) Cytochrome P$_{450}$ system. b) Rubredoxin system.

monoxygenases

The enzymes shown in Figure 6.2 are both monoxygenase because they introduce a single oxygen atom into the substrate molecule.

The introduction of a hydroxyl group enables cells to oxidise the compound further, first to form an aldehyde and then an acid group. Thus:

$$R-CH_3 \longrightarrow R-CH_2OH \longrightarrow R-CHO \longrightarrow R-COOH$$

NAD(P)H   NAD(P$^+$)   NAD(P$^+$)   NAD(P)H   NAD(P$^+$)   NAD(P)H
+ H$^+$ + H$^+$ + H$^+$

With linear alkanes, the carboxylic acid may be oxidised by the β-oxidation pathway leading to the production of acetyl CoA which can be further metabolised via central metabolic pathways such as the tricarboxylic and cycle.

With alicyclic hydrocarbons, the principle is the same. In Figure 6.3 we illustrate the initial breakdown of cyclohexane. You will notice that this pathway involves two oxygenases, one introduces an initial hydroxylic group into the ring, the second leads to the formation of an ester (in the form of a lactone) which is readily hydrolysed and the ring structure is opened to form a linear molecule.

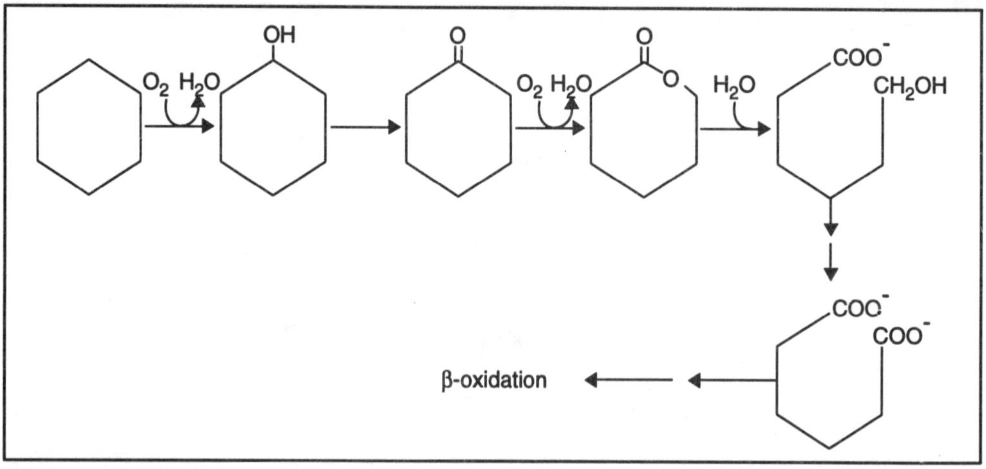

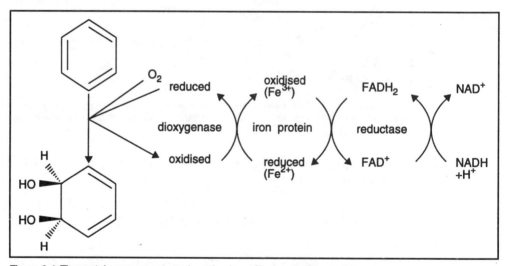

Figure 6.3 A proposed pathway for the microbial degradation of cyclohexane.

The same general principle applies to the oxidation of aromatic hydrocarbons. In this case, two oxygen atoms are added to the substrate and the enzymes are referred to as dioxygenases. We have illustrated this with the enzyme benzene dioxygenase in Figure 6.4.

dioxygenases

Figure 6.4 The protein components and mechanism of benzene dioxygenase.

Thus the general principle we can identify for the catabolism of hydrocarbons is that they are first activated by the addition of oxygen using one of a variety of oxygenases. The products are subsequently further oxidised leading to products which can be fed into central metabolic pathways.

∏ Will all monoxygenases have the same specificity?

You should have anticipated that the answer to this is no. A wide range of monoxygenases (and also dioxygenases) exist, some react best with short chain alkanes others with cyclic alkanes and so on. Thus, just because an organism contains a monooxygenase it does not mean that it will be able to metabolise all alkanes equally well. However, oxygenases tend not to be highly specific and each will oxidise a limited range of compounds. It should be self evident that the degradation of these xenobiotics in the environment will be carried out by a wide variety of organisms each capable of catabolising a limited range of compounds.

hydrocarbon-
degrading
pathways

Most of the priority pollutants are substituted aliphatic, alicyclic or aromatic hydrocarbons. These compounds, if biodegradable, are metabolised via conventional hydrocarbon-degrading pathways. The substituent group is usually removed or modified as one of the initial reactions though substituted aromatic compounds may exhibit substituent removal during a later part of the pathway.

Now that we have established some general principles let us examine some of the pathways in a little more detail.

6.5 Basic metabolic pathways of hydrocarbon degradation

6.5.1 C-1 compounds

Xenobiotic C-1 compounds include halogenated methanes and cyanides. The metabolism of C-1 compounds poses a number of metabolic constraints on micro-organisms but we will not go into these problems here. A number of micro-organisms, the methylotrophs, have been isolated that are capable of metabolising C-1 compounds such as methane, methanol, formaldehyde etc. The basic pathway is outlined below:

$$CH_4 \longrightarrow CH_3OH \longrightarrow CH_2O \longrightarrow HCOOH \longrightarrow CO_2 + H_2O$$

| methane | methanol | formaldehyde (methanal) | formate (methanoic acid) | |

The initial reaction, from methane to methanol, is catalysed by the enzyme methane monooxygenase which has been shown to have a relatively broad specificity in that it can transform a wide range of substrate analogues including halomethanes. Alternatively, oxidative dechlorination of halomethanes to the respective alcohol has been exhibited by a glutathione-dependent hydrolase, an anaerobic reaction in which the oxygen atom is derived from water.

halomethanes

cyanide

Cyanide (HCN) is found naturally at low concentrations, so it is not surprising that biological systems are present in the biosphere which will degrade it, albeit rather species limited. The basic pathways for cyanide metabolism are:

$$HCN \longrightarrow HCONH_2 \text{ by fungal hydratase}$$

$$HCN \longrightarrow NH_3 + CO_2 \text{ by bacteria eg } Pseudomonas \ fluorescens$$

The problems associated with cyanide which make it recalcitrant is that the biosphere is not adapted to coping with high concentrations of cyanide and it is toxic to biological systems at relatively low concentrations. Even the cyanide metabolising organisms are unable to cope with localised high concentrations.

∏ Although cyanides are only produced in low concentrations in natural systems, can you think of circumstances in which cyanides may be produced in high concentrations? What are the environmental consequences of these high concentrations?

Cyanides are produced and used in many chemical processes. Frequently, these processes produce high concentrations of cyanides in aqueous effluents or as solid residues. The release of these wastes into the environment are potentially very damaging because they are both recalcitrant and very toxic. Simply dumping cyanide residues in landfill is potentially hazardous because the cyanide may contaminate ground water in relatively high concentrations. Similarly, high concentrations of cyanides in aqueous effluents may have disasterous consequences on water treatment processes and much of the cyanide may pass, unchanged, through the water treatment. Furthermore, many cyanides (for example HCN and CH_3CN) are highly volatile and can pollute the atmosphere. It is not surprising, therefore, that the dispersal of cyanides is strictly controlled.

6.5.2 Aliphatic hydrocarbons

Saturated and unsaturated hydrocarbons occur naturally and thus it is relatively easy to isolate micro-organisms capable of biodegrading them. Problems occur due to their deposition in large amounts, for example, by oil spillage. The n-alkanes are the easiest of the aliphatic hydrocarbons to biodegrade.

n-alkanes

Certain generalisations can be made with respect to the persistent nature of aliphatic hydrocarbons. These are:

* long-chain n-alkanes are transformed more readily than are short-chain alkanes, probably due to the higher solubility and toxicity of the short-chain alkanes. However, once the alkane chain length exceeds a certain limit then the alkane will exist as a solid under normal climatic conditions and are difficult to degrade. Thus, it has been found that chain lengths of 10 to 24 carbon atoms are the most readily biodegraded;

* saturated aliphatics are transformed more readily than are unsaturated aliphatics;

* branching of the aliphatic chains is associated with lower rates of biodegradation.

As we have seen, the metabolism of aliphatic hydrocarbons usually involves an oxygen-requiring process initiated by monoxygenases ('mixed function oxidases') or dioxygenases to yield the primary alcohol which is then further oxidised to the aldehyde and then to the carboxylic acid.

In Figure 6.5, we have illustrated two different pathways for the oxidation of n-alkanes. Both involve monoxygenases. In one pathway a terminal methyl group is oxidised to form an n-alcohol, in the other the initial oxidation involves a β-methylene group. Despite such variations between organisms, the general principle of initial oxidation followed by conversion to products that can enter central metabolism are upheld.

Oxidation of both terminal methyl groups can occur to yield a dicarboxylic acid and is often used by micro-organisms to overcome metabolic blocks caused by branching of the carbon chain.

Once the carboxylic acid is formed, then this can be further catabolised by β-oxidation and the subsequent acetyl-CoA units oxidised to CO_2 via the tricarboxylic acid cycle.

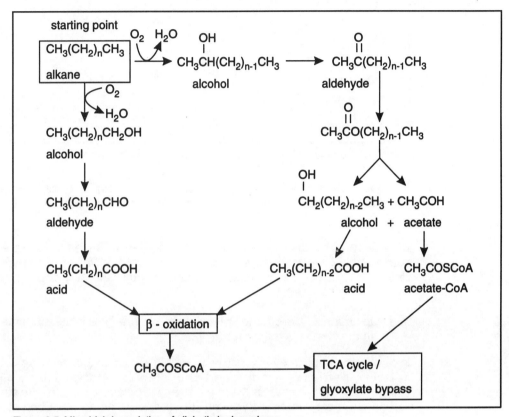

Figure 6.5 Microbial degradation of aliphatic hydrocarbons.

Alkene (ie unsaturated alkane) oxidation often involves monoxygenase-catalysed reactions oxygeneration to give oxides, diols, and primary and secondary alcohols which can be further metabolised by β-oxidation.

6.5.3 Alicyclic hydrocarbons

This group includes numerous natural products such as waxes from plants, crude oil and microbial lipids together with xenobiotic compounds including petroleum products and pesticides.

cyclohexane

Important general observations are that unsubstituted cyclohexane rings can be hydrolysed by monoxygenase-catalysed reactions. Then monoxygenases initiate ring cleavage by production of lactones from alicyclic ketones. Hydrolysis of the lactone results in ring cleavage (spontaneous or enzyme-mediated). The pathway was outlined in Figure 6.3.

6.5.4 Aromatic hydrocarbons

aromatic ring

The aromatic ring is a relatively stable structure. In order to overcome this stability certain micro-organisms have evolved which are capable of dihydroxylating the aromatic ring and then bringing about ring cleavage. Aerobically the dihydroxylation step involves a dioxygenase enzyme which catalyses the input of both atoms of molecular oxygen onto the aromatic ring as shown in Figure 6.6.

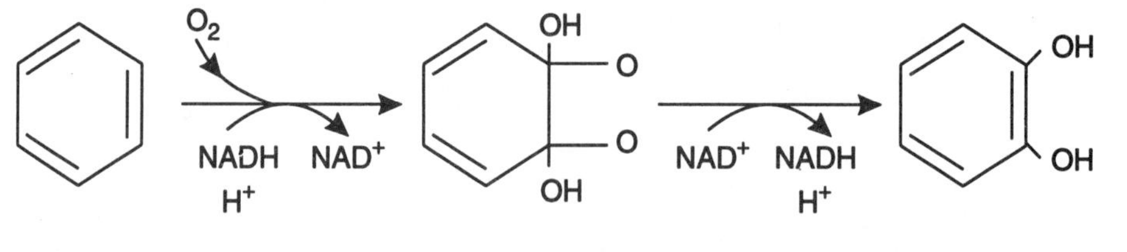

Figure 6.6 The initial oxidation reaction of aromatic ring cleavage: dihydroxylation.

catechols

The products of the initial metabolism of aromatic hydrocarbons are commonly catechols (or substituted catechols if the initial ring had substituent groups replacing the hydrogens on it).

Ring cleavage occurs by one of two pathways depending on the species and the initial compounds. These pathways are shown in Figure 6.6. One of these pathways we have called the *ortho* ring cleavage pathway and the other the *meta* ring cleavage pathway.

∏ Examine Figure 6.7 and see if you can put into words the fundamental difference in the cleavage of the ring.

ortho

meta

It can be seen that the *ortho* pathway involves cleavage of the ring between the two adjacent hydroxyl groups and this step is catalysed by a 1,2-dioxygenase (pyrocatechase) whereas the *meta* pathway involves cleavage of the ring between the carbon-atom containing a hydroxyl group and an adjacent carbon-atom without an hydroxyl group and this step is catalysed by a 2, 3-dioxygenase (*meta*pyrocatechase).

Both *ortho* and *meta* pathways are involved in aromatic hydrocarbon degradation but substituted catechols are generally degraded via the *meta* pathway. Benzene itself is degraded via *meta* ring cleavage. [Note most text books use the alternative terminology for 3-oxoadipate, that is β-ketoadipate. The *ortho* ring cleavage pathway is still generally known as the β-ketoadipate pathway].

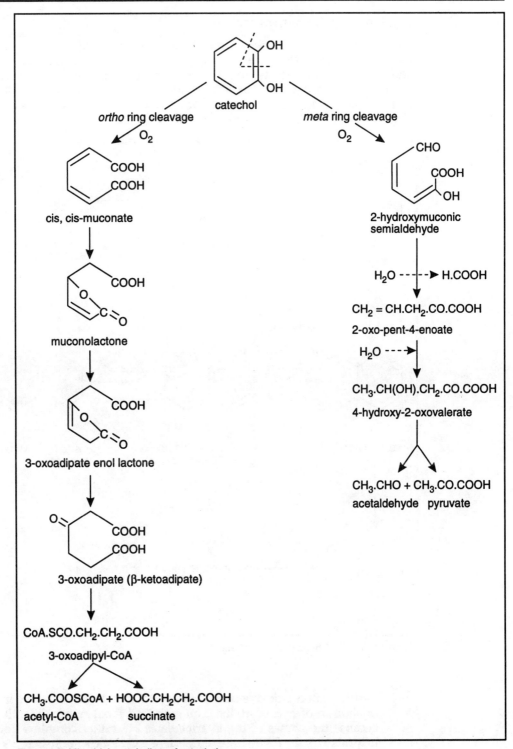

Figure 6.7 Microbial metabolism of catechols.

6.5.5 Polycyclic hydrocarbons

These compounds, with two or more ring structures, (for example naphthalene) are usually attacked at one of the terminal ring structures by dioxygenases to yield the diol and this step can then be followed by ring cleavage and further metabolised to ultimately yield a single substituted aromatic ring. This would then be further metabolised along a pathway similar to that previously described in Figure 6.6. Degradation of naphalene is shown in Figure 6.8.

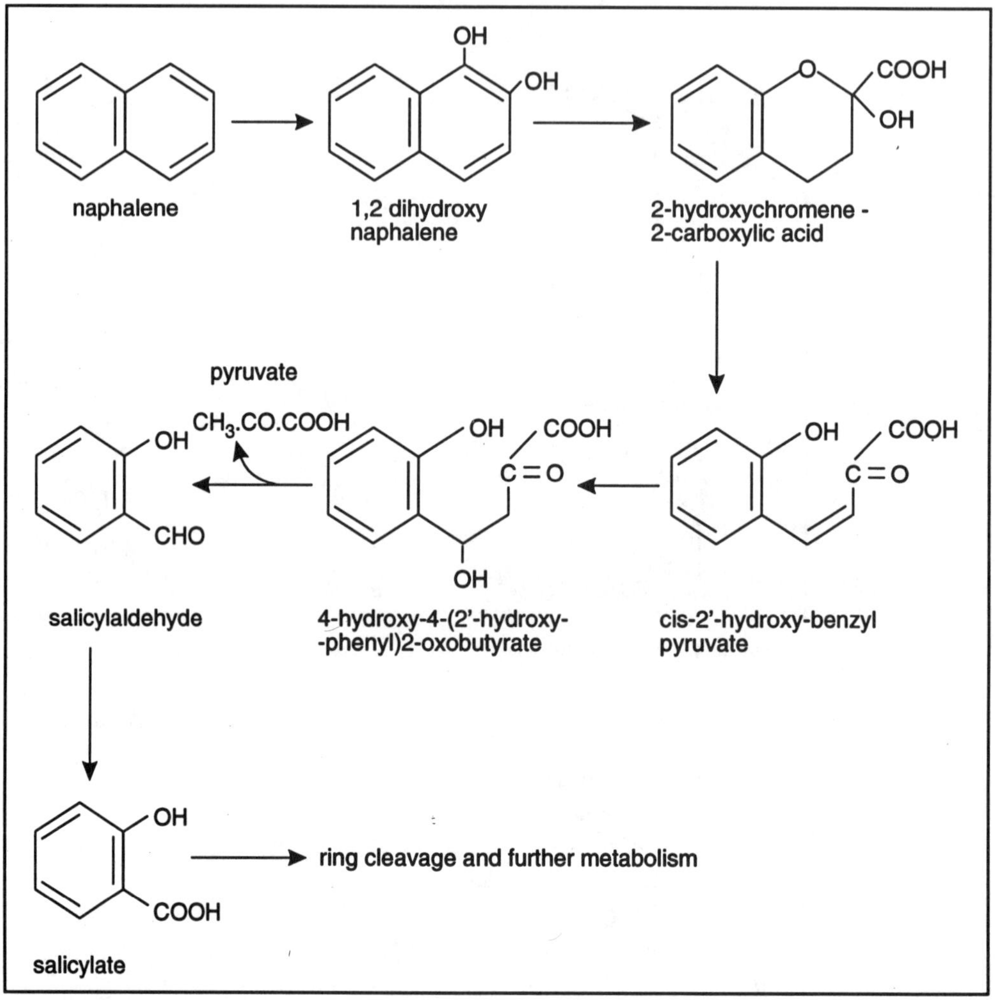

Figure 6.8 Microbial metabolism of naphalene.

∏ In this section we have described several metabolic pathways involved in the catabolism of hydrocarbons. You may find them rather difficult to remember because the names of the intermediates are rather complex. You may find it helpful to try the following exercise. Draw the following structures onto separate sheets of paper.

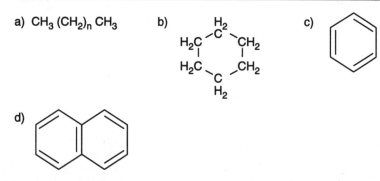

a) $CH_3 (CH_2)_n CH_3$

Then without looking at the text, try to write out the probable metabolic pathways involved in their degradation. When you have done this try to name the intermediates. You can check your pathways with the appropriate figures in the text. You may be surprised to find how quickly you can learn these pathways. The key is to remember that the first step involves molecular oxygen.

SAQ 6.2

Place the following processes occurring during the biodegradation of benzene in order of sequence of events . Omit those events that are not involved.

1)

Formation of acetyl CoA and succinate.

2)

Formation of an enoate.

3)

Reaction catalysed by a 1,2 dioxygenase.

4)

6.5.6 Factors affecting the biodegradation of hydrocarbons

complex
molecules

We have considered relatively simple organic molecules so far, for which certain generalisations have been made with respect to factors affecting their degradation. With more complex molecules containing aliphatic and aromatic, alicyclic, or heterocyclic components generalisations about their degradation become more difficult to make. However, if the portions are linked by amide-, ester- or ether bonds then these bonds are usually the first to be attacked and the products are metabolised as separate compounds. If attack on these linkage bonds cannot occur then, generally, the aliphatic portion is attacked first and if this is also prevented by extensive branching or other substituents then the initial attack may occur at the aromatic end. The site and mode of attack is determined not only by molecular structure but also by both the metabolic capabilities of the organisms involved and the environmental conditions. Such factors not only affect the rate of degradation but also the metabolic pathway used and hence the end products.

rate of
degradation

Certain compounds, particularly saturated ring structures such as cyclohexane, and halogenated aromatics, rarely support microbial growth yet can be degraded in the presence of another compound. The xenobiotic compound is then co-metabolised with the growth-supporting compound.

co-metabolism

Co-metabolism is a term used to describe the biochemical transformation of a compound (X) via a metabolic pathway that has been induced by the presence of another similar compound (Y). The organism, however, does not obtain any energy, carbon or other nutrient from the co-metabolised compound (X). Thus the degradation of the xenobiotic compound is dependent on a growth substrate for the provision of carbon, energy and reducing equivalents required for the transportation of the xenobiotic compound into the cell and for reactions required for transformation of the xenobiotic compound.

A similar situation can occur in which a xenobiotic compound may be metabolised by an existing pathway or sequence of enzyme reactions, as with co-metabolism, but from which the cell does gain energy and reducing equivalents from the metabolism of the

gratuitous metabolism

xenobiotic compound. This is termed gratuitous metabolism and as with co-metabolism, the xenobiotic compounds are structural analogs of naturally-occurring compounds and so act as substrates for the enzymes which are normally involved in the metabolism of the natural compound.

The extent of the biodegradation of xenobiotic compounds by gratuitous metabolism is often limited because one (or more) of the metabolites produced from the xenobiotic is not recognised by an enzyme later in the pathway. We can illustrate this in the following way:

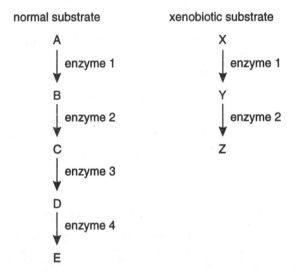

In this example enzymes 1 and 2 will catalyse reactions involving A or X and their products. Enzyme 3 on the other hand will only catalyse reactions involving C derived from its normal substrate. The equivalent product derived from the xenobiotic substrate (Z) is not used by enzyme 3.

In such a case, although the xenobiotic compound may not be completely degraded by the cell, it may be less of a pollutant problem or it may be metabolised further if it provides a beneficial substrate for other micro-organisms in the environment.

Since gratuitous metabolism and/or co-metabolism may play an important role in the degradation of certain xenobiotic compounds, it is important to consider how the structural differences between the normal substrate and the xenobiotic substrate and their metabolites influence the abilities of enzymes to use them.

In general we might anticipate that enzymes will work best with their natural substrates and that the rates of the reactions they catalyse will be lower with xenobiotic substrates.

We would argue that evolutionary selection will have led to the production of enzymes which have high affinities (low K_M's) and high turnover numbers (high k_{cat}) for their natural substrates. With the slight variations in the structure of the substrate (as represented by xenobiotics and their metabolites) it is likely that the enzymes will have lower affinities (higher K_M's) and lower turnover numbers (k_{cat}) with these substrates. This will mean that the natural substrates are likely to be catabolised (degraded) faster than their xenobiotic analogs. This effect can be observed by comparing the time needed for degrading various compounds in natural environments. We have illustrated this in Figure 6.9 in which we show the relative times needed for the degradation of various benezene derivatives in natural environments.

Compound	Structure	Relative biodegradation time
benzoate	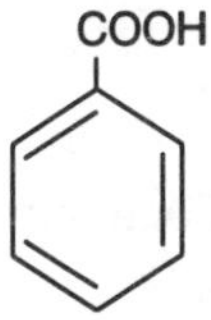 COOH	1
anisole (methoxybenzene)	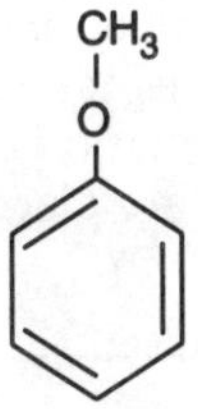 CH_3—O	8
nitrobenzene	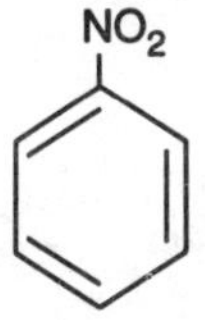 NO_2	> 64
aniline (aminobenzene)	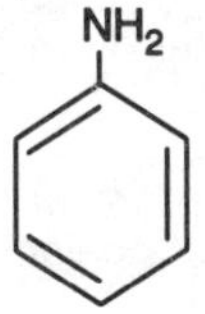 NH_2	4
benzene sulphonate	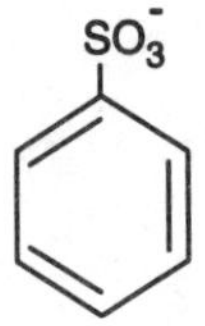 SO_3^-	16
phenol	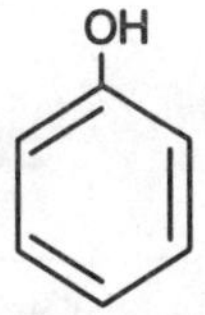 OH	1

Figure 6.9 The effect of substituent groups on the relative biodegradation time of various benzene derivatives.

∏ After studying Figure 6.9 place the compounds in order of increasing recalcitrance by writing the requisite substituent group (NH₂,-OH etc) in the relevant boxes below.

least
recalcitrant

most
recalcitrant

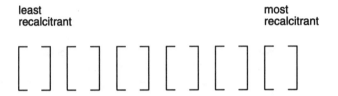

Phenol and benzoate have the shortest biodegradation times and are, therefore, the least recalcitrant whereas nitrobenzene has the longest biodegradation time and is, therefore, the most recalcitrant.

Similarly, substitution with halogens makes the resultant compounds more resistant to biodegradation.

We have seen that the type of substituent group on an aromatic ring affects the rate of biodegradation. Let us now consider the effect of the addition of a second substituent onto an already substituted aromatic. Look at the structures of the compounds depicted in Figure 6.10; the compounds are placed in order of increasing recalcitrance from left to right.

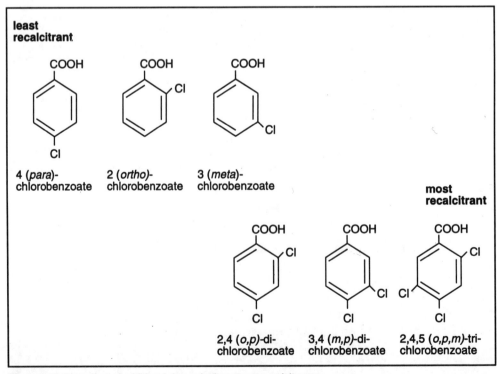

Figure 6.10 The effect of additional ring substituents on recalcitrance.

Π What general conclusions can you make from the information in Figure 6.10 about the effect of 1) the number, and 2) the position, of substituents on an aromatic ring? Briefly write down your answers before continuing.

It can be concluded that: 1) the greater the number of substituent groups on the aromatic ring, the greater the degree of recalcitrance; and 2) the position of the substituent groups affects the rate of biodegradation such that the degree of recalcitrance by substitution is in the order *meta > ortho > para*.

Therefore the type, number and position of substituent groups generally affect the rate of biodegradation of organic compounds. The position of substituent groups on the aromatic ring can be particularly important in that they can restrict the rate of, or partially block, the initial oxygenation reactions and hence the rate of ring cleavage and ultimately the rate of biodegradation.

With *meta* substituted aromatic rings we can rank the substitutions in order of their resistance to degradation. The general order is: $-NO_2 > CH_3CO- > Cl- > CH_3O- > -H$.

SAQ 6.3

The degree of recalcitrance of a compound is determined by the chemical nature of that compound. Look at the pairs of compounds given in Figure 6.11 and for each pair make a reasoned decision as to which compound ('a' or 'b') is the most recalcitrant. Write your reasons on a separate piece of paper and indicate in the relevant box which of the compounds ('a' or 'b') you think is the most recalcitrant.

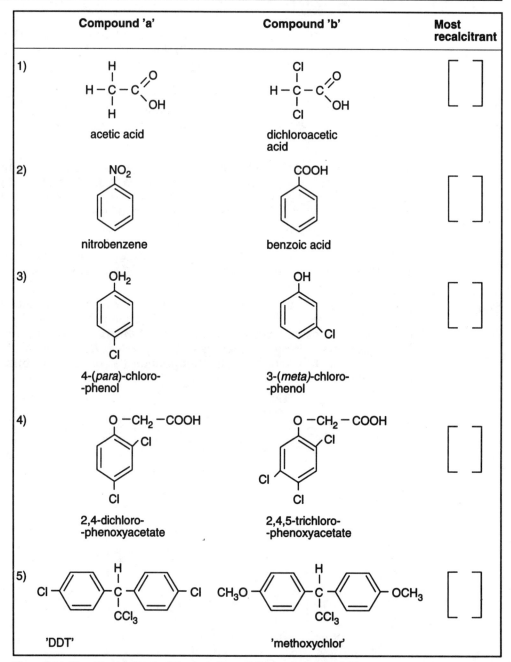

Figure 6.11 The effect of chemical structure on recalcitrance (see SAQ 6.3).

6.6 Biodegradation of halogenated compounds

removal of the
halogen

Complete mineralisation by biodegradation of organohalogen compounds involves
two stages: 1) removal of the halogen; and 2) degradation of the organic moiety of the

compound. Removal of the halogen is of course a requirement for mineralisation and this occurs by two possible basic mechanisms:

1) elimination of hydrogen halide:

$$\begin{array}{cc} \mathsf{H} \\ | \\ -\mathsf{C}-\mathsf{C}- \\ | \quad | \\ \mathsf{H} \quad \mathsf{Cl} \end{array} \longrightarrow \begin{array}{cc} \mathsf{H} \quad \mathsf{H} \\ | \quad | \\ -\mathsf{C}=\mathsf{C}- \end{array} + \mathsf{HCl}$$

or 2) substitution of the halogen by:

a) -H group (reductive reaction);

$$\begin{array}{cc} \mathsf{H} \quad \mathsf{H} \\ | \quad | \\ -\mathsf{C}-\mathsf{C}- \\ | \quad | \\ \mathsf{H} \quad \mathsf{Cl} \end{array} \longrightarrow \begin{array}{cc} \mathsf{H} \quad \mathsf{H} \\ | \quad | \\ -\mathsf{C}-\mathsf{C}- \\ | \quad | \\ \mathsf{H} \quad \mathsf{H} \end{array} + \mathsf{Cl}^-$$

b) -OH group (hydroxylation reaction);

$$\begin{array}{cc} \mathsf{H} \quad \mathsf{H} \\ | \quad | \\ -\mathsf{C}-\mathsf{C}- \\ | \quad | \\ \mathsf{H} \quad \mathsf{Cl} \end{array} \longrightarrow \begin{array}{cc} \mathsf{H} \quad \mathsf{H} \\ | \quad | \\ -\mathsf{C}-\mathsf{C}- \\ | \quad | \\ \mathsf{H} \quad \mathsf{OH} \end{array} + \mathsf{Cl}^-$$

c) -thio (S methyl, SCoA etc) group (replacement reaction).

$$\begin{array}{cc} \mathsf{H} \quad \mathsf{H} \\ | \quad | \\ -\mathsf{C}-\mathsf{C}- \\ | \quad | \\ \mathsf{H} \quad \mathsf{Cl} \end{array} \longrightarrow \begin{array}{cc} \mathsf{H} \quad \mathsf{H} \\ | \quad | \\ -\mathsf{C}-\mathsf{C}- \\ | \quad | \\ \mathsf{H} \quad \mathsf{SCoA} \end{array} + \mathsf{Cl}^-$$

1)   This direct elimination of the hydrogen halide (eg HCl) between two adjacent C-atoms to yield a double bond occurs rarely.

2)   This mechanism is more common, particularly displacement by hydroxyl groups which incorporate reactive oxygen groups into the molecule. This is particularly common for fully reduced aliphatics or aromatics.

6.6.1 Aerobic degradation of halogenated aromatic compounds

We have seen from the previous section how aromatic compounds may be aerobically biodegraded. Halogenated aromatic compounds are often biodegraded by the same metabolic pathway. In many cases, the pathway needs to be induced by a natural substrate (for example benzoate) and that the halogenated aromatic compound is co-metabolised along with this substrate. In other words, the enzymes induced by the co-metabolite do not recognise the structural differences of the xenobiotic compound and so catalyse its conversion into the corresponding halogenated products.

In order for complete mineralisation of halogenated aromatics to occur it is obviously necessary for the halogen(s) to be removed, usually by nucleophilic displacement with a hydroxyl group. Displacement of the halogen by a hydroxyl group generally occurs after ring cleavage due to the difficulty in displacing Cl⁻ from what is essentially a C-C double bond. Although Cl⁻ displacement after ring cleavage is usual, there are certain species of micro-organism which displace the Cl⁻ before ring cleavage.

We will now consider these possible routes in more detail.

6.6.2 Elimination of Cl⁻ after ring cleavage

The aerobic degradation of chlorinated aromatic compounds is usually achieved by a sequence of reactions. These are:

- hydroxylation or dioxygenation of the ring to form chlorinated catechols;

- cleavage of the aromatic ring;

- elimination of the Cl⁻ from the aliphatic intermediate;

- further metabolism of the dechlorinated aliphatic.

Examine Figure 6.12, as an example, which shows the *ortho* and *meta* pathways for the aerobic degradation of 4-chlorobenzoate during which ring cleavage occurs before chlorine release.

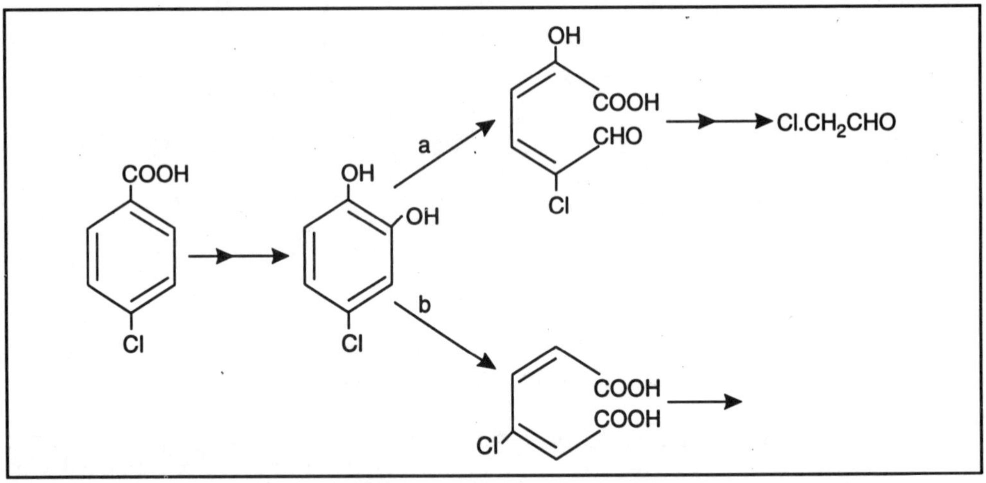

Figure 6.12 The metabolism of 4-chlorobenzoate showing ring cleavage before halogen release (a and b are referred to in the text).

As we have said, the first stage is the formation of chlorocatechols. Chlorocatechols are key intermediates in the degradative pathways of many chlorinated aromatics since they are produced not only from chlorobenzoate, as in Figure 6.12 but also from most other substituted chloroaromatics such as chlorophenol and chloroaniline. This initial hydroxylation reaction is mediated by dioxygenases (incorporate both atoms of O_2) or hydroxylases in the case of phenols in which one -OH group is already present.

The next stage is ring cleavage and, as for unsubstituted aromatics, this can occur either via the *ortho* route in which the ring is cleaved between the two adjacent -OH groups or via the *meta* route in which ring cleavage occurs between the C-2 -OH group and the adjacent C-atom.

∏ Which of the pathways shown in Figure 6.12 is the *ortho* route and which the *meta* route.

You should have identified route a) as the *meta* route and route b) as the *ortho*. If you got this wrong, re-examine the figure carefully.

Ortho cleavage of halocatechols is subsequently followed by elimination of the halogen and the non-halogenated product can then be further metabolised without any further problems. *Meta* cleavage metabolism often results in the halogen not being eliminated and this in turn may result in metabolites which are toxic and which inhibit further degradation of the original substrate (see Figure 6.12).

6.6.3 Elimination of Cl⁻ before ring cleavage

Elimination of the halogen prior to ring cleavage is not a common occurrence.

When the halogen is removed before ring cleavage then this usually occurs through replacement by hydrogen (reductive) or hydroxyl (hydroxylation) group.

Figure 6.13 shows the pathway for 4-chlorobenzoate.

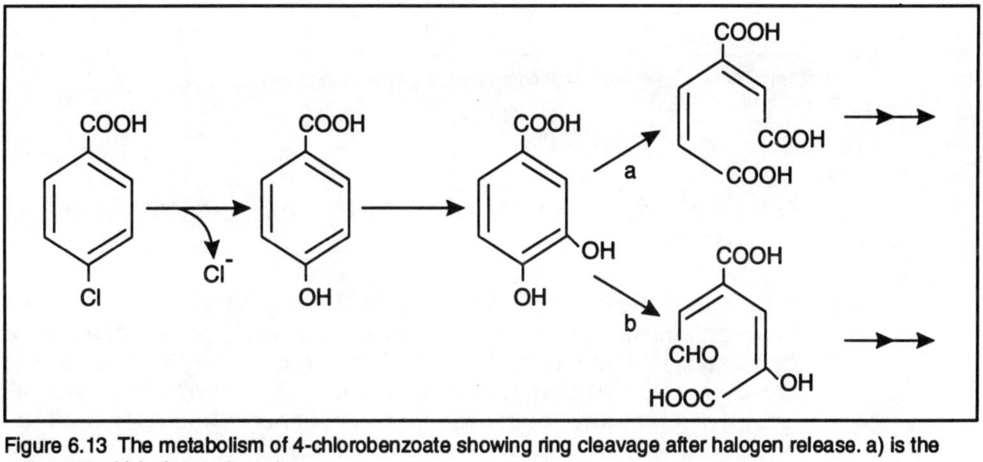

Figure 6.13 The metabolism of 4-chlorobenzoate showing ring cleavage after halogen release. a) is the *ortho* route, b) is the *meta* route.

Once the halogen(s) have been removed, then mineralisation of the compound can continue as for non-halogenated aromatics.

Figure 6.14 shows the dehalogenation of pentachlorophenol.

What types of dehalogenation reactions (substitution, elimination, replacement, reductive or hydroxylation) are involved at each stage of the complete dehalogenation of pentachlorophenol?

Type of dehalogenation reaction

Reaction 1

Reaction 2

Reaction 3

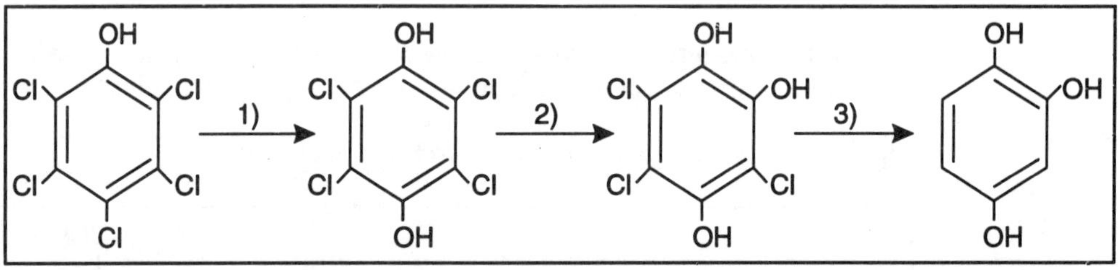

Figure 6.14 The microbial dehalogenation of pentachlorophenol.

6.7 The influence of plasmids and evolution on xenobiotic degradation

Mineralisation of any compound, including xenobiotics, by a single species of micro-organisms requires that species to possess all the necessary enzymes involved in the particular pathway and hence all of the necessary genetic information. We have already mentioned that certain xenobiotics can be wholly or partially degraded by pathways which have been induced by similar compounds (gratuitous metabolism) or require the co-metabolism of a second substrate.

genetic
information

Continual exposure of microbes to xenobiotics can result in the evolution of novel metabolic processes which enable the micro-organism to wholly or partially degrade the xenobiotic compound. Thus microbes are able to acquire the genetic information required for the production of the enzymes responsible for this.

How can microbes acquire this new information?

natural
mutation

There are two possible ways: firstly by natural mutation of existing genetic information and, secondly, from other living cells by natural genetic transfer.

We will now consider these two possibilities in more detail.

6.7.1 Natural mutation

Natural mutations occur at the rate of 1 mutation/100 cell divisions (in bacteria); the results of which are mostly lethal or non-beneficial to the cell. However, because the generation time of microbes is relatively short it does not take long for a cell to have passed through many generations (eg several hundred generations per year would not be an unreasonable value). In addition bacteria tend to exist in large populations (typically in the range $10^4 - 10^{10}$) and therefore there will be a significantly large number of cell divisions occurring at any one time period. Thus the chance of a beneficial mutation occurring over a certain time period is likely and this may allow a population of cells to evolve which has the capability of metabolising a particular xenobiotic compound.

Mutations of this sort normally only effect either slight conformational changes to the active site of an enzyme or the removal of regulatory control mechanisms. Both increase the activity of the enzyme towards its substrate rather than by creating an enzyme with totally new properties.

microbial infallibility

We should perhaps remind ourselves here of Martin Alexander's 'Principle of microbial infallibility', which we mentioned at the beginning of this chapter. It states that no natural organic compound is totally resistant to biodegradation given the right environmental conditions. Perhaps we should extend it to: 'the ability of microbes to degrade any organic compound given the right environmental conditions and a suitable period of time'.

6.7.2 Genetic transfer

Many genes that encode key enzymes important in the catabolism of recalcitrant compounds are plasmid encoded. (That is - the genetic information is present on a small, circular piece of double-stranded DNA which is separate from the chromosome and is capable of self-replication).

plasmids

Such plasmids can be capable of being transmitted from one cell to another by the processes of conjugation and transformation (or partially by transduction). You are perhaps more familiar with the transmission of plasmid encoded antibiotic resistance genes and the use of plasmids in genetic engineering. The transfer of genetic information between micro-organisms via plasmids is thought to have occurred throughout evolution. Just as in the cases we cited above, we must anticipate that genetic information required for the metabolism of xenobiotic compounds may be transferred between organisms in the form of plasmids.

Some plasmids encode for virtually the complete pathway to yield central metabolites; others encode for only a few enzymes. An example of a plasmid encoding for a complete degradation pathway is the TOL plasmid for toluene degradation. In contrast, the plasmids pAC21 and pAC25 encode for only a few enzymes in the degradation of p-chlorobiphenyl and 3-chlorobenzoate, respectively.

chromosomal genes

In cells possessing plasmids which only encode for a few enzymes of a functional xenobiotic degradation pathway, the plasmid genes must be complemented with chromosomal genes so that the plasmid encoded pathway is linked to an energy-yielding central metabolic pathway (chromosomally encoded).

Most of the plasmid-containing bacteria associated with xenobiotic degradation are Gram negative aerobes particularly of the genus *Pseudomonas*. The ability for genetic information to be transmitted from one cell to another, say of a different species, is important because:

- cells are able to acquire genetic information which enables them to complete a degradative pathway. For example, cells may be able to metabolise benzoate but not chlorobenzoate. The acquisition of a gene encoding for a dioxygenase with a broader substrate specificity may allow the formation of a chlorocatechol which could then be gratuitously metabolised along the benzoate pathway;

- cells may be able to acquire genes encoding for enzymes which allow them to improve the nature or rate of degradation. For example: acquisition of genes encoding for enzymes which catalyse the *ortho* cleavage of catechols as opposed to existing enzymes for *meta* cleavage will result in prevention or reduction of the formation of toxic *meta* cleavage metabolites.

genetic
engineering

The encoding of degradative genes on plasmids opens up the possibility for the construction of microbes by genetic engineering with the exact metabolic requirements for the treatment of xenobiotic industrial wastes. We will discuss this possibility in more detail later in this chapter.

6.8 Biodegradation in microbial communities

We have discussed the metabolic pathways involved in the degradation of selected xenobiotic compounds. In these cases we have considered the degradative pathway leading to complete mineralisation of the compound. In many cases complete mineralisation of xenobiotic compounds cannot be achieved by a single micro-organism because the micro-organism lacks (part of) the genetic information required to do this. Indeed, gratuitous metabolism and co-metabolism of xenobiotics by pure cultures of micro-organisms often results in the accumulation of metabolites, some of which can be more toxic than the original substrate. However, examination of the degradation of xenobiotic compounds by mixed microbial cultures which have adapted by selection and enrichment to the presence of the xenobiotic compound, often reveals that metabolites do not accumulate.

Π On the basis of what you have just read, why do you think xenobiotic metabolites often do not accumulate in the presence of mixed cultures?

sequential
metabolic
attack

There is a good chance that the metabolites of one organism (the primary utiliser) can be degraded by another species of micro-organism (the secondary utiliser) within the community. Thus, by a concerted, sequential metabolic attack of many micro-organisms, a xenobiotic compound can be completely degraded (mineralised) although no one species of micro-organism can totally utilise the substrate.

An example of sequential metabolism by two species of micro-organism is given in Figure 6.15.

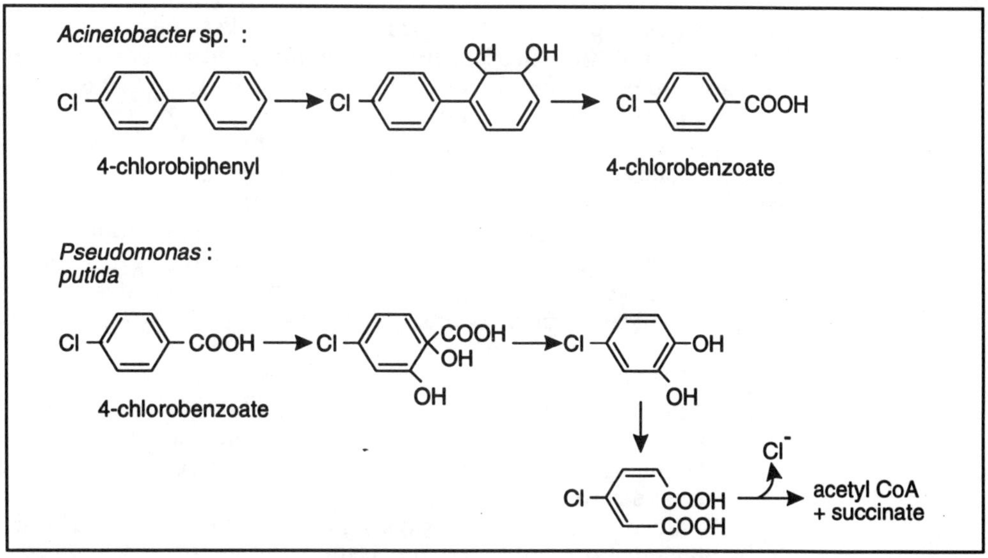

Figure 6.15 Biodegradation of 4-chlorobiphenyl by a mixed culture (see text for details).

The *Acinetobacter* sp. contains the plasmid pkF1 which encodes for the dihydroxylation of one of the rings of 4-chlorobiphenyl followed by *meta* cleavage of that ring and its subsequent degradation to yield 4-chlorobenzoate, a product which can not be further catabolised by the *Acinetobacter* sp. A *Pseudomonas putida* strain, which can not metabolise 4-chlorobiphenyl but is capable of 4-chlorobenzoate degradation (plasmid encoded), metabolises the 4-chlorobenzoate via *ortho* cleavage of the 4-chlorocatechol metabolite and subsequent dehalogenation. Thus, as a mixed culture, the *Acinetobacter* sp. and the *Pseudomonas putida* can together mineralise 4-chlorobenzoate, a compound which could not be mineralised by either species as a pure culture.

In certain cases, the lack of complete genetic information to mineralise a particular xenobiotic can result in the formation of toxic metabolites which restrict the rate of degradation or, if the concentration of the toxic metabolite gets high enough, can cause complete inhibition of cell metabolism or degradation of that compound. An example of such a situation is given in Figure 6.16.

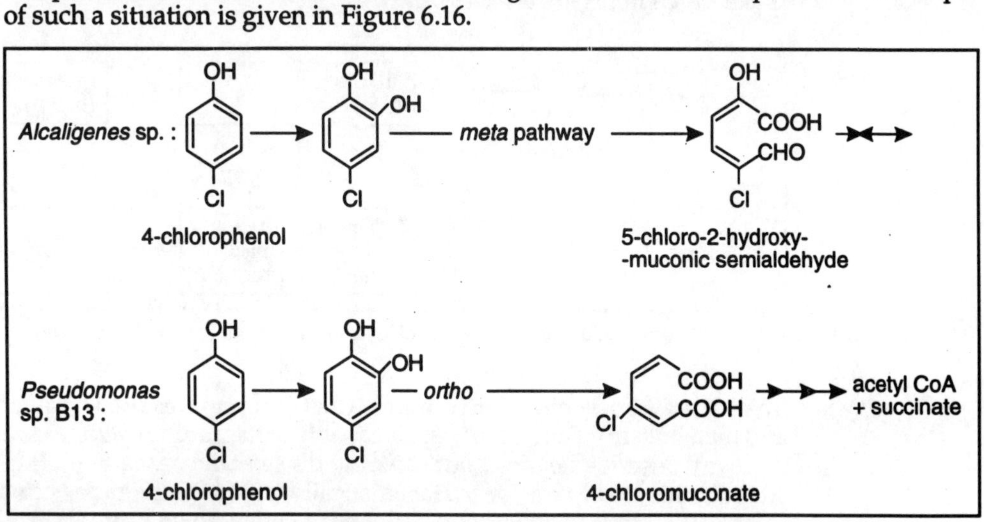

Figure 6.16 Biodegradation of 4-chlorophenol by a mixed culture (see text for details).

The *Alcaligenes* sp. metabolising phenol is capable of co-metabolising 2-chlorophenol, 3-chlorophenol and 4-chlorophenol via the 2,3,dioxygenase catalysed (*meta*) ring cleavage route in continuous culture. However, in the presence of high concentrations (>5mmol l^{-1}) 4-chlorophenol the concentration of the metabolite 5-chloro-2-hydroxymuconic semialdehyde increases to a toxic level and eventually terminates any further metabolism. *Pseudomonas* strain B13 contains the plasmid encoded enzyme catechol 1,2,dioxygenase expressed at high activity levels and can degrade 4-chlorophenol via the *ortho* ring cleavage route without the build up of toxic metabolites typical of *meta* cleavage of chlorocatechols. Thus, when this organism is present in co-culture with the *Alcaligenes* sp. an efficient treatment system for the removal of all chlorophenols under varying loading concentrations can be achieved.

∏ Continuous co-culture of the *Alcaligenes* sp. with *Pseudomonas* B13 under laboratory conditions resulted in the *Alcaligenes* sp. becoming dominant and eventually the *Pseudomonas* B13 was no longer detected in the continuous culture. Subsequent analyses of the effluent revealed no detectable levels of 5-chloro-2-hydroxymuconic semialdehyde despite the *Alcaligenes* sp. being the only micro-organism left in the culture. Can you explain what event(s) may have occurred? Briefly write down your explanations before continuing.

One possibility is that the *Alcaligenes* sp. has naturally mutated to a strain capable of 5-chloro-2-hydroxymuconic semialdehyde degradation, or was able to metabolise 4-chlorophenol degradation via the *ortho* route, and which resulted in the *Alcaligenes* having a faster growth rate than that of *Pseudomonas* B13. However, a more likely explanation would be that the plasmid encoding for catechol 1,2 dioxygenase was transmitted from *Pseudomanas* B13 to the *Alcaligenes* sp. thus enabling the *Alcaligenes* sp. to degrade 4-chlorophenol via the *ortho* route without the production of any toxic metabolites. This natural recombinant strain apparently had a higher growth rate than, and thus outgrew, *Pseudomanas* B13.

provision of specific nutrients

Another important interaction between micro-organisms which can effect the rate of degradation is that of the provision of specific nutrients or cofactors by one organism which are required for the growth or metabolism of another organism. An example of this is given in Figure 6.17.

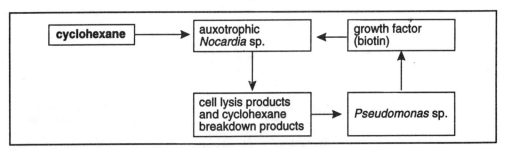

Figure 6.17 Biodegradation of cyclohexane by a mixed culture (see text for details).

The *Nocardia* sp. is biotin auxotrophic (that is it requires the vitamin biotin for growth and metabolism). The *Nocardia* sp. is capable of degrading cyclohexane, according to the pathway described earlier, but can only do so in presence of biotin. The *Pseudomonas* strain in this example is, uncharacteristically, unable to degrade cyclohexane but is able to utilise cell lysis products and cyclohexane breakdown products from the *Nocardia* sp.

and, in doing so, excretes biotin into the surrounding environment. The biotin produced by the *Pseudomonas* sp. is capable of supporting the degradation of cyclohexane by the *Nocardia* sp. which in turn provides the carbons and energy sources for the *Pseudomonas* which enables it to produce more biotin, and so on. Thus, in this system, neither organism is capable of degrading cyclohexane by itself but as a mixed culture can degrade cyclohexane.

∏ How might the above system be altered to achieve cyclohexane degradation by only one of the above named species of bacterium?

The incorporation of commercially available biotin into the surrounding medium would overcome the necessity for its production by the *Pseudomonas* and would thus enable the *Nocardia* sp. to degrade cyclohexane in pure culture.

The biodegradation potential of mixed cultures has distinct advantages over that of single species particularly in more realistic environments which contain mixed substrates and fluctuating loadings (concentrations). Overall these advantageous effects can be summarised as:

- there is a greater qualitative biodegradative capacity - the larger the diversity of species present the greater the 'gene pool' and hence the greater the chance of at least one organism being able to degrade the xenobiotic compound or the greater the chance of a compound being degraded by the concerted activity of two or more microbes;

- there is a greater quantitative biodegradative capacity - for example sequential metabolism of a compound involving primary and secondary utilisers etc thus allowing complete mineralisation;

- the biological treatment system is more stable due to a greater chance of microbes being present that can remove, or withstand, toxic metabolites which may be formed during shock loadings;

- improved biodegradation rates - the interactive provision of specific nutrients, more efficient removal of toxic intermediate metabolites etc can increase the metabolism and growth rate of the micro-organisms involved. In addition, single species of micro-organisms containing all the biodegradative capability for a particular xenobiotic compound may not be as good competitors because of the increased demand on biosynthesis will result in a slower growth rate. Sharing of genetic information and hence biosynthetic requirements, as in a mixed culture system, can result in these organisms having less restrictions on their growth rate.

Such communities of micro-organisms facilitate efficient catabolism of xenobiotic compounds due to interactions between the different species present.

Six strains of bacteria (A, B, C, D, E and F) were screened for their ability to degrade 1) polychlorobiphenyl (PCB) and 2) 4-chlorobenzoate as the sole sources of carbon in a mineral salts growth medium in pure culture and in co-culture with each of the other bacterial strains. The following results indicate the maximal percentage removal attained for each of the compounds by the bacterial strains A to F for all possible two strain combinations (for example: organism C with organism D achieves 58% removal of PCB). In every case of successful PCB degradation, as indicated by its removal from the growth medium, 4-chlorobenzoate was an intermediate. 4-chlorobenzoate was shown to be degraded via 4-chlorocatechol with subsequent ring cleavage either via the 'ortho' route for complete mineralisation, or by the 'meta' route in which case the product was 5-chloro-2-hydroxymuconic semialdehyde which was shown to be toxic to the bacteria at a critical concentration.

% PCB removal

	A	B	C	D	E	F
A	98	98	98	98	98	98
B		0	0	98	0	0
C			0	58	0	0
D				40	40	40
E					0	98
F						0

% 4-chlorobenzoate removal

	A	B	C	D	E	F
A	98	98	98	98	98	98
B		98	98	98	98	98
C			42	42	42	98
D				0	0	98
E					0	98
F						98

On the basis of the results presented above, which one of the bacterial strains (A, B, C, D, E or F) is best described by each of the following statements? Write down your answers and briefly give reasons for your choice.

1) The organism which can metabolise polychlorobiphenyl but is not capable of metabolising 4-chlorobenzoate.

2) The organism which cannot metabolise polychlorobiphenyl but is capable of metabolising 4-chlorobenzoate via the *ortho* pathway.

3) The organism which can only metabolise polychlorobiphenyl in the presence of a specific nutrient provided by a second organism and is capable of metabolising 4-chlorobenzoate via the *ortho* pathway.

4) The organism which cannot metabolise polychlorobiphenyl but is capable of metabolising 4-chlorobenzoate via the *meta* pathway.

5) The organism which cannot metabolise polychlorobiphenyl and is not capable of metabolising 4-chlorobenzoate.

6) The organism which completely mineralises the polychlorobiphenyl.

6.9 Practical approaches to the biodegradation of xenobiotics in the environment

The successful removal of xenobiotic compounds from the environment by biodegradation requires a detailed understanding of the factors affecting biodegradation. For example the nature of the environment containing the xenobiotic compounds to be removed is a determinative factor upon the success of xenobiotic degradation.

environmental factors

Let us consider the key environment factors involved in the biodegradation of xenobiotic compounds. These include:

1) xenobiotic (substrate) concentration - if the concentration is too high then toxicity problems may occur: if the concentration is too low then the rate of biodegradation may be limited by the affinity of the cells for these substrates;

2) the presence of the necessary genetic information and hence cells with the required metabolic capabilities;

3) the pH of the surrounding environment - since pH affects growth and metabolic rates;

4) the temperature of the surrounding environment - since temperature affects growth and metabolic rates;

5) water availability - the presence of water is essential for all life forms. Water may be present but not available to the cells due to the presence of high salt or high concentrations of organic molecules (eg sugars). This may result in the extracellular environment having a greater osmotic potential than the intracellular medium such that water would tend to leave the cell rather than to enter the cell;

6) availability of other nutrients required for growth - including oxygen or co-metabolites;

7) the presence of other organic material - organic compounds that may be metabolised in preference to the xenobiotic compound or act as inhibitors or co-metabolites.

The ideal environment is one in which all environmental conditions are constant or at least fairly predictably constant and are, of course, amenable to the growth and metabolism of micro-organisms. This ideal situation facilitates laboratory-based studies to allow the determination of the optimal treatment regime for successful xenobiotic removal. Such a situation of constant conditions rarely occurs naturally but is most likely to occur for treatment of industrial effluents at source for which the physical and environmental conditions can be controlled.

waste water
treatment
processes

Treatment of xenobiotic compounds at municipal waste water treatment works is feasible, and usually successful, but often depends on a constant supply of the xenobiotic compound to the system as opposed to intermittent loading of the xenobiotic.

∏ See if you can list reasons why intermittent loading with xenobiotics may reduce the efficacy of a system to remove these xenobiotics.

Intermittent loading of a xenobiotic to a treatment system causes problems due to loss of microbial capabilities since:

1)   the system continually naturally selects and enriches micro-organisms on the basis of their metabolic capabilities according to the substrates present. The absence of a particular xenobiotic compound could result in the loss, or depletion, of the micro-organisms capable of degrading it;

2)   shock loadings of xenobiotic compounds to the system can result in elevated concentrations of the xenobiotic compound or its metabolites. They may reach toxic levels which will reduce or destroy the effectiveness of the microbial population in degrading organic compounds.

Waste water treatment systems may never be able to treat certain xenobiotic compounds successfully due to:

1)   the inability of the system to sustain a suitably high enough population of cells capable of degrading the compound because the xenobiotic degraders are unable to successfully compete for nutrients, other than the carbon source, with the normal flora present;

2)   other waste compounds present are metabolised in preference to the xenobiotic, or their presence inhibits the degradation of the xenobiotic compound. An example of the latter is the degradation of chlorophenols in the presence of phenol. Chlorophenols can be successfully degraded via the *ortho*-cleavage pathway but phenol induces the *meta*-cleavage pathway. If the latter is utilised by the organism for the metabolism of any chlorophenols present, subsequent build-up to toxic levels of *meta* cleavage metabolites of chlorophenol, will occur.

The successful biodegradation of xenobiotics due to spillage on land is less predictable. This is due to the complexity of the environmental system and the constant changes occurring particularly in temperature and water availability.

6.9.1 Novel approaches to xenobiotic degradation

We should realise that there are key fundamental requirements for xenobiotic degradation to occur. As we indicated above, the physical factors, such as pH and temperature should be monitored and controlled so that they are within the limits required for microbial activity. When the physical parameters are acceptable for microbial activity there are several key factors which can be manipulated to induce or improve degradation of xenobiotic compounds. These are:

• provision of sufficient nutrients or co-metabolites;

• manipulation of xenobiotic substrate concentrations to non-toxic levels;

- provision of microbial capability.

Let us briefly consider each of these approaches in more detail.

Provision of nutrients and or co-metabolites

addition of nutrients to ensure these do not become rate limiting

Growth and metabolism of cells require the presence of the elements: carbon, hydrogen, oxygen, nitrogen, phosphorus, and sulphur in some form which is capable of being utilised by the cell. In addition certain trace elements are also required. In most situations involving xenobiotic degradation, the limiting elements are often nitrogen and phosphorus and sometimes sulphur, the carbon and hydrogen being obtainable from the xenobiotic molecule and oxygen from water or from the atmosphere as molecular oxygen. In certain cases, the limiting element can be oxygen and this is often easily resolved by implementing a forced aeration system. It is thus logical to realise that the addition of suitable nutrients can stimulate the growth and metabolism of the organisms involved and thus effect a more rapid xenobiotic degradation rate.

The addition of compounds that act as co-metabolites may be required to induce the requisite metabolic pathway or provide the necessary energy source in order for the cells to be able to degrade the particular xenobiotic compound.

Manipulation of the xenobiotic concentration

concentrations of xenobiotics may be manipulated by dilution

We know that certain xenobiotic compounds, when present at high enough concentrations, are toxic to the cells; they cause cell death or inhibition of growth and metabolism. If this is the case, then usually the problem can be resolved by initially diluting the xenobiotic compound with a suitable diluent, usually water. In a continuous-flow system for the degradation of xenobiotic-containing effluents relatively high concentrations of xenobiotics in the effluent can be dealt with as long as the amount of xenobiotic entering the treatment system does not exceed the amount being degraded within the system. Thus the concentration of the xenobiotic in the treatment system can be kept relatively low to maintain the continuous metabolic activity of the microbes present.

On the other hand, if the xenobiotic concentration is relatively low then this may have effect of lowering the rate of biodegradation according to Monod kinetics. This may result in longer periods of time being required to reduce the xenobiotic concentration to the accepted level. One approach to overcome this problem is to attempt to increase the number of catalytic sites ie increase the number of cells with which the substrate can interact.

Provision of microbial capability

It is likely that environment systems that have never been exposed to the xenobiotic compound of interest will not immediately contain microbial cells capable of rapidly metabolising that compound when first presented with it. In addition, environments which have been exposed to the xenobiotic compound do not always contain a sufficiently high population of suitable cells required for rapid degradation of the compound. In both cases, microbes will eventually be selected for or suitable naturally-occurring cells can be induced to proliferate by addition of nutrients, but in both cases a substantial lag period is likely. During such a lag time substantial environmental damage may have occurred.

bioaugmentation

This problem of not having a large enough population of metabolically capable cells can be overcome by the process of bioaugmentation. Bioaugmentation is the technique of increasing the biological efficiency of the treatment system through the programmed additions of specific bacterial formulations.

microbial formulations or inoculants

There are a number of these microbial formulations or inoculants on the market that have been shown to be effective in degrading a mixture of xenobiotic compounds. Such inoculants are usually mixed cultures based upon naturally occurring micro-organisms. The individual microbial species are selected on the basis of their metabolic capabilities. They are grown in pure culture, harvested and then preserved by either air drying or freeze drying techniques. These pure cultures are then carefully matched and blended in order to achieve pre-defined biodegradative capabilities. Often bulking agents, dispersant chemicals, wetting agents and nutrients are added to produce a product which is a dry powder that can be transported and stored easily and reconstituted on site by the addition of water.

genetically engineered micro-organisms

The prospects for microbial inoculants containing genetically engineered micro-organisms are still being evaluated. Certainly the capabilities are available for the construction of recombinant pathways into organisms for certain xenobiotic pollutants for which the biochemistry is fully understood.

Realistic strategies are to:

• manipulate genes encoding for broad-specificity enzymes to expand existing catabolic pathways thus producing organisms which have a greater diversity of capabilities;

• modify, by genetic manipulation, selected regulatory and structural genes involved in existing catabolic pathways in order to overcome problems associated with feedback repression or inhibition of catabolic enzymes, to promote enzyme synthesis, or to promote assimilation of the xenobiotic compound into the cell.

∏ Why might the vision of a 'super bug' capable of degrading a wide variety of xenobiotic compounds at high rates, under a variety of conditions, and in stable association with indigenous micro-organisms, be an unrealistic proposition?

The construction of such a 'super bug' would most likely involve the insertion of additional genetic information. We know that cells which have received substantial amounts of additional DNA are subject to additional constraints with respect to their rate of growth and proliferation. One may therefore hypothesise that engineered organisms would not be good competitors against a community of micro-organisms which collectively can degrade the same compounds but which have the required genetic information more widely distributed.

In addition, the public concern over the release of genetically engineered organisms into the environment is such that, at present, in many countries the release of such organisms is illegal.

| **SAQ 6.6** | Which of the following methods could be considered as being the most unsuitable for increasing the degradation rate of a halogenated aromatic compound present in a mixture of other organic molecules? Comment on the relative merits of each method. |

1) Increase the number of cells present.

2) Implementation of a pH-control system

3) Implementation of an aeration system.

4) Addition of a source of nitrogen, phosphorus and sulphur.

5) Increase the concentration of the compound to be degraded.

6) Decrease the concentration of the compound to be degraded.

7) Addition of benzoate.

8) Introduction of a heating system or heat retention system.

9) Addition of a genetically engineered bacterium with improved capabilities for halogenated aromatic degradation.

| **SAQ 6.7** | A pesticide-producing company has developed a new halogenated aromatic compound for use as a pesticide and intermittently discharges high concentrations of halogenated aromatics into the waste water it dispenses to the municipal water treatment plant. Which of the following effects may occur due to such an action? |

Comment briefly on your reasoning.

1) Observable reduction in the biodegradative activity of the micro-organisms in the average treatment system.

2) Appearance of high levels of halogenated aromatic compounds in the discharge water leaving the treatment plant.

3) An increase in the COD (Chemical Oxygen Demand) value of the discharge water leaving the treatment plant.

4) An increase in the BOD (Biological Oxygen Demand) value of the discharge water leaving the treatment plant.

SAQ 6.8

How might the water treatment plant remedy the problem described in SAQ 6.7? Consider each of the proposals below and write down any reasons for or against each proposal.

1) Impose a ban on the company from discharging halogenated aromatic compounds into the sewage line and suggest the company has it tankered away and buried.

2) Develop a separate treatment facility or persuade the company to develop its own, on-site facility using selected bacteria.

3) Persuade the company to hold the waste and allow it to be slowly released over a longer time period.

4) Increase the aeration rate to the activated sludge tank at the water treatment plant.

5) Alter the operative pH of the effluent being treated at the water treatment plant.

Summary and objectives

This has been a long chapter in which we have examined the biodegradation of xenobiotic compounds. We began by describing the range and sources of xenobiotics and discussed the problems which arise from the release of xenobiotics into the environment. We then provided an overview of the metabolic pathways used to degrade xenobiotics and introduced the concepts of co-metabolism and gratuitous metabolism. We also considered the factors which influenced the recalcitrance of organic compounds and the importance of mixed populations in xenobiotic degradation. In the final part of the chapter, we considered some practical issues relating to the utilisation of biodegradation to reduce the environmental problems associated with the production and release of xenobiotics.

Now that you have completed this chapter you should be able to:

- define and explain the terms xenobiotic, recalcitrant, persistant, mineralisation, co-metabolism, gratuitous metabolism etc in relation to known pollutant compounds;

- list the sources of the major types of compounds considered as xenobiotic;

- understand how the chemical nature of compounds can determine their recalcitrant nature;

- list the factors used to assess the seriousness of xenobiotic pollutants and describe the impact of certain xenobiotic compounds on the environment;

- demonstrate an understanding of the metabolic capabilities of microbial cells in the biodegradation of xenobiotic compounds by describing in detail the biochemical mechanisms involved in the biodegradation of selected xenobiotic compounds;

- identify and discuss the factors affecting the rate of biodegradation/mineralisation of selected xenobiotic compounds;

- understand the microbial capabilities of xenobiotic biodegradation at the gene level, the involvement of plasmids, and the evolutionary consequences;

- explain the importance of microbial communities in the biodegradation/mineralisation of xenobiotic compounds by understanding and describing selected examples;

- understand and describe the current practical approaches to the biodegradation of xenobiotic compounds in the environment.

Water analysis and quality control

Water analysis and quality control

7.1 Introduction

In the previous four chapters, we considered the nature of waste and pollution and the major ways these materials are processed to render them less harmful to mankind and the environment. Special attention was given to the application of biological processes used in the disposal of solid and water-borne wastes. Implicit in these processes is the need to assess the diversity and concentrations of substances in both the inputs and outputs of the treatment processes. We have, for example, briefly alluded to such measures as the biological and chemical oxygen demand (BOD, COD) of waste and treated waters. In this chapter, we extend the discussion of this aspect of waste treatment and environmental management. We will focus mainly on the standards that are applied to the control of water quality and the methods used to monitor water quality. Particular emphasis well be placed at the end of the chapter on how recent biotechnological developments may facilitate the processes of water analysis and quality control.

7.2 The importance of water analysis

∏ Before reading on, see if you can list 3 or 4 ways in which the analysis of water may play an important role in the water management.

There are, of course many ways in which water analysis plays a vital role in water management. Here we list six of the major ones, you may have thought of many more. Water analysis is used to:

* assess the suitability of water for a particular use (for example, drinking water abstraction, fisheries, amenity and recreational use, reception of effluents);

* provide data which can be used to set water quality standards and objectives;

* ensure that quality standards are met;

* assess the polluting potential of effluents and other wastes thus enabling the setting of discharge standards;

* ensure discharge standards are maintained;

* develop and monitor the efficiency of treatment processes.

objectives of
water analysis

We can therefore identify that the primary objective of water analysis is to maintain water quality. This is essential to:

* safeguard public health;

- protect water sources used by industry and the community;

- maintain the amenity and recreational value of waters;

- maintain (and improve) fisheries;

- conserve aquatic wildlife.

multiple use of water

Because the demand for water is so high, it is often necessary to use the same water sources for drinking water abstraction and effluent discharges as well as using it for recreational and conservation purposes. This inevitably demands a continual and thorough monitoring of the water and effluent (treated or untreated) discharges and the setting of legal quality standards.

7.3 The legal framework of quality standards in water management

The quality standards that are applied to the management of water are part of an evolving system and are the subject of numerous supra-national and national regulations. It is not our purpose here to review these regulations in detail. We will, however, outline the situation within the European Union and give some illustrative examples of national regulations.

7.3.1 EC - Directives relating to water quality

In Table 7.1 we have listed the main EC - Directives relating to water management and quality.

Π Scan this list to get some idea of the range of regulations which are in force. (We would not expect you to remember them all - but you should be aware of their existence).

The list has been arranged in chronological order. You should, however, be able to identify some recurrent themes. For example, you may have noticed Directives relating to dangerous substances in water.

Some of the Directives relate to specific compounds, others are more generic. We do not intend to review all of these Directives in detail. Instead we will discuss some of the main ones directly relevant to our discussion of water analysis. For convenience we can divide these into two groups:

- those concerned with the quality standards that are applied to waters used for particular purposes;

- those concerned with the presence of dangerous substances in water.

Π Before we go on to discuss each of those groups, it would be helpful to you to re-read Table 7.1 and to identify the Directives which relate to these two groups. (Construct a table of these for yourself). You will have an opportunity to check whether or not you have correctly identified the appropriate Directives as you read the following sections.

Directive	Reference
EEC Council Regulation applying rules of competition to transport by rail, road and inland water ways	OJ EEC 1968 NoL 175/1-1017/68/EEC
EEC Council Directive on the approximation of the laws of Member States relating to detergents	OJ EEC 1973 NoL 347/51-73/404/EEC
EEC Council Directive on the approximation of the laws of the Member States relating to methods of testing the biodegradabity of anionic surfactants	OJ EEC 1973 NoL 347/53-73/405/EEC
EEC Council Directive on the disposal of waste oils	OJ EEC 1975 NoL 194/23-75/439/EEC
EEC Council Directive concerning the quality required of surface water intended for the abstraction of drinking water in the Member States	OJ EEC 1975 NoL 194/26-75/440/EEC
EEC Council Directive on reciprocal recognition of navigability licences for inland water vessels	OJ EEC 1976 NoL 21/10-76/135/EEC
EEC Council Directive concerning the quality of bathing water	OJ EEC 1976 NoL 31/1-76/160/EEC
EEC Council Directive on the disposal of polychlorinated biphenyls and polychlorinated terphenyls	OJ EEC 1976 NoL 108/41-76/403/EEC
EEC Council Directive on pollution caused by certain dangerous substances discharged into the aquatic environment of the community	OJ EEC 1976 NoL 129/23-76/464/EEC
EEC Council Directive on waste from the titanium dioxide industry	OJ EEC 1978 NoL 54/19-78/176/EEC
EEC Council Directive on the quality of fresh waters needing protection or improvement in order to support fish life	OJ EEC 1978 NoL 222/1-78/659/EEC
EEC Council Directive on the conservation of wild birds	OJ EEC 1979 NoL 103/1-79/409/EEC
EEC Council Directive concerning the measurement and frequency of sampling and analysis of surface water intended for the abstraction of drinking water in the Member States	OJ EEC 1979 NoL 271/44-79/869/EEC
EEC Council Directive on the quality required of shellfish waters	OJ EEC 1979 NoL 281/47-79/923/EEC
EEC Council Directive on the protection of ground water against pollution caused by certain dangerous substances	OJ EEC 1980 NoL 20/43-80/68/EEC
EEC Council Directive on the approximation of the laws of the Member States relating to the exploitation and marketing of natural mineral waters	OJ EEC 1980 NoL 229/1-80/777/EEC
EEC Council Directive relating to the quality of water intended for human consumption	OJ EEC 1980 NoL 229/11-80/778/EEC
EEC Council Directive amending the Directives laying down the basic safety standards for the health protection of the general public and workers against the dangers of ionising radiation	OJ EEC 1980 NoL 246/1- 80/836/Euratom

Table 7.1 A selection of EEC - Directives and Regulations relating to the management and quality of water (up to December 1993). Cont/d.............

Directive	Reference
EEC Council Directive on limit values and quality objectives for mercury discharges by the chlor-alkali electrolysis industry	OJ EEC 1982 NoL 81/29-82/176/EEC
EEC Council Directive on the approximation of the laws of the Member States relating to method of testing the biodegradability of non-ionic surfactants	OJ EEC 1982 NoL 109/-82/242/EEC
EEC Council Directive on the conservation of migratory species of wild animals	OJ EEC 1982 NoL 210/11
EEC Council Directive laying down technical requirements for inland waterway vessels	OJ EEC 1982 NoL 301/1-82/714/EEC
EEC Council Directive on the procedures for the surveillance and monitoring of environments concerned by waste from the titanium dioxide industry	OJ EEC 1982 NoL 378/1-82/883/EEC
EEC Council Directive amending Directive 78/176/EEC on waste from the titanium dioxide industry	OJ EEC 1983 NoL 32/28-83/29/EEC
EEC Council Directive on limit values and quality objectives for mercury discharges by sectors other than the chlor-alkali electrolysis industry	OJ EEC 1984 NoL 74/49-84/156/EEC
EEC Council Directive on limit values and quality objectives for discharges of hexachlorocyclohexane	OJ EEC 1984 NoL 274/11-84/491/EEC
EEC Council Directive on limit values and quality objective for discharges of certain dangerous substances included on List 1 of the Annex to Directive 76/464/EEC	OJ EEC 1986 NoL 181/16-86/280/EEC
EEC Council Directive on the protection of the environment and in particular of the soil, when sewage sludge is used in agriculture	OJ EEC 1986 NoL 181/6-86/278 EEC
EEC Council Directive amending Directive 75/439/EEC on the disposal of waste oils	OJ EEC 1987 NoL 42/43-87/101/EEC
EEC Council Directive amending Annex II to Directive 86/280/EEC on the limit values and quality objective for discharges of certain dangerous substances included in List 1 of the Annex to Directive 76/464/EEC	OJ EEC 1988 NoL 158-88/347/EEC
EEC Council Directive on procedures for harmonising the programme for the reduction and eventual elimination of pollution caused by waste from the titanium dioxide industry	OJ EEC 1989 NoL 201/56-89/428/EEC
EEC Council Directive on urban waste water treatment	OJ EEC 1991 NoL 135/40-91/271/EEC
EEC Council Directive on laying down health conditions for the production and placing on the market of live bivalve molluses	OJ EEC 1991 NoL 268/1-91/492/EEC
EEC Council Directive on laying down the health conditions for the production and placing on the market of fishery products	OJ EEC 1991 NoL 268/15-91/493/EEC
EEC Council Directive concerning the protection of waters against pollution caused by nitrates from agricultural sources	OJ EEC 1991 NoL 375/1-91/676/EEC
EEC Council Directive standardising and rationalising reports on the implementation of certain Directives relating to the environment	OJ EEC 1991 NoL 76/14-91/692/EEC

Table 7.1 A selection of EEC - Directives and Regulations relating to the management and quality of water (up to December 1993).

7.3.2 EC - Directives relating to the quality of drinking water

Directive
80/778/EEC

The main Directive is Directive 80/778/EEC (OJ L229/11/80/778/EEC) relating to the quality of water intended for human consumption. The purpose of this Directive is to set standards for the quality of water intended for drinking or for use in food and drinks manufacture in order to protect human health. The Directive has some bearing on the protection of the environment as the drinking water must be sufficiently free from contamination to allow inexpensive water treatment. This Directive is also linked to Directives 75/440/EEC and 79/869/EEC which are concerned with surface water. Natural mineral waters are covered by a separate Directive (80/777/EEC).

62 water
quality
standards

Directive 80/778/EEC includes 62 water quality standards and guidelines for water quality monitoring. The water quality standards are listed in Annex I of the Directive. It includes three types of standards:

- the Guide Levels (GL);

- the Maximum Admissible Concentrations (MAC);

- the Minimum Required Concentrations (MRC).

MAC or MRC
values

Where MAC or MRC values are included in the Directive, Member States must set values which fall within these MAC or MRC values. They must also ensure that waters used for consumption meets these standards.

GL standards

When the Directive gives only a GL standard, Member States may use their discretion as to whether or not to set a standard. You should also note that the Directive sets no standards for some parameters. They were included in the Directive to indicate that they may be subject to subsequent regulatory control.

The Guide Levels and Maximum Admissible Concentrations of some key parameters set by Directive 80/778/EEC are shown in Table 7.2. Notice that some parameters are given both Guide Levels (GLs) and Maximum Admissible Concentrations (MACs). Where both are given, the Guide Level sets a useful target concentration to aim for. Notice also that for some parameters only MACs are given, whilst for others only GL's are given.

Π Examine the microbiological standards given in Table 7.2. Why are standards set for the number of coliforms present in the water?

Coliforms are organisms that are commonly found in the alimentary tracts of mammals. Thus the presence of coliforms is indicative of contamination by mammalian excreta. Waters contaminated by excreta are also potentially contaminated by pathogens

coliforms are
used as
indicator
organisms

derived from the alimentary tract (for example *Salmonella* spp., *Shigella* spp. and *Vibrio cholera*). The coliforms are therefore used as 'indicator' organisms, indicating that the water may harbour various pathogens. Microbiological tests are available to distinguish between faecal and non-faecal coliforms. These tests are usually based on a statistical approach and allow determination of the most probable number (MPN) of coliforms present.

The general microbiological quality of a water is evaluated by inoculating samples of rich media and incubating them at 37°C and 22°C. Saprophytic organisms tend to grow at 22°C whilst organisms capable of colonising humans tend to grow better at 37°C.

Parameter	Expressed as	Guide level	Maximum admissible concentration
Colours	mg l^{-1} (Pt/Co scale)	1	20
Turbidity	mg l^{-1} (Sii O_2 scale)	1	10
Odour	Dilution number	0	2 (12°C) 3 (25°C)
pH		6.5 ≤ ≥ 8.5	
Total dissolved solids	mg l^{-1}		1500
Conductivity	μS cm (20°C)	400	
Chloride	mg l^{-1}	25	
Sulphate	mg l^{-1}	25	250
Calcium	mg l^{-1}	100	
Magnesium	mg l^{-1}	30	50
Sodium	mg l^{-1}	20	175
Nitrate	mg l^{-1}	25	50
Ammonia	mg l^{-1}	0.05	0.5
Phenols	μg l^{-1}		0.5
Boron	μg l^{-1}	1000	
Iron	μg l^{-1}	50	200
Manganese	μg l^{-1}	20	50
Phosphate	μg l^{-1}	400	5000
Fluoride	μg l^{-1}		1500 (12°C) -700 (25°C)
Arsenic	μg l^{-1}		50
Cadmium	μg l^{-1}		5
Cyanide	μg l^{-1}		50
Mercury	μg l^{-1}		1
Lead	μg l^{-1}		50
Pesticides	μg l^{-1}		0.5
Microbiological standards			
Total coliforms	MPN/100 ml		< 1
Faecal coliforms	MPN/100 ml		< 1
Total colonies 37°	ml^{-1}	10	
Total colonies	ml^{-1}	100	

Table 7.2 Some key EC - Drinking water standards as described by Directive 80/778/EEC. MPN = most probable number.

| **SAQ 7.1** | Which of the following statements are false and which are true? |

1) EC Member States must set standards that are consistent with all Guide Levels (GL) stated in Directive 80/778/EEC relating to the quality of water intended for consumption.

2) It is safe to assume that if the MPN of faecal coliforms in a water is 0 per 100 ml, the water is safe to drink.

3) In cases where Directive 80/778/EEC sets both Guide Levels and Maximum Admissible Concentrations for a parameter, the Guide Levels are always lower than the Maximum Admissible Levels.

7.3.3 EC - Directives relating to amenity value of water

Here we will focus on just one example, the use of water for bathing. EEC - Directive 76/160 (OJ L31/1-76/160/EEC) is concerned with the quality of bathing water. The quality of bathing water is to be raised over time largely by ensuring that sewage is not present or has been adequately destroyed or diluted.

sampling procedures are specified

I and G values

The Directive specifies minimum sampling frequencies and specifies where and how samples are to be taken. It lists 19 physical, chemical and microbiological parameters which need to be analysed. Standards are specified by I (= imperative) values or by G (= guide) values. The most critical of these standards are the coliform counts. The I values set by the Directive should have been met by Member States by December 1985, although, in some cases, derogations were allowed providing they were justified by a management plan.

The Directive also included 'reference methods' of how the analysis was to be conducted, although other comparable methods are allowed.

Analogous Directives are also in force for the waters which are to be used, for example, as fisheries.

<table>
<tr><td>SAQ 7.2</td></tr>
</table>

Below is a table of water quality standards relating to waters used in fisheries. It is often said that coarse fish are more tolerant to pollution than salmonids. In what ways do the EC quality standards reflect this?

Some EC fresh water fish water quality standards as described in Directive 80/778/EEC.

Parameter (mg l^{-1} except where otherwise stated)	Average dissolved concentration ions	
	Coarse fish	Salmonid fish
Arsenic	0.05	0.05
Cadmium	0.005	0.005
Chromium	0.15-0.25	0.005-0.05
Cooper	0.001-0.028	0.001-0.028
Lead	0.05-0.25	0.004-0.02
Mercury	0.001	0.001
Zinc	0.075-0.5	0.01-0.125
Phosphate	131	65
Ammonium (total)	0.16	0.031
BOD	6	3
pH (units)	6-9	6-9

7.3.4 EC - Directives relating to dangerous substances in water

Framework Directive 76/464/EEC

Directives relating to dangerous substances in water are built around the Framework Directive (76/464/EEC) on pollution caused by certain dangerous substances discharged into the aquatic environment. The purpose of this Directive is to provide a framework for the reduction (elimination) of pollution from waters by various dangerous substances.

A key element of this Framework Directive is the inclusion of two lists (List I and List II) of groups of dangerous substances. List I includes substances which, because of their toxicity, persistence and bioaccumulation (biomagnification) are regarded as particularly damaging. It is not surprising that this list has become known as the 'Black List'.

Black List

Π From your knowledge of chemistry and biology and from what you have read earlier in this text, see if you can suggest some examples of the types of compounds you might expect to be included in List I.

The sorts of compounds that you should have anticipated as being in this list are many of the highly toxic, recalcitrant man-made organohalogens and organophosphorus compounds. You may also have included carcinogenic materials and the highly toxic metals such as cadmium and mercury.

List II includes compounds that are considered to be less dangerous because they are either less toxic or are less persistent (more readily biodegraded). This list includes such compounds as cyanide and ammonia and compounds of metals such zinc, copper and lead. Although regarded as less dangerous than List I materials, they still present substantial threats. List II is sometimes referred to as the 'Grey List'.

Grey List

Discharges of substances included in these two lists are subject to prior authorisation by a competant authority within each Member State. The mechanisms of authorisation are different for List I and II substances and there are some variations between Member States.

This Framework Directive has been subsequently added to by a variety of secondary Directives which set standards for particular substances. Some examples of these are given in Table 7.2.

∏ Are the secondary Directives listed in Table 7.3 related only to List I (Black List) or List II (Grey List) substances?

The Directives we have listed are mainly confined to List I substances. It is clear that the Community has prioritised these. You should, however, anticipate that, increasingly, Directives will be issued which specify new quality standards for a wider group of substances including those currently placed in List II.

Substance	Directive	OJ Reference
Mercury	Directive on limit values and quality objectives for mercury discharges by the chlor-alkali electrolysis industry (82/176/EEC)	L81 27.03.82
	Directive on limit values and quality objectives for mercury by sectors other than the chlor-alkali electrolysis industry (84/156/EEC)	L74 17.03.84
Cadmium	Directive on limit values and quality objectives for cadmium discharges (83/513/EEC)	L291 24.10.83
Hexachlorocyclohexane	Directive on limit values and quality objectives for discharges of hexachlorocyclohexane (84/491/EEC)	L274 17.10.84
Carbon tetrachloride DDT, pentachlorophenol	Directive on limit values and quality objectives for discharges of certain dangerous substances included in List I of the Annex to Directive 76/464/EEC (86/280/EEC)	L81 04.07.86
'Drins' (aldrin, dieldrin, endrin, isodrin) hexachlorobenzene, hexachlorobutadiene, chloroform	Directive amending Annex II to Directive 86/280/EEC on limit discharges of certain dangerous substance included in List I of the Annex to Directive 76/464/EEC (88/347/EEC)	L158 25.06.88
1,2-dichloroethane, perchloroethane, trichloroethane, trichlorobenzene	Directive amending Annex II to Directive 86/280/EEC on limit discharges of certain dangerous substance included in List I of the Annex to Directive 76/464/EEC (88/347/EEC)	L219 14.08.90

Table 7.3 Examples of secondary Directives to Framework Directive 76/464/EEC on dangerous substances in water.

7.3.5 implementation of regulations within Member States

Member States of the EC are under an obligation to meet the quality standards set within the Directives described above. They are also obliged to put into place a management and monitoring regime that ensures that these quality standards are maintained. The situation is an evolutionary one and the details of how these obligations are met differs between Member States. In Britain, for example, the Water Act 1989 and subsequent legislation established the National Rivers Authority as an independent statutory body with the responsibility for the control of water quality and water pollution. Previously, these responsibilities were divided between a variety of bodies. As a result of its obligations arising from the Directives, the British Government has introduced legally defined water quality objectives. The National Rivers Authority has a statutory duty to ensure that these objectives are met. It, therefore, organises the monitoring of water quality at sampling points throughout its jurisdiction. The same authority has responsibility for authorising and monitoring discharges.

National regulations

7.4 The parameters measured to maintain or improve water quality

Our attention has been predominantly focussed on the physical and chemical parameters of water, but you should realise that the ability of water to support a diverse fauna and flora are indicative of its quality. There is, therefore, a strong case, especially for surface waters, to use ecological parameters (biotic indices) as a means of determining water quality. These ecological parameters are of significance to the objective of using waters to enhance the conservation of wild life as well as to the value of these waters as a resource for human use. Ecological measures are, however, predominantly influenced by the physical and chemical parameters and it is on these that we will concentrate.

ecological parameters

∏ See how many parameters you can list that you think should be monitored for water that is to be abstracted for human consumption. (Do this without looking back to Table 7.2).

There are many parameters that you could have listed. Let us see if we can bring some order to these.

Some parameters measure some general properties of the water. These included physical parameters such as:

- pH;
- colour/turbidity;
- suspended solids;
- temperature (°C);
- conductivity ($\mu S\ cm^{-1}$);
- odour.

Also included in these general parameters are those that relate to the chemicals present in the water. These include such parameters as:

- chemical oxygen demand (COD);
- biological oxygen demand (BOD);
- total nitrogen (Kjeldhal);
- total phosphate;
- total pesticides;
- dissolved oxygen (DO).

We can also identify some microbiological parameters such as:

- total coliform counts;
- faecal coliform counts;
- faecal streptococci counts;
- Salmonella counts.

Our list of parameters does not, however, end there. The Directives we discussed earlier imposes quality standards for a wide range of specific substances. Therefore, these also need to be monitored. We have provided a list of some of these specific materials in Table 7.4. Again, this is an indicative list, we have not attempted to be fully comprehensive.

Inorganic	Organic
Ammonia	Specific pesticides
Arsenic	Hydrocarbons
Barium	Organohalogens
Boron	Phenolics
Cadmium	
Calcium	
Chloride	
Chromium	
Copper	
Cyanide	
Iron	
Lead	
Manganese	
Mercury	
Nitrate	
Selenium	
Zinc	

Table 7.4 Examples of specific parameters that are measured for waters used for abstraction and consumption.

Π What is done with the data produced from measuring these parameters?

The data are used for a multitude of purposes. First, of course, it enables a decision to be taken whether or not the water is suitable for its intended use. EEC - Directive 76/464/, for example, specifies the type of treatment that needs to be undertaken if a surface water is to be used for consumption depending upon the quality of the water (see Table 7.5, do not attempt to remember all the figures).

Π Which parameters are the most variable that can be tolerated for using surface waters for abstraction and consumption?

You should have concluded that the presence of micro-organisms (especially coliforms and streptococci) is most tolerated. This is because they can be readily removed by treatment processes. As you might expect, however, highly contaminated surface waters have to be treated more intensively than are clean waters.

normal treatment processes do not remove metal ions
In contrast, normal treatment processes do not remove many of the metal ions. Thus the limits that are set for these are similar irrespective of the treatment process. They are also very similar to those that are set for drinking water (compare the standards in Table 7.5 with those in Table 7.2).

A second purpose for including such a wide variety of parameters is to detect discharges. A progressive rise in, for example, the levels of mercury in a body of water would alert the water authorities to seek out the source and to put into place measures that will avert or alleviate the problem.

relevance of analysis to the authorisation of discharge
A third purpose for measuring these parameters is to monitor the consequences of authorised discharges and to evaluate proposals to allow new discharges. For example, during the planning stage of new industrial developments, the proposed industries may seek to discharge certain materials into a body of water. The authorising authorities need to evaluate the likely outcome of such discharges. The outcome may be that they authorise discharge or impose certain conditions (for example they may require some pre-treatment before discharge) or they may completely prevent discharge. The measurement of these parameters also enables evaluation of the efficacy of existing treatment processes and to plan for future developments of waste water treatment.

Treatment type	A1		A2		A3	
Parameter (mg/1 except where noted)	Guide limited	Mandatory limit 1)	Guide limit	Mandatory limit 1)	Guide limit	Mandatory limit 1)
pH units	6.5-8.5		5.5-9.0		5.5-9.0	
Colour units	10	20	50	100	50	200
Suspended solids	25					
Temperature, °C	22	25	22	25	22	25
Conductivity (µS/cm)	1000		1000		1000	
Odour (TON)	3		10		20	
Nitrate (as NO_3^-)		50		50		50
Fluoride	0.7-1.0	1.5	0.7-1.7		0.7-1.7	
Iron (soluble)	0.1	0.3	1.0	2.0	1.0	
Manganese	0.05		0.1		1.0	
Copper	0.02	0.05	0.05		1.0	
Zinc	0.5	3.0	1.0	5.0	1.0	5.0
Boron	1.0		1.0		1.0	
Arsenic	0.01	0.05		0.05	0.05	0.1
Cadmium	0.001	0.005	0.001	0.005	0.001	0.005
Chromium (total)		0.05		0.05		0.05
Lead		0.05		0.05		0.05
Selenium		0.01		0.01		0.01
Mercury	0.0005	0.001	0.0005	0.0001	0.0005	0.001
Barium		0.1		1.0		1.0
Cyanide		0.05		0.05		0.05
Sulphate	150	250	150	250	200	250
Chloride	200		200		200	
Phosphate (as P_2O_5)	0.4		0.7		0.7	
Phenol		0.001	0.001	0.005	0.01	0.1
Hydrocarbons (ether soluble)		0.05		0.2	0.5	1.0
PAH (polycyclic aromatic hydrocarbons)		0.0002		0.0002		0.001
Pesticides		0.001		0.0025		0.005
COD					30	
BOD (with ATU)	< 3		< 5		< 7	

Table 7.5 EC standards for surface waters used for human consumption (Directive 76/464/EEC).
Treatment types: A1 = Simple physical treatment and disinfection. A2 = Normal full physical and chemical treatment with disinfection. A3 = Intensive physical and chemical treatment with disinfection.
1) Mandatory levels 95% compliance, 5% not complying should not exceed 150% of mandatory level.

Cont/d.............

Treatment type	A1		A2		A3	
Parameter (mg/1 except where noted)	Guide limited	Mandatory limit 1)	Guide limited	Mandatory limit 1)	Guide limited	Mandatory limit 1)
DO (percent saturation)	> 70		> 50		> 30	
Nitrogen (Kjeldahl)	1		2		3	
Ammonia (as NH_4^+)	0.05			1.5	2	4
Total coliforms/ 100 ml	50		5000		50 000	
Faecal coliforms/ 100 ml	20		2000		20 000	
Faecal streptococci/ 100 ml	20		1000		10 000	
Salmonella	absent in 5 litres		absent in 1 litre			

Table 7.5 EC standards for surface waters used for human consumption (Directive 76/464/EEC). Treatment types: A1 = Simple physical treatment and disinfection. A2 = Normal full physical and chemical treatment with disinfection. A3 = Intensive physical and chemical treatment with disinfection.
1) Mandatory levels 95% compliance, 5% not complying should not exceed 150% of mandatory level.

Simply for illustrative purposes, we have reproduced the prescribed concentrations (or values) that are applied to water used for consumption in England and Wales (see Table 7.6). These are specified within the Water Supply (Water Quality) Regulations 1989. Although different values for these parameters may be used in different Member States, they must all set values which fall within the limits set by the relevant Directives. Do not attempt to memorise all these values. You should, however, be impressed by the extensive range of parameters that are legally controlled.

∏ It would be a helpful learning activity for you to compare the quality standards given in Table 7.6 with those shown in Table 7.2.

Basically, what you should have discovered is that the quality standards of the UK (1989) regulations are consistant with the standards of EC - Directive 80/778/EEC.

PART A			
Item	Parameters	Units of measurement	Concentration or value (maximum unless otherwise stated)
1)	Colour	mg/1 Pt/Co scale	20
2)	Turbidity (including suspended solids)	Formazin turbidity units	4
3)	Odour (including hydrogen sulphide)	Dilution number	3 at 25°C
4)	Taste	Dilution number	3 at 25°C
5)	Temperature	°C	25
6)	Hydrogen ion	pH value	9.5 5.5 (minimum)
7)	Sulphate	mg SO_4/l	250
8)	Magnesium	mg Mg/l	50
9)	Sodium	mg Na/l	150 (i)
10)	Potassium	mg K/l	12
11)	Dry residues	mg/l	1500 (after drying at 180°C)
12)	Nitrate	mg NO_3/l	50
13)	Nitrite	mg No_2/l	0.1
14)	Ammonium (ammonia and ammonium ions)	mg NH_4/l	0.5
15)	Kjeldahl nitrogen	mg N/l	1
16)	Oxidizability (permanganate value)	mg Os/l	5
17)	Total organic carbon	mg C/l	No significant increase over that normally observed
18)	Dissolved or emulsified hydrocarbons (after extraction with petroleum ether); mineral oils	µg/l	10
19)	Phenols	µg C_6H_5OH/l	0.5
20)	Surfactants	µg/l (as lauryl sulphate)	200
21)	Aluminium	µg A1/l	200
22)	Iron	µg Fe/l	200
23)	Manganese	µg Mn/l	50
24)	Copper	µg Cu/l	3000
25)	Zinc	µg Zn/l	5000
26)	Phosphorus	µg P/l	2200
27)	Fluoride	µg F/l	1500
28)	Silver	µg Ag/l	10 (i)

(i) If silver is used in a water treatment process, 80 may be substituted for 10.

Table 7.6 Prescribed concentrations or values from the Water Supply (Water Quality) Regulations 1989.
Cont/d.............

		PART B	
Item	**Parameters**	**Units of measurement**	**Maximum concentration**
1)	Arsenic	µg As/l	50
2)	Cadmium	µg Cd/l	5
3)	Cyanide	µg CN/l	50
4)	Chromium	µg Cr/l	50
5)	Mercury	µg Hg/l	1
6)	Nickel	µg Ni/l	50
7)	Lead	µg Pb/l	50
8)	Antimony	µg Sb/l	10
9)	Selenium	µg Se/l	10
10)	Pesticides and related products:		
	a) individual substances	µg/l	0.1
	b) total substances (i)	µg/l	0.5
11)	Polycyclic aromatic hydrocarbons (ii)	µg/l	0.2

(i) The sum of the detected concentrations of individual substances.
(ii) The sum of the detected concentrations of fluoranthene, benzo 3,4 fluoranthene, benzo 11.12 fluoranthene, benzo 3,4 pyrene, benzo 1,12 perylene and indeno (1,2,3-cd) pyrene.

		PART C	
Item	**Parameters**	**Units of measurement**	**Maximum concentration**
1)	Total coliforms	number/100 ml	0 (i)
2)	Faecal coliforms	number/100 ml	0
3)	Faecal streptococci	number/100 ml	0
4)	Sulphite-reducing clostridia	number/20 ml	≤ 1 (ii)
5)	Colony counts	number/1 ml at 22°C or 37°C	No significant increase over that normally observed

(i) Analysis by multiple tube method.

		PART D	
Item	**Parameters**	**Units of measurement**	**Maximum concentration or value**
1)	Conductivity	µS/cm	1500 at 20°C
2)	Chloride	mg Cl/l	400
3)	Calcium	mg Ca/l	250
4)	Substances extractable in chloroform	mg/l dry residue	1
5)	Boron	µg B/l	2000
6)	Barium	µg Ba/l	1000
7)	Benzo 3,4 pyrene	ng/l	10
8)	Tetrachloromethane	µg/l	3
9)	Trichloroethene	µg/l	30
10)	Tetrachloroethene	µg/l	10

Table 7.6 Prescribed concentrations or values from the Water Supply (Water Quality) Regulations 1989.
Cont/d.............

PART E			
Item	Parameters	Units of measurement	Minimum concentration
1)	Total hardness	mg Ca/l	60
2)	Alkalinity	mg HCO₃/l	30

Table 7.6 Prescribed concentrations or values from the Water Supply (Water Quality) Regulations 1989.

7.5 The measurement of parameters important in water quality management

high costs of monitoring

the principle of the polluter pays

The great variety of parameters that need to be measured, the frequency of sampling and the administration involved in the monitoring of discharges and consideration, and review, of consents to discharge involve considerable costs. Although the principles of 'the polluter pays' is increasingly being adopted, the recurrent costs are high. There is a need, therefore, to use cost effective, but reliable, methods to measure the desired parameters. In many, but not all, cases the methods to be used are specified or recommended by EC - Directives and other Regulations and Guidelines. It is not our purpose to give details of these procedures here. However, we will draw attention to some general features of some of these procedures.

The measurement of specific inorganic parameters generally rely on specific colourmetric reactions or upon the use of ion-specific electrodes. In many instances, specific kits may be purchased which conform to the specified/recommended procedures. Thus, for most inorganic species, suitable procedures are readily available. These procedures are predominantly designed to require a minimum of pre-treatment, high reliability and specificity, give the required degree of precision and have a low labour input. Also important is the speed with which data can be generated after sampling.

In the case of complex organic materials, the situation is usually not so easily solved. They are often present in very low concentrations and these compounds often do not possess an accessible, distinctive chemical reactions. Analysis often involves a concentrating step followed by chromatographic characterisation (for example by High Performance Liquid Chromatography). These types of analysis are, therefore, costly in terms of labour and equipment and there is often substantial delay in obtaining data. It also means that the analysis cannot be done on site but have to be conducted in a centralised laboratory.

biosensors

There are however other options. Much development work is being undertaken to use the specificity of biological systems (exhibited by enzymes, antibodies and receptors) to generate analytical tools suitable for, amongst other purposes, monitoring water quality and discharges. The most widely held concept is the production of biosensors which are capable of detecting and measuring specific molecules at low concentrations. It is on these that we will concentrate our attention.

7.6 Biosensors and the measurement of water quality parameters

7.6.1 General principle of biosensors

The field of biosensor technology is developing very rapidly and involves the use of biological molecules to specifically interact with substances and to use this interaction as a method of detecting and measuring substances of interest. In most biosensors, the biomolecules that are used are enzymes although biosensors using antibodies, receptors and nucleic acids have been designed. Here we will outline the principles of enzyme-based biosensors. If you wish to follow up this in greater detail, we would recommend the BIOTOL text 'Technological Applications of Biocatalysts'.

A generalised scheme for a biosensor is shown in Figure 7.1.

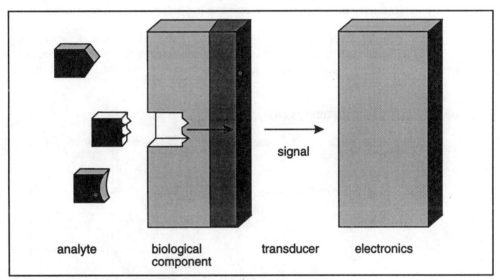

Figure 7.1 A generalised scheme for a biosensor (see text for details).

The substance to be measured (the analyte) specifically interacts with the biological component (the enzyme) on the surface of the sensor. This interaction leads to some change in the biological component (or other components in its vicinity) and this change is transformed into an electronic signal by a transducer. This electronic signal is processed to give a measurement.

7.6.2 Generation of an electronic signal

Transducers are used in biosensors to convert the interaction of the analyte and the enzyme into an electronic response. In order to be successful, this transducer must also be amenable to the immobilisation of the enzyme on its surface so that the two are in intimate contact. Many different types of transducers have been used in the construction of biosensors. A list is given in Table 7.7.

role of
transducers in
biosensors

Transducer	Examples
Electrochemical	
a) Amperometric	Clark oxygen electrode, mediated electrode systems.
b) Potentiometric	Redox electrodes, ion-selective electrodes, field effect transistors, light addressable potentiometric sensors.
c) Conductimetric	Platinum or gold electrodes for the measurement of change in conductivity of the solution due to the generation of ions.
Optical	Photodiodes, waveguide systems, integrated optical sensors.
Acoustic	Piezoelectric crystals, surface acoustic wave devices.
Calorimetric	Thermistor or thermopile.

Table 7.7 Transducers commonly used in biosensors.

We do not intend to examine all of these transducers in detail. We will, however, illustrate how transducers work using just one example, amperometric transduction.

Consider the following system. We have an enzyme attached to the surface of a transducer. This enzyme is of the oxido-reductase type. If the reduced substrate of this enzyme is present, the enzyme may catalyse the oxidation of the substrate releasing electrons.

Thus, diagrammatically:

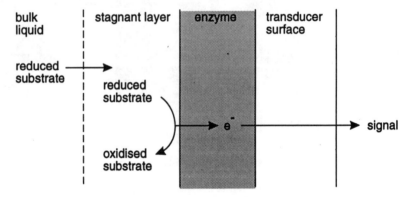

Π   What factors will influence the rate of electron generation by the enzyme?

The main factor will be the concentration of reduced substrate in the vicinity of the enzyme (providing the enzyme is present in excess).

Π   What factors influence the concentration of the substrate in the vicinity of the enzyme?

The main factors are:

• the concentration of the substrate in the bulk liquid;

- the thickness of the unmixed (stagnant) layer around the enzyme (this depends on the mixing of the bulk liquid);

- the properties of the liquid (for example its viscosity);

- the properties of the substrate (for example its molecular mass) which influence its rate of diffusion;

- the temperature of the liquid (which influences diffusion rates).

For any particular substrate, providing other factors such as mixing rates, viscosity of the medium and temperature are kept constant, the rate of transfer of the substrate should relate simply to the concentration of the substrate in the bulk liquid.

Π What factors other than mass transfer rates may influence the size of the signal produced in such a system?

Obviously the activity of the enzyme is important. If it was completely denatured, the signal would fall to zero. Except under conditions in which the enzyme is present in vaste excess, the signal is dependent upon the enzyme reaction kinetics. Also importance are the kinetics of the electrode (eg efficiency of transfer of electrons to the transducer) and the presence of inhibitors.

use of mediators This picture of biosensors based upon immobilised enzymes is greatly simplified. Frequently other agents (mediators) are used to transfer electrons from the enzyme to the electrode. $Fe(CN)_6^{3-}$ and ferrocene are two good example of such mediators.

In coulometric biosensors, the total amount of electrons transferred rather than the rate of electron transferred is measured.

The transfer of electrons from substrate to electrode is not the only method of generating a signal. In optical transducers the enzyme is held close to a photodiode which is illuminated. In these systems catalysis of a reaction involving the substrate **use of photodiodes** (analyte) leads to a change in absorbance, either of the substrate or a co-reactant. The change in absorbance is detected by the photodiode. Diagrammatically these systems look like this:

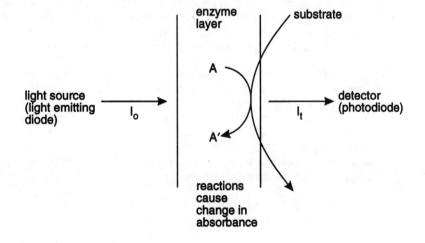

use of
thermopiles

Alternatively, the heat generated during the reaction of the substrate may be used as the signal generating mechanism. In these cases, the transducer consists of a thermopile or some other heat responsive detector. (For details of the physical principles of these and other detection devices we recommend the BIOTOL text 'Technological Applications of Biocatalysts').

7.6.3 Limitations to developing usable biosensors

Π From your knowledge of enzymes, see if you can list some difficulties that need to be overcome in order to use biosensors to detect and measure pollutants in water.

There are many practical difficulties in converting the concept of biosensors into practical instruments. First we need to find biomolecules that show specific interactions with the compounds we wish to measure. In the case of enzymes biosensors, this means finding an enzyme that will either:

- specifically catalyse a reaction involving the compound of interest, and which shows minimal interfence by other compounds;

- is specifically inhibited by the compound of interest.

In the first example, higher concentrations of the compound of interest leads to a greater signal whilst in the second, the signal is reduced by higher concentrations of the compound of interest.

affinity is
important

Assuming that we have found an enzyme with the desired degree of specificity, it must also have an appropriate affinity for the compound of interest. For example, if the anticipated concentration of the compound to be measured is of the order of 10^{-13} mol l^{-1}, then if the K_M (or K_I) of the enzyme is of the order of 10^{-5} mol l^{-1}, the enzyme will barely be sensitive enough to detect the compound. Although this type of difficulty may be overcome in some instances by using highly sensitive detection systems, the system is prone to imprecision. You should, however, note that biosensors have been produced which have lower detection limits as low as 10^{-21} moles.

In addition to finding an enzyme of appropriate specificity and sensitivity, we also need to devise a way of converting the interaction between the substrate and enzyme into a detectable signal. Not all reactions and detection systems are as sensitive as each other. For example, the sensitivity of thermal detectors depends upon the amount of heat released by the reaction whilst optical detectors depend upon the changes in optical properties brought by the interaction of substrate and enzyme.

stability is
important

Also of great importance is the availability of the enzyme and its stability. Enzymes are generally sensitive to a wide range of physical (temperature, pH) and chemical parameters. Ideally, we need to use enzymes which are relatively stable and robust so that the biosensor has a long life. A major difficulty encountered in biosensor production and use is the question of stability. Although searches for stable enzymes (for example from thermopiles) and the techniques of contemporary genetic engineering (such as site-directed mutagenesis) facilitates the production of more stable enzymes, success is not guaranteed.

SAQ 7.3

1) Two enzymes have been isolated which catalyse the reaction.

$$X_{reduced} \rightarrow X_{oxidised} + 2e^- + 2H^+$$

These enzymes are called enzyme A and enzyme B. Solely on the basis of the specificity data shown below and assuming that a suitable transducer was available, which enzyme would you recommend to use for the construction of a biosensor to detect and measure $X_{reduced}$.

Substrate	Relative activities	
	Enzyme A	Enzyme B
$X_{reduced}$	100	100
$CH_3\text{-}X_{reduced}$	0.2	15
$CH_3\text{-}O\text{-}X_{reduced}$	0.8	5.5
$C_6H_5\text{-}X_{reduced}$	8	5
$C_6H_{11}O_6\text{-}X_{reduced}$	0.5	3

2) What other parameters would influence your choice of enzyme?

7.6.4 Other types of biosensors

Microbial biosensors

use of whole cells in biosensors

So far we have focused on using specific enzymes as the sensor part of the biosensor. Whole cells can also be held near to a transducer to form what are usually called microbial biosensors. We learnt, in the previous chapter, that particular micro-organisms have a capacity to metabolise a wide range of recalcitrant organic molecules. In principle, we can use these abilities instead of using isolated enzymes. This approach may be advantageous in situations where it may be impractical to use a purified enzyme. This may arise if the enzyme is unstable, requires a soluble cofactor, or is simply difficult to purify.

The variability and multireceptor behaviour of whole cells has advantages as well as disadvantages. The low specificity may be undesirable if a highly specific probe is required. On the other, in some circumstances, this lower specificity is desirable as it enables the use of the same probe to measure a range of compounds.

Whole cell sensors are generally more stable as the labile enzymes are retained in the protective milieu of the cells. This makes them more suitable for continuous operation. However, response times are generally slow and frequent recalibration is usually necessary.

Substrate specificity can often be improved by pre-incubation of the cells used in microbial sensors with the substrate we wish to determine.

∏ See if you can explain why pre-incubation of cells with the desired substrate may lead to greater substrate specificity.

importance of pre-incubation

The answer is that the production of enzymes by microbial cells is, at least in part, controlled by exogenously available substrates. For example, in the presence of glucose, the bacterium *Escherichia coli* does not express the enzyme β-galactosidase needed to metabolise lactose. On the other hand, in the absence of glucose and in the presence of lactose, β-galactosidase is produced. Thus if we used lactose grown cells, we could use these cells to detect lactose and glucose since the cells would metabolise either sugar. If, on the other hand, we used glucose grown cells, they could not be used to detect lactose. The point we are making is that by pre-conditioning cells, we influence their metabolic capabilities.

use of mutants
in biosensors

We could also consider the use of mutants which have changed metabolic capabilities. For example, we may improve selectivity by using mutants in which undesirable metabolic pathways are blocked. Alternatively we may consider the use of metabolic inhibitors to improve selectivity.

Electrochemical transduction is the favoured technique. Amperometric and potentiometric devices are most common. To date, about fifty electrochemical biosensors based on micro-organisms have been developed for the detection of alcohols, ammonia, antibiotics, biochemical oxygen demand, enzyme activities, mutagenicity, nitrate, organic acids, peptides, phosphate, sugars and vitamins.

SAQ 7.4

You have a strain of bacteria that carry a plasmid which enables the bacteria to catabolise phenol. Plasmid-carrying bacteria are capable of using phenol as their sole source of carbon and energy.

Biochemical analysis shows, however, that the cells, even when cultivated in phenol-containing media, produce the enzymes of glycolysis.

You would like to use this bacterium to produce a phenol biosensor, using a thermal detection system. Unfortunately, the samples you need to test contain both phenol and glucose. Glucose is known to switch off phenol-metabolising genes. Explain what you should do to make a successful phenol biosensor.

Immunosensors

Biosensors may be based on affinity reactions of biological molecules. In affinity biosensors, rather than relying on an initial binding of a molecule followed by a biocatalytically-induced chemical change for detection, the binding event itself is detected. Although a range of biological molecules with specific binding properties could be used in affinity biosensors, the most versatile and sensitive molecules are antibodies. Antibody-based biosensors are commonly termed immunosensors.

use of
ELISA-based
systems

The basic principles of all immunosensors are based on classical immunoassay methods which have been routinely used in clinical laboratories and diagnostic tests for a number of years. One of the most common immunoassay methods is the enzyme-linked immunosorbant assay, or ELISA method. This involves the use of a solid phase for the separation of free and bound immunobiological complexes. There are two main types of ELISA; 'competitive' and 'sandwich', both of which most commonly use colorimetric detection.

competitive
ELISA

Competitive ELISA is used mainly for small molecules. Here, the antigen of interest is bound to an inert surface, such as polystyrene, in a microtitre plate. The sample (which may, or may not contain antigen) is then added in the presence of antibody-enzyme conjugate. If no antigen is present in the sample, then the antibody-enzyme conjugate forms a complex with the bound antigen and after washing, addition of substrate will lead to colour formation. If the sample contains antigen, then this competes with the bound antigen for the antibody-enzyme conjugate and reduces the amount of bound complex formation (and hence colour) in the microtitre wells. Therefore, the final measured product concentration is inversely proportional to the concentration of the analyte. A similar, but alternative competitive ELISA technique involves antibody bound to the solid phase and in this situation free antigen in the sample competes with antigen-enzyme conjugate.

sandwich
ELISA

The sandwich ELISA approach is used for larger antigen molecules such as proteins or whole cells. The antigen is simply 'sandwiched' between bound and free antibodies. For this, the sample is added along with antibody-enzyme conjugate, to wells containing bound antibody and then washed. Bound complex is subsequently determined by adding a chromogenic substrate for the enzyme. The final product concentration is therefore directly proportional to the antigen present in the sample. In immunosensors these reactions take place on the surface of a transducer.

A wide range of monoclonal antibodies are now commercially available for immunoassay, allowing immunological detection of a huge number of molecules. Analytes with molecular masses ranging from 100 up to over 10^6 Daltons, can be detected by immunoassay techniques.

The two transduction methodologies most commonly used in immunosensors are electrochemical and optical. Further details of this technology can be found in the BIOTOL text 'Technological Applications of Biocatalysts'.

7.7 Miscellaneous contributions of biotechnology to water analysis

7.7.1 Immune assays

Biotechnological developments in analytical procedures are not entirely dependent upon the development of biosensors. The ability to produce large quantities of monoclonal antibodies enables the production of specific analytical procedures for a wide variety of molecules, especially macromolecules.

Although antibodies have long been used to detect and measure specific molecules, a difficulty that was often encountered centred on the quantification of the interaction between antibody and the target compound.

The development of enzyme-linked immunosorbent assays (ELISA) have largely overcome this difficulty and ELISA techniques have become a method of choice. We outlined these techniques above. (If you need further details we recommend the BIOTOL texts 'Technological Application of Biocatalysts' and 'Technological Application of Immunochemicals').

Π The development of these techniques enables us to rapidly detect and assay a wide variety of molecular species. See if you can suggest how this type of technique might be used in water analysis.

It is rather unusual for us to seek to assay for specific macromolecules in water. We are, however, often interested in the presence of specific organisms (especially certain bacteria and viruses) particularly those derived from human excreta. Bacteria are traditional detected by incubating water samples with suitable media and to carry out biochemical characteristion of the cultivated organisms. Such techniques are costly and slow and are not readily usable for the detection of viruses. ELISA assays which detect specific antigens associated with particular organisms offer an alternative route to the monitoring of the microbial quality of water. Thus we might visualise using ELISA techniques to monitor the microbiological quality of water.

7.7.2 Assays based on the use of enzymes

The biosensors described earlier are only one way in which enzymes may be used to detect and measure compounds of interest. They differ from conventional assays which rely on the activities enzymes in that they are usually produced in the form of a portable probe. You should not forget, however, that more conventional assay formats involving enzymes may also be employed. In most cases, such assays are used to measure particular molecules in biologically-derived materials and find particular application in the food and health industries. However, they can be used to detect compounds such as the organophosphate insecticides/nerve poisons. The enzyme used for this is acetyl cholinesterase which is inhibited by these compounds. (For a detailed discussion of the use of enzymes in analysis we recommend the BIOTOL text 'Technological Applications of Enzymes').

∏ From what we said about the production of suitable enzyme-based biosensors for use in monitoring water quality, what do you think has mainly restricted the wider application of enzymes for analysing water-borne substances?

The main problems with using enzyme-based assays for analysing components in water samples relates to the:

*   sensitivity of the assays;

*   instability of enzymes;

*   need to conduct the assay under poorly-defined conditions;

*   the cost of producing enzymes.

∏ Can biotechnology alleviate these problems?

The answer is that, in principle, biotechnological techniques may reduce these problems. For example, genetic engineering techniques may provide routes to producing enzymes with altered characteristics (lower K_M and therefore greater sensitivity; altered specificities; greater stability). Here, we would cite the techniques of enzyme modelling and site-directed mutagenesis as being of particular importance. Genetic engineering techniques may also be used to increase the production of enzymes and to simplify the purification of these enzymes thereby reducing the cost of producing the desired enzymes. (If you would like to learn more about these techniques, we would again recommend the BIOTOL text 'Technological Applications of Biocatalysts'). Because of the advent of these techniques we may anticipate that water analysis based on the use of enzymes (either in conventional forms or as biosensors) will play an increasing role in water quality control in the future.

7.8 Concluding remarks

In Chapters 4, 5 and 6 we predominantly considered the application of biotechnological processes to the treatment of water-borne contaminants. From what we have described in this chapter you should have become more acquainted with the quality standards these processes help to achieve. You should also be aware of the complexity of the analysis that need to be undertaken to ensure waters are of good quality and the importance of this analysis in monitoring and regulating discharges. You should also be aware of the potential of applying biotechnological products and processes to the analysis of water.

You should, however, be aware that biotechnological products and processes are also being applied to the analysis and purification of gases and soils.

Summary and objectives

In this chapter, we have examined the importance of water analysis in the control of water quality and on the regulatory framework in which water quality standards are set. We have also examined how recent biotechnological developments may facilitate water analysis thereby contributing to the achievement of quality control in water management.

Now that you have completed this chapter you should be able to:

• list the roles water analysis plays in water management;

• give examples of EC - Directives which set quality standards in water supplying and use;

• explain why microbiological as well as chemical quality standards are applied to waters;

• interpret data relating to water quality standards;

• demonstrate an awareness of the existence of List I and List II dangerous substances and give examples of each;

• list examples of parameters which are measured for waters used for abstraction and consumption;

• explain the potential of using enzyme-based biosensors in water analysis;

• list criteria for the selection of enzymes suitable for use in bioprobes;

• explain how biosensors with high degrees of specificity may be produced using whole cells;

• list ways in which molecular genetic engineering may lead to increases in the use of biological systems as analytical reagents in water analysis.

Biofuels

Biofuels

8.1 Introduction

In our discussion of the use of energy resources in Chapter 2, we pointed out that the use of fossil fuels is, in the longer terms, unsustainable and that there is a need to use renewable resources. In essence, this means using solar energy either directly or through the use of biomass and its derivatives. Historically mankind was dependent upon biomass to supply energy a fact that is reflected in the energy sources used in developing countries. Figure 8.1 compares the energy sources used in developed and developing countries. You will notice that in developed countries only about 1% of the energy consumed is derived from biomass whereas, in developing countries, over 40% is obtained either directly or indirectly from biomass.

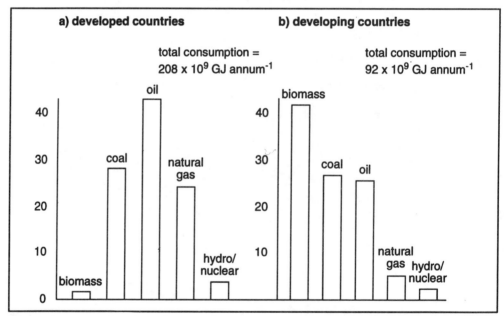

Figure 8.1 Comparison of energy sources used by developed and developing countries. (Data derived from Parikh J.K. Energy Systems and Development, Oxford University Press, 1980).

∏ In addition to the differences in the sources of energy used by developed and developing countries, what other major difference is there between the two types of countries in terms of their use of energy?

The feature we hope you will have identified is that the developed countries consumed much more energy then the developing countries. This is reflected by the amount of energy consumed per person per year. This varies enormously from community to community but, from numerous surveys, typical figures for developing countries can be as low as 5-10 GJ per person per year (see, for example, data collated by Hall D.O., Barnard G.W. and Moss P.A. 'Biomass for Energy in Developing Countries'). In the developed countries energy consumption per person may be 100-500 times greater.

social and
economic
pressures for
using fossil
fuels

The lower dependence on non-renewable energy sources by the developing countries, although environmentally more satisfactory, is, however, under threat. As these countries strive for economic and social party with their richer, developed counterparts, there is enormous pressures to adopt the practices of the developed world. It is in this context that we need to consider how biological resources may be more fully employed to meet the energy demands of human society.

In part, the move towards using renewable energy resources will only be achieved through the development of a political and social climate conducive to its application. The development of this politico-social morality is largely outside the scope of this text. Here we mainly confine ourselves to a discussion of the technical possibilities of using biological processes and resources to provide the energy that human society needs.

We will begin by providing an overview of the biological sources we might use and the processes by which these sources may be converted to more usable forms. We will then go on to discuss the conversion processes in more detail.

8.2 The biological energy options

In order to consider the options available to use, we need to:

- identify sources of biomass, or derivatives of biomass, that may be used to provide energy;
- identify processes by which these sources may be converted into a usable form (that is their conversion into fuels);
- identify how these fuels will be used.

∏ See if you can list 3 or 4 sources of biological materials that may be used to provide energy.

The sorts of items we anticipate you may have listed are wood, crop residues such as straw and bagasse, crops specifically grown to provide energy (energy crops) and organic wastes such as sewage.

∏ Now see if you can list some ways in which these sources may be used to provide energy.

burning to
provide heat

methane
production

The most obvious way is simply to burn the biomass to provide heat. From what you read in earlier chapters you should have also included the production of methane from anaerobic digestion by, for example landfill and waste water treatment processes. You could have also included fermentation processes leading to, for example, ethanol production. We should also include thermo-chemical processes which lead to the production of intermediary fuel forms such as charcoal that can be produced by the pyrolysis of wood. Of course, these conversion processes are not applicable to all bio-energy sources.

∏   How may these energy sources be utilised?

In broad terms they may be used directly to provide heat or, indirectly, to provide electricity or mechanical power.

In Figure 8.2, we have linked the biological energy sources to the energy conversion processes and to their end use. This figure provides a useful summary so it is worthwhile examining it carefully.

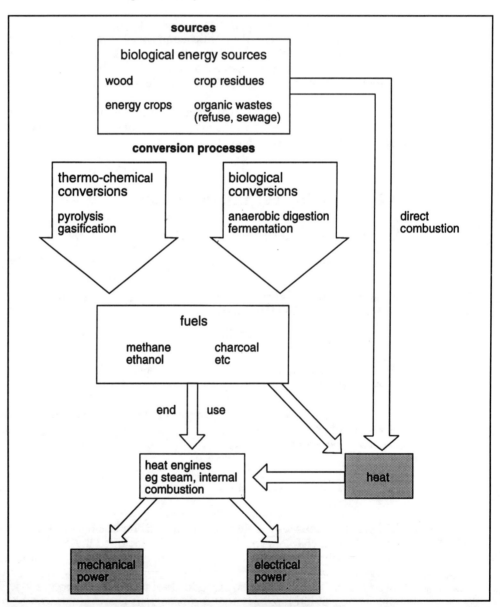

Figure 8.2 Generalised schemes for the use of biological energy sources.

ways in which
biotechnology
may increase
biologically-
generated
energy
resources
Since we are mainly concerned with the application of biotechnology, we will not discuss the thermo-chemical conversion processes further in this text. Nor will we discuss, in any detail heat engines, and the production of mechanical and electrical power. We can visualise biotechnology as making a number of major contributions to the realisation of using biologically-generated energy resources. By improving crop productivity, biotechnology may enhance the amount of biomass that can be made available for providing energy. It may also lead to improvements in the characteristics and yields of crops grown specifically for energy purposes. The main alternatives of energy crops are:

- fast growing trees;

- starch/sugar crops for producing ethanol;

- oil and hydrocarbon producing crops.

The development of new strains of plants using biotechnological techniques are described in detail in the BIOTOL text 'Biotechnological Innovations in Crop Production' so we will not re-examine this aspect here. However, we give some indications of how biotechnology may lead to improved crop production in the next chapter.

In this text, we will predominantly focus on the application of biotechnology in the conversion of biomass and biomass derived materials into fuels.

8.3 Biomass as fuel

Biomass is organic material of biological origin that has ultimately been derived from the fixation of carbon dioxide and the trapping of solar energy during photosynthesis. If it is to be used as a fuel, the biomass can either be burned, releasing the chemical energy directly, or it can be converted to a liquid or gaseous fuel that is higher in energy per kilogram than the original biomass. Although unconverted biomass can be used as a fuel, and indeed plays an important role in countries such as India where the collection and burning of firewood is important to the rural economy, the large scale use of such fuel suffers from a number of disadvantages over fossil fuels such as coal and oil.

∏ There are several potential disadvantages of using biomass as a fuel; list any that you can think of.

disadvantage
of using
biomass
directly as a
fuel
Firstly, biomass usually has a lower thermal content than fossil fuels. Secondly, it often has a high moisture content, thus inhibiting easy combustion, resulting in a large energy loss on burning - mainly as latent heat of steam. The high moisture content also causes the material to be biodegradable so that it cannot readily be stored. Thirdly, biomass tends to have a low density, in particular a low bulk density, which increases the size of equipment necessary for handling, storage and burning. Finally, the material is rarely in a homogenous physical form and is not free-flowing, making automatic feeding of combustion plants difficult.

In comparison, the geoconversion processes which have formed fossil fuels have increased the thermal content per unit weight, have reduced the moisture content, have substantially increased the density and bulk density and have fluidised the material (eg oil and natural gas) or produced a readily handled solid, as in the case of coal. It seems logical therefore, that if biofuels are to compete with, or replace fossil fuels some kind of conversion must take place, rendering the energy contained within the biomass more usable. This is particularly the case for transport fuel, where the portability and thermal energy content of the fuel rapidly become limiting factors. Therefore, while the burning of biomass will continue on a local scale and the generation of heat and electricity from waste biomass may become important, it is the processed biofuels that are most likely to be the major developments in the immediate future.

8.4 Fuels extracted directly from biomass

An intermediate stage to chemically or biologically converting the energy of biomass to a higher energy fuel is to extract the fuel from the biomass, usually as an oil or other hydrocarbon. There are many plants which accumulate oil, usually in their seeds and these are often already exploited commercially as sources of edible oil. Such oils include sunflower, rape-seed, olive, peanut, linseed, soyabean and safflower. All of these products are high in energy value and all are theoretically capable of being burned in boilers or used as diesel engine fuel. The main problem with the latter use is that often the oils are rather viscous and so do not inject easily into the combustion chamber of the engine. This problem can be overcome by esterifying the oils, although this will add a processing cost.

naturally occurring plant oils as fuels

In addition to the plants used at present for edible oils, a number of species of plant have shown promise as producers of hydrocarbons for fuel use. These include the Euphorbias (spurges), the milkweeds (*Asclepias spp.*) and a tropical tree *Copaifera*. An important feature of the first two examples is that they need little water to grow and hence can be grown in relatively dry environments where they will not compete for land used for food production. The Euphorbias are related to the plants used to commercially produce rubber, and produce a latex which is an emulsion containing 30% hydrocarbon in water. After the water has been removed, the resulting product is a liquid containing hydrocarbons of a lower molecular weight than present in petrol. The common Milkweed (*Ascepsias speciosa*) also contains around 30% hydrocarbons. Perhaps the most promise in this field is offered by the plant *Copaifera multijuga* which is found in Brazil. This plant is a legume and therefore is able to fix nitrogen in its root nodules, which makes cultivation more economical as less fertiliser is needed. It grows to a tree of some 30 metres in height, and when tapped like a rubber plant or maple produces a large volume of a liquid which has properties very similar to diesel oil. The tapping of the plant can take place twice a year, and so the potential for commercial exploitation is good. Another possible source of large scale production of hydrocarbons are certain freshwater and marine algae which are known to accumulate similar substances.

the use of Euphorbias to produce hydrocarbons

Although there is clearly potential for the production of biofuels by direct extraction from plants, at the present time (apart from small scale trials) these are only research projects or theoretical possibilities. There are, however, biofuels currently being produced by conversion of biomass to gaseous or liquid fuels and these will be considered in the following sections.

8.5 Biogas

Biogas is a mixture of gases, primarily methane and carbon dioxide, which is produced when organic material is degraded under anaerobic conditions.

∏ It is important that you remember the significance of anaerobic conditions. Briefly distinguish the nature of the products produced by micro-organisms growing aerobically from those of micro-organisms growing anaerobically.

Under strictly aerobic conditions the principle products are cellular material, carbon dioxide and water. The cellular material will itself be recycled when the cell dies generating more carbon dioxide and water. Under anaerobic conditions the end products are far more varied. They include carbon dioxide and water but also many reduced organic compounds such as acids, alcohols and so on.

biogas
produced in
landfill and
anaerobic
digesters

Biogas production is dependent on the development of a community of different bacterial species, and occurs naturally in environments such as lake and pond sediments, and in the rumen of cattle. As we learn, in Chapters 3 and 5, the process has now been applied in man-made environments such as in landfill sites and anaerobic digesters for solid and liquid waste treatment. Although, in each environment, the relative numbers and species of bacteria involved in the production of biogas varies, they all have some common features. We remind you of the general scheme of mass and energy flow through an anaerobic system in Figure 8.3.

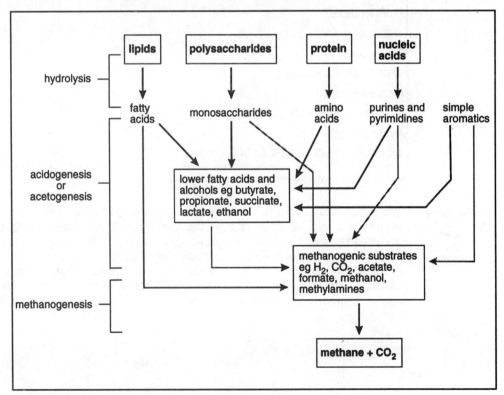

Figure 8.3 Carbon and energy flow through an anaerobic system.

∏ Look at Figure 8.3 and identify those compounds entering the anaerobic system. What would be the predominant types of compound entering naturally into a lake sediment (away from human influence) and how would these differ from those entering a landfill site?

In a lake sediment, the compounds would primarily be decaying plant material, principally cellulose, but other plant material such as xylans, lipids and proteins would be present. The compounds entering landfill sites would be much more variable - almost any compound could be represented. You should also remember that not all of the material deposited in landfill sites is biodegradable.

products of hydrolysis are converted to acids, alcohols and simple gases — Polymeric materials are hydrolysed by micro-organisms, such as the cellulolytic and proteolytic bacteria, to smaller molecules which then act as substrates for a second group of microbes, the acetogens. These fermentative bacteria further degrade the higher fatty acids to low molecular weight fatty acids such as acetate and formate, alcohols such as methanol, methylamines and gases such as hydrogen and carbon dioxide. Finally, these then act as substrates for the methanogenic bacteria, which convert them to methane and carbon dioxide. The methanogenic bacteria are a group of strictly anaerobic bacteria which requirement highly reducing conditions (redox potential (Eh) < - 400 mV). The types of energy yielding reactions which methanogenic bacteria as a group are capable of are shown in Table 8.1.

From hydrogen and carbon dioxide:

$$4H_2 + CO_2 \rightarrow CH_4 + 2H_2O$$

From formate:

$$4HCOOH \rightarrow CH_4 + 3CO_2 + 2H_2O$$

From acetate:

$$CH_3COOH \rightarrow CH_4 + CO_2$$

From methanol:

$$4CH_3OH \rightarrow 3CH_4 + CO_2 + 2H_2O$$

From trimethylamine:

$$4(CH_3)_3N + 6H_2O \rightarrow 9CH_4 + 3CO_2 + 4NH_3$$

Table 8.1 Energy yielding reactions in methanogenesis.

methanogens use CO_2 as the terminal electron acceptor — These microbes, unlike the fermentative acetogenic bacteria, are capable of anaerobic respiration using carbon dioxide as the terminal electron acceptor and, in the process, produce methane. The resulting gas which is generated by the system, biogas, contains roughly 60% methane, 40% carbon dioxide and traces of hydrogen. Biogas burns readily, although it does not have as high an energy value as natural gas which is almost pure methane. It is suitable as a fuel for boilers and kilns, and for internal combustion engines, although impurities in the gas can cause corrosion problems in the latter case.

<table>
<tr><td>

SAQ 8.1

</td><td>

1) Which one of the following substrates used to support an anaerobic digestion is likely to lead to the greatest yield of methane per kg of substrate? (Try to do this without looking at Table 7.1)

   a) CH_3OH b) HCOOH   c) CH_3COOH

2) What will be the proportions (v:v) of methane and carbon dioxide in the gases produced using these substrates?

</td></tr>
</table>

8.5.1 Anaerobic digesters

In Chapter 5, we described anaerobic digesters in some detail so we will not repeat a description of them here. We remind you however that they range from simple septic tanks through to highly technical fast flow systems. We suggest that you check your knowledge by attempting the following SAQ. If you are unable to correctly answer these questions, we suggest you re-read the appropriate sections in Chapter 5.

<table>
<tr><td>

SAQ 8.2

</td><td>

Indicate if each of the following is true or false.

1) Thermophilic digestion processes are more stable than mesophilic digestion processes.

2) Septic tanks retain the sludge for a long time and generate copious amounts of methane.

3) The upward feed flow in fluidised bed anaerobic digesters is usually greater than that used in packed bed and expanded bed reactors.

4) The average size of the microbial flocs produced in the treatment of munipal waste waters and from wastes derived from food processing units and piggeries is the same.

5) Thermophilic digestion is a faster process than mesophilic digestion.

</td></tr>
</table>

The methane generated from the anaerobic digestion of wastes is increasingly being used. The main motivations are:

- to reduce the costs of waste treatment process;

- to reduce the risk of explosions arising from the production of the flammable methane.

economics of methane production are important
Although the consequences of generating methane from waste is to reduce dependence on the use of fossil fuels, the amount of usable energy generated by this process represents only a tiny fraction of the total energy needs of contemporary society. This is reflected by the fact that it is predominantly economic considerations which govern the actual use of the methane. In many instances, it is simply flared to remove the risk of explosion. However, in larger municipal processes, the methane is now usually collected and burnt to produce steam to drive electric generators. The electricity generated may be sufficient to power the pumps and motors used in the treatment process but rarely produces sufficient for export. The treatment of concentrated wastes from large pig and chicken farms also generates significant quantities of methane which

can be used to generate electricity. The application of this technology to these circumstances is still not universal because of the high capital costs involved in installation and the limited economic (at least in the farmer's perception) return on this investment. Whether or not the use of anaerobic digesters to produce usable energy from wastes will be extended in the next decades depends more on the political, social and economic perceptions of environmental costs than it does on technological development.

8.5.2 Landfill gas

development of landfill to produce biogas

A further development of the anaerobic digester concept is the landfill site, which can be considered to be a very large batch anaerobic bioreactor. You will recall from Chapter 3, that landfill sites are typically sites into which domestic refuse is tipped as a means of disposal. When full, the landfill site is covered with soil and landscaped, then left. Within the landfill the conditions rapidly become anaerobic due to the activities of aerobic micro-organisms which quickly deplete any oxygen originally present or which diffuses into the site. There then follows the development of communities of anaerobic bacteria as described above, which, after a time, produce significant quantities of biogas. This methane/CO_2 mixture accumulates within the landfill and then diffuses out through the surface and sides of the site. When landfill was first used as a method of waste disposal, this biogas leakage was considered to be a problem, as indeed it was, since there was a risk of fires or explosions from accumulating methane as well as an odour problem. During the 1960's it was realised by scientists and engineers in both the USA and Germany that this landfill gas had potential as a fuel if it could be captured and effectively managed. There then followed, during the 1970's and 80's, a period of research and development culminating in the production of managed landfill sites designed from the start to produce biogas - so-called 'biofilling'. By 1987 there were at least 146 commercial landfill gas schemes in operation worldwide, with the majority in the USA, Germany and the United Kingdom.

SAQ 8.3

For anyone to seriously consider harvesting biogas from a landfill site the major criterion after safety is obviously economic viability. List and discuss some of the factors which have to be considered before embarking on commercial exploitation. Read our response carefully.

use of gas wells

The process of gas abstraction at a landfill site is shown in Figure 8.4. The gas is initially collected by sinking a number of 'gas wells' into the top of the landfill. These are constructed of perforated or slotted piping which is then interconnected via a network of pipes to a gas pump or compressor. If an existing landfill is being harvested for gas then there will be no internal structure within the landfill. With new sites which have been planned from the start for gas production then the gas wells are more complex and may have internal walls made of clay in order to trap the gas more effectively. The shape of these wells is the subject of research, but a number of different types have been tried including vertical wells, short horizontal, short vertical and bell-shaped. From the compressor, the gas is passed to a chiller, which cools the biogas in order to remove water from it, which would render it less combustible and cause corrosion problems. There is also a flare stack which is used to burn off excess gas when production exceeds demand, or when there is a problem further down the line. Finally, the gas is filtered and its composition determined by monitoring prior to end use.

As we indicated in our response to SAQ 8.3, landfill gas technology is currently more prevalent in industrialised Western countries though small scale conventional biogas reactors can, and will increasingly, play a part in the Third World.

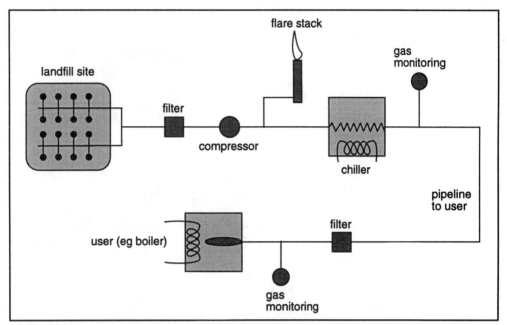

Figure 8.4 Landfill gas abstraction plant.

There are three major uses for landfill gas and these are illustrated in Figure 8.5. These end uses are:

- direct combustion in boilers, kilns or furnaces;

- generation of electricity;

- upgrading and export or compression or liquification.

direct combustion is economically the most viable

Of these alternatives, the first is the most economically feasible. The most effective use of the landfill gas is burning at a site close to the landfill itself, since building gas pipelines is expensive. One of the most successful landfill gas schemes to date in the UK is that of the London Brick Company who are ideally positioned to exploit this technology since they have disused clay pits which can be used for landfill, and, they have a major use for the biogas in firing brick kilns produced near to the landfill. Other examples of direct combustion of landfill gas include the heating of greenhouses, the firing of cement kilns and the production of steam for various industrial uses.

use of biogas to generate electricity

If a local end-user of the gas is not available, a second alternative is to generate electricity from the biogas. This can then be exported relatively cheaply by building electricity lines to connect in with the National Grid. At small scale plants generating up to 1-2 MW this usually involves the use of converted internal combustion gas engines, although larger schemes using gas turbines also exist. The main problem with this approach is that landfill gas is not pure a methane/carbon dioxide mixture, but contains traces of many other gases some of which are corrosive to metal parts of engines. This problem can be partially overcome by gas cleaning and by modifications to the generating engines and their lubricating oils. In the UK in 1987 six out of the 20 landfill gas schemes in operation generated electricity, while in the USA the figure was 22 out of 54, with most new schemes opting for this approach.

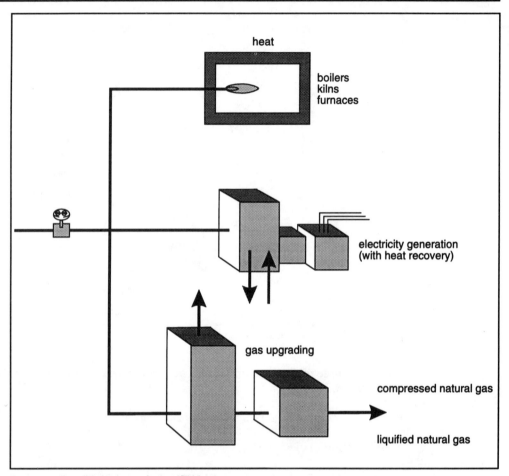

Figure 8.5 Possible end uses for landfill biogas.

The third alternative for landfill gas use is purification followed by export, either in a pipeline to a gas utility network or by bottling as compressed or liquified gas. The process of gas cleanup to provide a suitably pure methane gas is complex and expensive and so this option is the least favoured of the three. There are, however, a number of purification plants operating, mainly in the USA, and one scheme in Santiago, Chile where landfill gas is used to supplement the city's existing gas distribution system.

SAQ 8.4	

Decide whether the following statements are true or false giving reasons for your decisions.

1) Biogas contains methane with significant amounts of carbon dioxide and hydrogen.

2) Biogas has an energy value virtually equivalent to that of natural gas.

3) Methane is produced biologically by bacteria.

4) The predominant substrate in lake sediments is decaying plant material.

5) The predominant substrate in landfill sites is decaying plant material.

6) Anaerobically virtually all biopolymers are metabolised to simple fatty acids and alcohols.

8.6 Bioethanol

An alternative to conversion of biomass to a gas is to convert it to a liquid fuel, ethanol. Because it is a liquid, ethanol clearly has potential as a possible replacement for petrol. This possibility, as well as the methods of ethanol production, will now be considered.

8.6.1 Bioethanol as a fuel for internal combustion engines

Transport in the developed world relies largely on internal combustion engines. It therefore follows that any biofuel that is going to have a major impact on transport as it is today must comply with the requirements of these engines. If we consider that petrol is at present the most commonly used fuel for cars then a replacement biofuel must substitute for the qualities of petrol in the following respects:

- portability: it must be possible to carry the fuel in the vehicle in sufficient quantity to give an equivalent range to petrol;

- it must be possible to burn the fuel by an internal combustion engine;

- the fuel must have a roughly equivalent energy value to petrol.

Table 8.2 gives some of the chemical and physical properties of petrol and ethanol.

	Petrol	Ethanol
Freezing point (°C)	< -130	-117
Boiling point (°C)	35-200	78
Energy value (MJ kg^{-1})	44.0	27.2
Density (kg litre^{-1})	0.74	0.79
Flashpoint (°C)	13	45
Latent heat of vaporisation (MJ kg^{-1})	293	855
Octane number	80-100	99

Table 8.2 Chemical and physical properties of petrol and ethanol.

Examination of Table 8.2 shows that ethanol and petrol have broadly similar properties. It appears that ethanol is not as high in energy value as petrol (and hence theoretically would produce less 'kilometers per litre') but it is not quite as simple as this. Ethanol, in fact has some advantages over petrol, and these are connected with the latent heat of vaporisation and the octane number. A spark ignition internal combustion engine works by mixing the liquid fuel with air in the correct proportions before it enters the cylinder. This mixing occurs in the carburettor and involves the vaporisation of fuel in the air stream. This vaporisation requires energy (in the form of heat) which is supplied by the air, and is measured as the latent heat of vaporisation, which is expressed in MJoules of heat required to vaporise one kilogram of fuel. For ethanol, this latent heat of vaporisation is higher, which means that the mixture of ethanol vapour and air entering the engine is cooler than it would be for a petrol/air mixture, and this causes it to be denser. This in turn means that the volume of the ethanol/air mixture entering the cylinder at each induction stroke contains a greater mass of ethanol and thus more fuel enters the cylinder than it would in the case of using petrol. The result of this is that although the energy value per gram of ethanol is lower than that of petrol, the energy released during each combustion stroke is only slightly lower than with petrol.

The other advantage of ethanol relates to the octane number. Broadly speaking, this is a measure of the fuel's capacity to prevent pre-ignition, that is ignition before the piston has reached the correct position on the compression stroke. If pre-ignition occurs then the engine loses power and damage may occur to valves and piston and a 'pinking' noise can be heard. Since ethanol has a higher octane rating then petrol, pre-ignition does not occur in engines set up for petrol combustion, and the ethanol is effectively equivalent to a higher 'star rating' of petrol. Additionally, this also means that the compression ratio of ethanol burning engines can be raised with a concomitant increase in power, and this, together with the previously discussed density factor means that ethanol burning engines have a fuel consumption which is approximately only 10% higher than the equivalent petrol engine.

Table 8.2 also shows one other advantage of ethanol as a fuel for cars with respect to safety - the flashpoint (the temperature at which a substance ignites) of ethanol is more than three times higher than that of petrol. This means that in a fuel leak situation, say in a collision, the ethanol is far less likely to ignite or explode than petrol. Ethanol also burns cleaner than petrol resulting in less hydrocarbon emissions.

Ethanol does, however, have some disadvantages over petrol. The higher latent heat of vaporisation means that if the air temperature is cold, engine starting may be difficult. A more serious problem is that ethanol is miscible with water, so that fuel stored in tanks could pick up water from the air or from accidental additions. The resulting ethanol/water mixture will not burn readily and will cause corrosion problems in tanks and engines. Ethanol also reacts with some metals such as magnesium and aluminium alloys. These problems are not insurmountable; carburetters can be heated electrically to aid starting, and other metals such as nickel can be used for engine alloys.

An alternative to using pure ethanol as a fuel is to mix it with petrol in which case it not only becomes part of the energy of the fuel but also increases the octane rating of the petrol. Such a mixture, containing 20% ethanol, is currently marketed in the USA as 'Gasohol'. The drawback with this approach is that the ethanol has to be distilled to 100% purity before it is mixed with the petrol, otherwise separation will occur.

The foregoing discussion has not considered the diesel engine, which is used for most heavy haulage throughout the world, either by lorry or railway locomotive. Unfortunately, due to its chemical and physical characteristics, ethanol is not a good replacement for diesel oil. There are, however, other possible biofuels that could be

advantages of ethanol

importance of the latent heat of vaporisation

octane number of ethanol

importance of the higher flash point of ethanol

disadvantages of ethanol

gasohol

used in diesel engines including vegetable oils, plant derived hydrocarbons and hydrogen gas.

8.6.2 Production of bioethanol

The technology for the biological production of ethanol is as old as biotechnology itself since the production of alcoholic beverages by Man has been recorded for thousands, of years. However, although the production of fuel ethanol has undoubtedly drawn upon this older technology, the processes involved are more complex. They are, however, less constrained since the product does not have to be potable. The basic process involves conversion of the feedstock to a fermentable form, followed by fermentation and then distillation. With some feedstocks, such as sugar cane, the initial stage can be omitted since the extracted substrate is readily fermentable. The unit operations involved in the production of ethanol from feedstocks such as maize and sugar cane are shown in Figure 8.6.

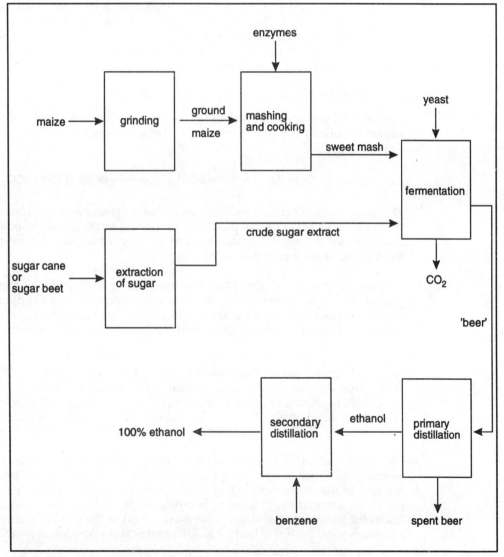

Figure 8.6 Process flowsheet for the production of bioethanol. In the examples shown, maize, sugar cane and/or sugar beet serve as the primary feedstock.

8.6.3 Feedstocks for the ethanol process

The feedstocks for the ethanol process are usually high in starch, for example maize, other grains such as cassava or millet, and potatoes; or are high in sugars, such as sugar cane and sugar beet. Additionally, less biodegradable substrates such as wood have been considered as feedstocks. The micro-organism that is used for the fermentation process (usually the yeast *Saccharomyces cerevisiae*) is incapable of hydrolysing starch to sugar and so if a starchy feedstock is used then enzymatic hydrolysis of the starch to sugars must first occur. To perform this, the grain is mixed with water and heated to the optimum temperature for the starch hydrolysing enzymes (around 66°C), the enzymes are added and the mash, as it is termed, is stirred until saccharification is complete. With feedstocks such as sugar cane, the process is simpler and involves breaking open the cane and extracting the sugar with water. One potential problem with these feedstocks is that they are also utilisable as food products and so the economic (and political) clash of biofuels with food is particularly important. A possible solution is to use wood as a feedstock, although this is very difficult to process to a form that can be fermented. There has been research into the use of steam treatment coupled with enzymic degradation of lignin and cellulose in order to produce a sugar solution from wood which can then be fermented to ethanol. Pilot plants are already operating with some success in North America.

economic and political clash between producing food and producing fuel

8.6.4 Fermentation

The fermentation process involves the conversion of sucrose (or glucose) to ethanol and carbon dioxide by the yeast according to the stoichiometric equation:

$$C_{12}H_{22}O_6 \xrightarrow{\ H_2O\ } 2C_6H_{12}O_6 \longrightarrow 4C_2H_5OH + 4CO_2$$

This is an anaerobic process which means that the fermenters used can be simpler than those use in say baker's yeast production in that no aeration and less mixing is required. The yeast, *Saccharomyces cerevisiae* is a facultative anaerobe, which means that it has both aerobic and anaerobic metabolic pathways.

Aerobically yeasts use glycolysis and the TCA cycle to produce some 38 ATP per glucose molecule. Anaerobically the yeasts use glycolysis to yield pyruvate and then, in order to maintain redox (reduction/oxidation) balance, convert pyruvate to carbon dioxide and ethanol. In this system, only 2 ATP per glucose are produced.

∏ Consider a yeast culture having a finite amount of glucose available. Fill in the gaps with either the word aerobic or the word anaerobic. a) [   ] conditions lead to production of more biomass. b) [   ] conditions lead to production of more ethanol. c) [   ] conditions are preferable for producing single cell protein. d) [   ] conditions are preferable for producing biofuels.

The answers are: a) aerobic because more energy is derived from each glucose molecule yielding, all other things being equal, 19 times more growth. b) anaerobic because ethanol is the main product of fermentation whereas relatively little ethanol is produced aerobically. c) aerobic because single cell protein is largely biomass. d) anaerobic because the ethanol is the most useful of the products in terms of biofuel. Although yeast itself will burn, it is not a practical proposition as a fuel.

⫪ Why is more ethanol produced under anaerobic conditions?

Because the cell needs to survive and, if possible, grow and divide. All of these processes require energy and the only way that yeast can obtain energy whilst maintaining its redox balance under anaerobic conditions is, as described above, by glycolysis and reduction of pyruvate to ethanol.

importance of catabolite repression

From the discussion above it appears obligatory to maintain anaerobic conditions. This is in practice expensive and fortunately there is a cheaper alternative. When the substrate level in the fermenter is above a certain concentration (around 20 mmol l^{-1}) catabolite repression occurs and the aerobic pathway is inhibited, even if air is present, and so the fermentation does not have to be kept as strictly anaerobic as when an organism such as *Clostridium* or the methanogenic bacteria are being grown. When the yeast is growing anaerobically, the flux of substrate through the metabolic pathway to ethanol is greater than that to carbon dioxide under aerobic conditions. The yield of biomass grown aerobically to produce enzymes or to produce baker's yeast is greater than that under anaerobic conditions, but for ethanol production anaerobic conditions are favourable.

⫪ What is the real limiting factor in the production of large quantities of ethanol?

You may have suggested reactor size, lack of balanced media etc. but all of these things can be manipulated. The problem is in fact ethanol toxicity. Above certain concentrations ethanol inhibits growth and further increase causes cell inactivation and death.

For most beer brewing strains of yeast, this inhibition occurs at ethanol concentrations above about 10%. The strains of yeast used in fuel ethanol production have been selected for ethanol tolerance, but even these strains cannot produce a fermentation broth with a greater ethanol concentration than approximately 18-20%. Since fuel ethanol must have a concentration of greater than 96% clearly extensive downstream processing is required, and this, in turn, will require energy.

8.6.5 Downstream processing in bioethanol production

The standard technique currently used to process ethanol from the liquor leaving the fermenter (termed 'beer' in the industry) to the purity required for fuel ethanol is distillation. This is usually performed on a continuous basis, although batch distillation, as is performed in the whisky industry, is possible. The process of distillation relies on the differing boiling points of ethanol and water (87°C and 100°C respectively in the pure state). When a mixture of water and ethanol containing less that 95% ethanol is heated to boiling the vapour produced has a higher concentration of ethanol than the liquid phase, and if cooled will condense to produce a distillate which is enriched in ethanol. If a single volume of an ethanol/water mixture was heated the initial concentration of ethanol in the vapour would be relatively high, but the liquid phase would now be depleted in ethanol and so the vapour would gradually become richer in water and poorer in ethanol. This means that a single batch distillation can only produce a limited quantity of enriched ethanol/water mixture. If, however, the initial distillate is then re-distilled a further ethanol distillate could be obtained, although this would be a smaller volume than the first distillate.

continuous and batch distillation

sequential re-distillation

This principle of a sequential re-distillation is employed in the batch distillation column (Figure 8.7), in which steam is used to boil an ethanol/water mixture in a vessel mounted below a cylindrical column divided into a series of chambers by perforated plates. The boiling liquid in the vessel gives off a vapour which is richer in ethanol than the original liquid and this passes into the first chamber where it mixes with the cooler liquid held on the plate by the pressure of the vapour. As the vapour mixes it gives off heat to the liquid, causing it to re-boil and give off a vapour which is still richer in ethanol. Some of the vapour/liquid mixture condenses and accumulates on the plate until it overflows from a weir back into the vessel. This process of re-boiling, or re-fluxing, takes place in exactly the same way on the next plate up, and so on to the top of the column where, if the column is designed properly, 95% ethanol will be produced.

continuous distillation

The continuous distillation column uses the same principle except that the fermenter 'beer' is fed into a point on the column where the concentration of ethanol is the same as the feed. The 'beer' is then boiled by condensing vapour from lower in the column, causing enriched vapour to pass up the column, and more watery 'beer' to fall down the column. In this way, 95% ethanol is continuously removed from the top of the column and water (known as stillage) is removed from the bottom.

azeotrope

The reason why the concentration of ethanol is limited to 95% under these conditions is that at this concentration the water in the liquid and vapour phase has the same concentration, and so no further separation is possible. This state is termed an azeotrope, and in order to obtain higher concentrations of ethanol, a small quantity of a substance such as benzene must be mixed with the ethanol/water. This effectively 'breaks' the azeotrope and allows nearly 100% ethanol to be obtained, while the benzene can be recovered.

energy consumption to produce steam is high

The consumption of steam during distillation of fuel ethanol can amount to a large proportion of the energy value of the ethanol obtained, and it is this perhaps more than any other consideration which limits the economic feasibility of bioethanol as a real alternative to petrol. The distillation process can be made more energy efficient by recovering the heat lost from boiling the liquid, but there are limits to this. Another approach is to burn waste biomass such as straw or bagasse in order to generate heat for the stills. This approach can be taken a step further by using the totally integrated grain-ethanol feedlot system (Figure 8.8) in which a biogas plant is integrated with an ethanol plant and a cattle (beef) farm such that nothing is wasted from the system.

For the future, a number of possibilities exist to make the bioethanol process more energetically feasible:

- recombinant DNA technology could be used to increase both the yield and ethanol tolerance of the yeast;

- procedures can be adopted to remove the ethanol from the fermenter before inhibition occurs. This could involve separating the yeast from the effluent by centrifugation and recycling the biomass to the fermenter. Alternatively ethanol could be removed continuously from the fermenter by vacuum distillation in which a partial vacuum is maintained in the headspace above the fermentation broth so that the ethanol evaporates rapidly at the operating temperature of the fermenter;

- the separation of ethanol from water can be made more energy efficient by using alternatives to distillation. These include reverse osmosis, where the ethanol is separated from the water by filtration through extremely small pores under high pressure; selective adsorption using solid adsorbants and the use of supercritical carbon dioxide to selectively extract the ethanol. (Details of these techniques are given in the BIOTOL text 'Product Recovery in Bioprocess Technology').

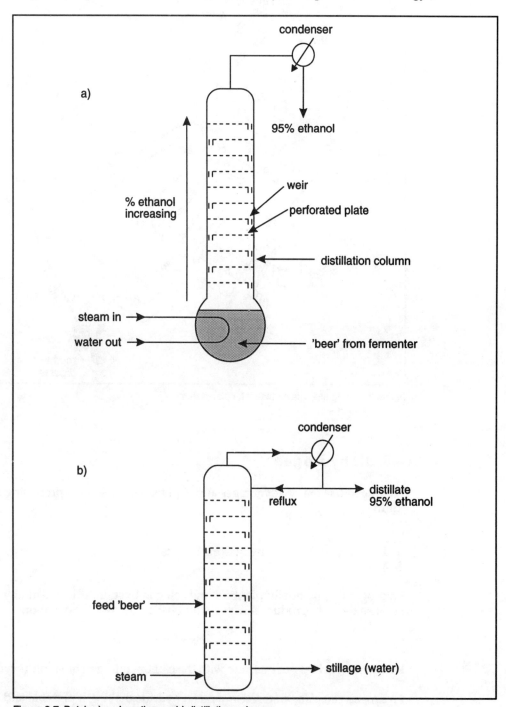

Figure 8.7 Batch a) and continuous b) distillation columns.

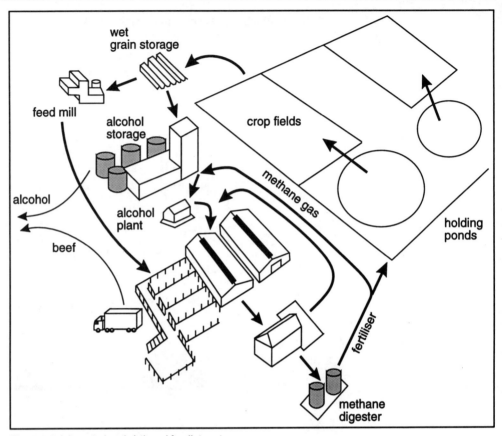

Figure 8.8 Integrated grain/ethanol feedlot system.

8.7 Biohydrogen

Hydrogen has many advantages and is potentially the most promising biofuel for the future.

Ⅱ Can you think of any reasons for this?

Hydrogen is potentially plentiful and, on combustion, it is essentially non-polluting because the only product is water. Combustion follows the equation:

$$2H_2 + O_2 \rightarrow 2H_2O$$

It is therefore does not interfere with the carbon, nitrogen or sulphur cycles.

Hydrogen can also be stored as a metal hydride making it portable and safe for use as an automobile fuel.

Hydrogen gas can be produced by a number of methods including the electrolysis of water, reaction of biomass with steam (gasification), and by biological processes. The latter method will be considered here.

hydrogen can
be used as an
electron sink Hydrogen is an important energy carrier in many biological systems, particularly in anaerobic fermentation processes. Hydrogen can be produced by anaerobic fermentative bacteria such as Clostridia as an 'electron sink', thereby providing a method of removing NADH from the system. When a substrate is oxidised during fermentative metabolism there must be a subsequent reduction, and this usually generates NADH from NAD^+. The accumulation of NADH would upset the equilibrium and cause these oxidation reactions to stop, and so the surplus NADH is channelled to hydrogen ions with the production of hydrogen, which is released from the cells:

$$NADH + H^+ \rightarrow NAD^+ + H_2$$

In anaerobic environments, this hydrogen is an important source of energy. In Chapter 5, for example, we described the methanogenic bacteria which can oxidise the hydrogen by using carbon dioxide as an electron acceptor to produce methane:

$$4H_2 + CO_2 \rightarrow CH_4 + 2H_2O$$

If the hydrogen-producing bacteria are grown in appropriate conditions in the absence of hydrogen-utilising bacteria then the hydrogen accumulates and can be collected. The feedstock for this fermentation could be cellulose, which can be hydrolysed to sugars and then fatty acids by certain anaerobic bacteria providing a substrate for the hydrogen-generating Clostridia. Some microscopic algae also produce hydrogen, and there has been research into growing these organisms under sunlight in 'photobioreactors', with the additional advantage of biomass production which could be used as animal feed. A more direct approach has also been suggested in which the photosynthetic apparatus of plants is linked biochemically to the hydrogen producing system. This possibility is shown in Figure 8.9. During photosynthesis the water molecule is split using light energy to yield hydrogen ions (protons), electrons and oxygen:

$$H_2O \xrightarrow{\text{light}} 2H^+ + 2e^{-1} + 1/2\ O_2$$

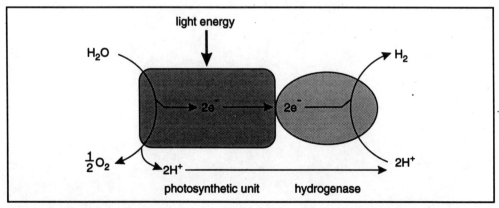

Figure 8.9 Possible mechanism for production of hydrogen by linking photosynthesis with hydrogenase.

Hydrogen generating microbes such *Clostridium spp.* contain an enzyme called hydrogenase which catalyses the reaction:

$$2H^+ + 2e^- \rightarrow H_2$$

The idea is to link the photosynthetic apparatus isolated from plants to hydrogenase in an *'in vitro'* situation thus generating hydrogen from water using solar energy. This system has been made to work but, unfortunately, is unstable, partly due to the sensitivity of hydrogenase to the oxygen generated by the photosystem. However, it does have potential for further investigation.

Summary and objectives

This chapter considers the use of biofuels as a necessary alternative to the use of fossil fuels. It investigates the limitations of biomass as a fuel and discusses its possible future exploitation. Production of biogas by anaerobic bacteria is described drawing upon information provided by earlier chapters. The potential value of bioethanol is described and its production and processing are outlined. In the final part of the chapter, we briefly examine the production of hydrogen gas by biological systems.

Now that you have completed this chapter you should be able to:

- recognise the need for alternatives to fossil fuels;

- outline the current use and status of biofuels;

- list the limitations of biomass as a potential fuel;

- describe the production of biogas (methane) and compare its commercial potential when produced in anaerobic digesters to that obtained from landfill sites;

- list the advantages of bioethanol as a fuel and describe its production and purification from the process in which yeast ferments glucose;

- demonstrate the value of biohydrogen as probably the potentially most important of all biofuels.

Alternative strategies in agriculture

Alternative strategies in agriculture

9.1 Introduction

Traditional agricultural practices largely depended upon harnessing natural, biologically-based processes to produce crops to service mankind's needs. In these traditional practices, there was little use of fossil fuels and no generation of recalcitrant chemicals that would resist biodegradation. The varieties of crop plants used were produced by the naturally accessible process of cross pollination; muscle power supplied the energy needed to till the soil, remove weeds and harvest the crop and soil fertility was maintained by the use of organic 'manures'.

population and economic pressures encourage the use of high technology

high energy consuming agriculture

Such agricultural practices represent merely an extension of the natural cycling of material described in Chapter 2. The social and economic pressures to increase crop productivity, together with developments in mechanical engineering and the chemical technologies has changed the complexion of agriculture. No longer is it free from fossil fuel consumption or from the production of man made, potentially environmentally damaging, chemicals for use as crop protectants. Continuing economic pressures and the need to feed a growing population encourages further extension of the 'high' technology, high energy demanding approach to agriculture. Nevertheless, there is an increasing body of opinion that predicts that such an approach is, in the longer term, unsustainable.

The future challenges of agriculture are not, however, confined to fulfilling the food requirements of mankind without causing extensive environmental damage. Arguments were given in Chapter 2 that mankind needs to become less reliant upon the consumption of fossil fuels in order to establish sustainable development. This calls for agriculture, and related sectors, are not only to provide food but also to be providers of energy. We have discussed the production of biofuels in Chapter 8 and have described how hitherto 'waste' biomass may be used to generate usable energy sources. In this chapter, we will focus on other aspects of the application of biotechnology to agriculture. In principle, we can foresee biotechnology providing novel approaches in the following main areas:

- plant strain development;

- crop protection;

- soil fertility;

- processing agri-products.

development of plant strains with desirable properties

Plant strain development involves the production of varieties with 'desirable' features such as increase yield, better shelf life and disease resistance. In most cases, the motivation for producing these strains is largely economic. Nevertheless it is becoming increasingly apparent that the application of biotechnology to strain improvement may have many, potentially important, environmental consequences. For example, the production of disease resistant varieties may reduce the use of recalcitrant crop protectants. Similarly, by modifying the morphology of root crops they may be 'designed' to be easier (costing less energy) to harvest or easier to extract particular products from. The techniques and application of biotechnology to crop improvement is described in detail in the BIOTOL text 'Biotechnological Innovations in Crop Improvements', so we will not examine this aspect further here.

The yield of crops is largely depended upon soil fertility, climate and the presence of pests and diseases. Biotechnology can do little to change climate, but it has much to offer in terms of improving soil fertility and reducing the ravages of pests and diseases.

use of case studies

Soils, crops and their associated pathogens and pests are very diverse. We do not propose to review the whole topic here. Instead we have selected two case studies to illustrate the principles involved. In the first, we will discuss a biological alternative to using inorganic fertilisers to improve soil fertility. In the second case study, we will examine the use of genetic engineering to develop potent biological crop protectants. In the final part of the chapter, we will provide a brief overview of some future perspectives in agricultural practices.

9.2 Nitrogen fertilisers - their role, importance and alternative sources

Crop productivity is dependant upon many factors including the availability of mineral nutrients and water, climatic conditions, the occurrence and prevalence of pathogens and pests and upon the physiological capabilities of the plants being cultivated. In this section, we will concentrate on the availability of mineral nutrients, especially nitrogen.

factors affecting crop productivity

The availability of nitrogen for crop production is a very large topic indeed. It encompasses studies on the biochemical and biological aspects of nitrogen turnover, soil chemistry, plant physiology, ecological and agronomic aspects. It should be realised that each crop, soil type and region imposes its own specific features. Here, we will examine in broad terms, the main issues involved and show, by using specific examples, how biotechnology may lead to an approach to agriculture which is environmentally more sustainable. The main emphasis is environmental rather than economic. On completing this section, the reader should be able to apply the principles described in a wide variety of situations. For those who wish to follow up specific aspects, some suggestions for further reading are given at the end of the text.

9.2.1 The mineral requirement of crops

The nutrition of plants is complex and, as we might anticipate, the exact requirements vary, in quantitative terms, from crop to crop. Nevertheless, all plants require at least 16 different elements. Some (the macro-elements) are required in substantial quantities while others (the micro-elements) are required in only trace amounts. Table 9.1 lists these elements and gives quantities required by a 'typical' crop.

macro- and micro-elements

Element	Chemical symbol	Form available to plants	Concentration in dry tissue	Relative No. of atoms compared to molybdenum
			ppm	
molybdenum	Mo	MoO_4^{2-}	0.001	1
copper	Cu	Cu^+, Cu^{2+}	0.01	100
zinc	Zn	Zn^{2+}	0.3	300
manganese	Mn	Mn^{2+}	1.0	1 000
iron	Fe	Fe^{3+}, Fe^{2+}	2.0	2 000
boron	B	$BO_3^{2-}, B_4O_7^{2-}$	2.0	2 000
chlorine	Cl	Cl^-	3.0	3 000
sulphur	S	SO_4^{2-}	30	30 000
phosphorus	P	HPO_4^{2-}	60	60 000
magnesium	Mg	Mg^{2+}	80	8000
calcium	Ca	Ca^{2+}	125	125 000
potassium	K	K^+	250	250 000
nitrogen	N	NO_3^-, NH_4^+	1000	1 000 000
oxygen	O	O_2, H_2O	30 000	30 000 000
carbon	C	CO_2	40 000	35 000 000
hydrogen	H	H_2O	60 000	60 000 000

Table 9.1 Form of mineral absorbed and concentration considered adequate. (After Epstein, E. Mineral Nutrition of Plants, Wiley, Chichester, 1972).

The essential point to bear in mind when considering the mineral requirement of plants is that if any one of these is not available in sufficient quantities, then growth and development of the crop will be restricted. It is, therefore, essential to provide plants with a balanced source of these elements in forms that may be utilised by them. An added complication is that excess of certain elements may cause toxicity problems.

∏   Which of the elements listed in Table 9.1 most frequently limit crop production?

K, P and N are
most
commonly the
limiting nutrient
in crop
production

We anticipate that you would have concluded that it is likely to be one of the macro-elements (especially sulphur, phosphorus, magnesium, calcium, potassium and nitrogen) that most frequently limit crop production. This is true. However, of this group, potassium, phosphorus and nitrogen are most commonly limiting and are, in contemporary agricultural practices, frequently supplemented using inorganic fertilisers. Of these limiting nutrients, nitrogen availability is most commonly the nutritional factor which limits crop production. We will examine why this is so in the next section.

9.2.2 Nitrogen as the limiting nutrient

nitrogen is only
a small portion
of terrestrial
matter

Various estimates of the elemental composition of the Earth have been made. All of these emphasise that nitrogen is only a very small proportion of total terrestrial matter. A generally accepted figure is that nitrogen represents only 0.046% (w/w) of the total matter present on Earth. (Compare this with the nitrogen content of plants in Table 9.1).

nitrogen cycle

This low level of nitrogen is further accentuated by the distribution of nitrogen and by its geocycling. In Figure 9.1, we have provided a simplified version of the nitrogen cycle. Of central importance to plant growth are the levels of NO_3^- and NH_4^+ in soils. These, especially NO_3^-, are the forms in which nitrogen can be taken up and used by plants. We will, therefore, begin our discussion of the cycle with NO_3^- and NH_4^+.

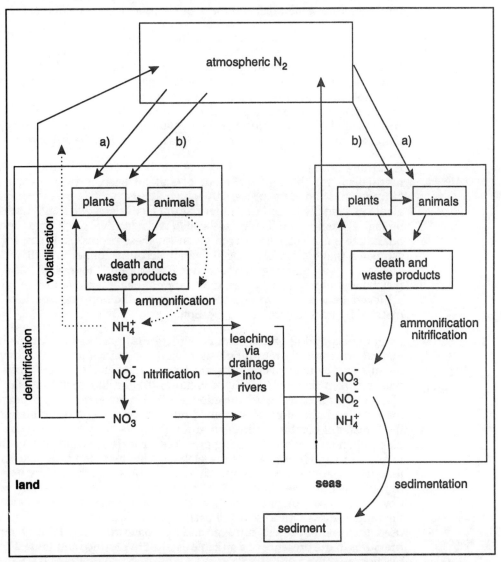

Figure 9.1 Simplified version of the natural global cycling of nitrogenous materials. a) biological fixation. b) lightning fixation.

leaching

denitrification

volatilisation

Four important processes tend to remove NO_3^- and NH_4^+ from soils. Salts containing NO_3^- and NH_4^+ ions are water soluble and these salts will tend to leach out of soils with drainage. These salts will eventually reach the oceans where they may be taken up by phytoplankton or sedimented. The second important process, denitrification, arises in anaerobic conditions. In this, some micro-organisms use NO_3^- as the electron acceptor in their respiratory processes. The product (mainly N_2) is lost to the atmosphere. A further loss of inorganic nitrogen from soils may occur through the volatilisation of NH_3. The formation of NH_3 from NH_4^+ occurs under alkaline conditions so we must anticipate that this process will be predominantly encountered in alkaline soils. The three processes described above may be regarded as natural processes. The extent to which they occur will, of course, dependent upon climatic conditions (for example heavy rainfall will increase leaching), the tendency of the soil to become waterlogged (and therefore, anaerobic) and the general mineral composition of the soil.

crop harvesting removes nitrogen

The fourth process contributing to the diminution of the availability of inorganic nitrogen is even more significant to agricultural productivity. The removal of plant parts during crop harvesting means that organic nitrogen is being removed from soils and will not replenish the available inorganic nitrogen through the processes of decomposition and mineralisation. Thus, if crops are removed from the land without the inorganic nitrogen being replaced, nitrogen availability will soon become reduced and the ability of the land to support plant growth will be seriously restricted. It is, therefore, important to consider how this nitrogen is to be replaced.

nitrogen fixation

Returning once more to the nitrogen cycle illustrated in Figure 9.1, the main natural mechanism for the replenishment of soil nitrogen is by the fixation of nitrogen. Nitrogen fixation is the process by which atmospheric N_2 gas is combined with other elements (usually O_2 or H_2). Under natural conditions two main mechanisms operate; biological fixation and fixation associated with electrical discharges (lightning). Various quantitative estimates have been given for these processes. It has, for example, been estimated that biological nitrogen fixation accounts for perhaps 10^9 tonnes N fixed per annum, whilst lightning may account for about 10^7 tonnes N fixed per annum. In natural environments in which crops are not being removed some kind of balance is achieved between the processes of nitrogen fixation and the loss of nitrogenous material by leaching and denitrification.

use of manure and crop rotation to maintain N levels

Until the beginning of this century, agricultural practices in Europe harnessed these natural processes. By returning waste plant and animal materials to soil in the form of manure and by using a crop rotation system in which one of the crops was associated with nitrogen fixation (usually a legume) the fertility of soils could be maintained. However, the increasing urbanisation resulted in increased rates of net removal of plant material (and, therefore of combined nitrogen) from agricultural lands and increased the transfer of combined nitrogen, through rivers, to the oceans as a result of sewage disposal. In other words, farming practices, urbanisation and sewage dispersal tended to accelerate the lower part of the global cycle illustrated in Figure 9.1. The ability to maintain or even increase productivity could only be achieved by using processes for replenishing the levels of combined nitrogen in soils. The key to this was the development of industrial processes to fix atmospheric dinitrogen gas. First, electrical discharges (the Arc process) were used and subsequently, the Haber Bosch process was used. In the latter process, nitrogen and hydrogen are reacted at high temperatures and pressures in the presence of a suitable catalyst to produce ammonia. We may write this reaction as:

Arc and Haber Bosch processes

$$N_2 + 3H_2 \rightarrow 2NH_3$$

The process is an energy demanding one since the $N \equiv N$ bond is extremely stable.

Table 9.2 gives figures for the world production of nitrogen fertilisers for the period 1913 - 1985.

Year	N (tonnes x 10^6 annum^{-1})
1913	1.27
1938	2.5
1962	9.3
1964	10.6
1975	30.3
1985	42.4

Table 9.2 Estimated world production of nitrogen fertilisers (data from Food and Agriculture Organisation Production year books).

We can now, therefore, represent the nitrogen cycle in a slightly different way (see Figure 9.2). You will notice that the two new elements in this cycle compared to that in Figure 9.1 are industrial nitrogen fixation and the removal of combined nitrogen via sewage disposal.

What, in effect, the change to using industrially fixed nitrogen has done is to accelerate the cycle. In this way we have achieved a greater production and turnover of food plants needed to feed the growing human population.

∏ What source (or sources) of energy is driving this increase in cycling?

In part, the cycle is being driven by increased rates of photosynthesis made possible by providing the minerals needed by plants. In other words, solar energy is, in part, being harnessed to drive the cycle. Perhaps more important is the consumption of fossil fuels. These are used not only for the industrial fixation of nitrogen, but also for its transport from its sites of production to its site of use and for its dispersal at the site of use.

cost of commercial nitrogen fertilisers is high

This use of fossil fuels has important consequences both in the shorter and the longer term. In the short term, the high energy costs and therefore commercial costs of the product, places this form of fertiliser out of reach of farmers in poorer regions of the world. These farmers and their communities are those who would benefit most from the use of these products. In the longer term, as we indicated in Chapter 2, the maintenance of human societies dependent upon the consumption of fossil fuels is unsustainable. In other words, if we are to maintain or even enhance agricultural productivity to meet increased demands of food, we need to find a mechanism for maintaining soil nitrogen levels which is dependent upon solar rather than fossil energy sources. This will, inevitably, mean that we need to find ways to better harness biological nitrogen fixation.

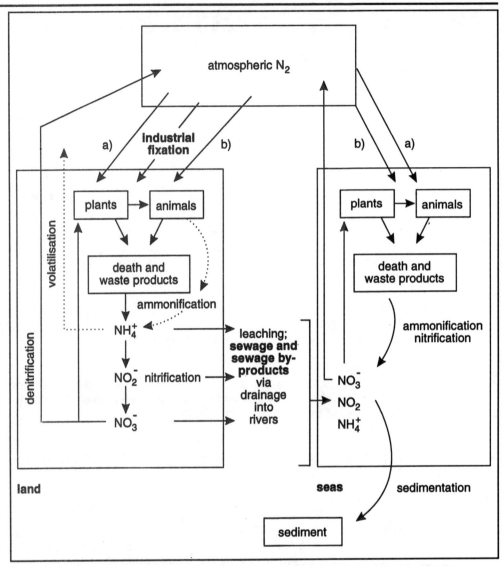

Figure 9.2 The imposition of human activities on the global cycling of nitrogen. a) biological fixation
b) lightning fixation.

environmental
consequences
of using NH_4^+
fertilisers

Before we turn our attention to this, we make one further point. Although the use of industrial fertilisers has enabled greater crop production, the application of ammonium salts to soils has other important environmental consequences. Ammonium ions (or NO_3^- ions produced from ammonium by nitrification) are readily water soluble. Additions of nitrogen fertilisers in this form means that much of the added material is quickly leached from the soil. In other words, the increase in soil fertility is short lived and much of the added nitrogen ends up in rivers and impounded water. This may lead to increases in algal blooms in such waters causing the water to become anaerobic when the algae die and decompose. The water also becomes discoloured, thus the water loses its amenity value as a potable water supply or as a recreational resource. Furthermore, increased NO_3^- levels in drinking water have been associated with important clinical problems. There are, therefore, environmental, energy and economic costs to rectify the problems which arise from the use of industrially-generated nitrogen fertilisers.

SAQ 9.1

The amount of nitrogen fixed by industrial processes represents only a small % of the total nitrogen fixed per annum. Why has industrially fixed nitrogen been so important to improving crop yields (read our response carefully)?

9.2.3 General principles of biological nitrogen fixation

Nitrogen fixation is a process carried out by a limited number of prokaryotic organisms. These organisms, however, belong to physiologically diverse groups.

Table 9.3 summarises these groups. They include free-living anaerobic, aerobic and facultative anaerobic heterotrophs; free-living photo-autotrophs; symbiotic forms and species which are in part free-living but usually grow in association with other, particular, non-nitrogen fixing species.

	Phototrophic	Chemo-autotrophic	Heterotrophic
free-living aerobic	cyanobacteria	methane oxidisers thiobacillus	*Azotobacter* group *Mycobacterium*
free-living anaerobes	purple bacteria green bacteria	-	*Clostridium* *Klebsiella** *Desulphovibrio* *Desulphotomaculum* Methanogenic bacteria
symbiotic aerobes (including those which grow in the rhizosphere)	cyanobacteria (with fungi, ferns)	-	*Rhizobium spp.* (and legumes) *Azospirillum* (and grass) *Frankia* (and alder, hawthorn)
symbiotic anaerobes	-	-	*Citrobacter* (and termites)

Table 9.3 Types of nitrogen fixing organisms. Note that all of the known nitrogen fixers are prokaryotes. No attempt has been made to catalogue all known groups. * Genera such as *Klebsiella* are facultative anaerobes. Commonly such organisms only fix nitrogen when they are cultivated under anaerobic conditions.

nitrogenase

Much is known about the biochemistry of nitrogen fixation. The core process is very similar in each case. The key enzyme system is nitrogenase which uses strong reductants and ATP generated by the cells to reduce N_2 gas. We can represent this as:

$$N_2 + \begin{array}{c}\text{strong}\\\text{reductant}\end{array} \xrightarrow[\text{nitrogenase}]{\text{ATP}} 2NH_3$$

production of ATP and strong reductants

The different physiological groups of nitrogen fixing organisms differ in the mechanisms by which they supply the necessary reductant and ATP. The photosynthetic types use the energy of solar radiation to generate both ATP and reductant. The aerobic heterotrophs use respiration coupled to oxidative phosphorylation to generate the necessary ATP and the reductant is generated from the oxidation of the organic substrates. (Note these organisms contain high levels of dehydrogenase activities). Anaerobic heterotrophs use a rather specialised intermediary metabolism to generate these reactants. Of particular importance in these micro-organisms in the phosphoroclastic reaction in which the oxidation of pyruvate is coupled to both the generation of reducing potential (in the form of reduced ferredoxin) and ATP production.

This process can be represented as:

We will not examine the biochemistry of nitrogen fixation in detail here. (For those interested in following up this aspect, we recommend the BIOTOL text 'Biosynthesis and the Integration of Cell Metabolism'). We should however point out that it is generally estimated that the stoichiometry of nitrogen fixation can be written as:

$$N_2 + 8H^+ + 8e^- + 16ATP \rightarrow 2NH_3 + H_2 + 16ADP + 16Pi$$

The ammonia produced by this process is assimilated into cellular material through amination reactions particularly those involving glutamine and glutamate.

A common feature of aerobic nitrogen fixing organisms is the need to protect the nitrogenase system against O_2. Nitrogen fixation requires strong reductants which may be auto-oxidised by molecular oxygen. Nitrogenase is also inactivated by O_2. Three types of protective mechanisms are used. These are:

- respiratory protection;
- nodule formation;
- heterocyst production.

Respiratory protection

Respiratory protection is found in many free-living aerobic heterotrophic nitrogen fixers. A good example is *Azotobacter vinelandii*. In these organisms there are two types of electron transport chains. Type 1 operates at both high and low O_2 tensions and is coupled to ATP production. This is 'normal' aerobic respiration. Type 2 operates at high oxygen tensions only and is not coupled to oxidative phosphorylation. This system dramatically reduces the O_2 tension in the vicinity of the nitrogenase.

Root nodules

Formation of root nodules is a feature of symbiotic nitrogen fixers which live in association with higher plants. It is especially associated with rhizobia which live in association with legumes and *Frankia spp.* which grow in association with alder.

rhizobial-legume associations

creation of a micro-aerophilic environment

In the rhizobial-legume associations, the infection of the roots of legumes by rhizobia leads to the formation of nodules. The host plants produce a protein called leghaemoglobin (or legumohaemoglobin). This is similar to the haemoglobin of mammalian blood except that it has a higher affinity for oxygen. The leghaemoglobin prevents the accumulation of high concentration of free oxygen but at the same time provides the oxygen necessary for the metabolism of *Rhizobium spp.* In effect, the presence of leghaemoglobin produces a micro-aerophilic (low O_2 tension) environment.

Heterocysts

nitrogen
fixation is
compart-
mentalised into
heterocysts in
some
cyanobacteria

Heterocysts are produced by certain species of nitrogen fixing cyanobacteria. These photosynthetic organisms generate O_2 during photosynthesis. Protection of nitrogenase is achieved by physically separating photosynthesis and nitrogen fixation by confining nitrogen fixation to specialised cells called heterocysts. Within these cells, the photosystem involved with O_2 evolution (photosystem 2) is absent. Thus photosynthetic activity is confined to producing ATP. The reducing potential needed is generated from normal photosynthesis involving both photosystems in normal cells. Figure 9.3 shows the metabolic exchange between vegative cells and heterocysts.

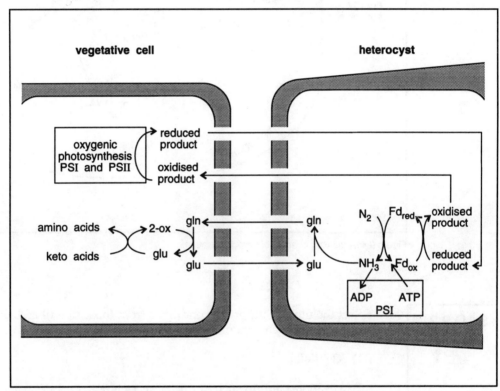

Figure 9.3 Metabolic exchange between vegetative cells and heterocysts. (Abbreviations: PSI, photsystem 1; PSII, photosystem 2; glu, glutamate; gln, glutamine; 2-ox, 2-oxoglutarate).

Despite these differences between systems, we can nevertheless, produce a unifying scheme for biological nitrogen fixation (see Figure 9.4).

Π There is one other aspect of nitrogen fixation which we have not mentioned in the text but which is illustrated in Figure 9.4. What is it?

regulation of
nitrogen fixation

You should have noted that ammonia leads to the repression of the genes which code for the proteins involved in nitrogen fixation. This is a common feature of nitrogen fixers. Nitrogen fixation is an energy consuming process. Switching off nitrogen fixation when usable combined nitrogen is available makes biological sense.

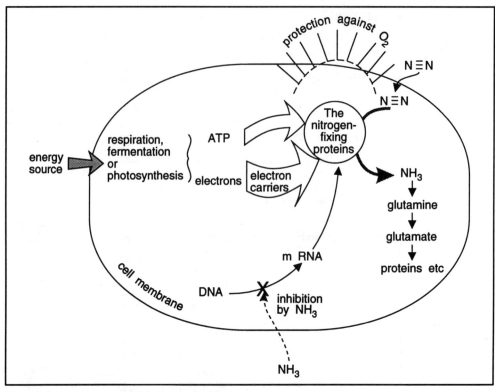

Figure 9.4 A unifying concept of nitrogen fixation.

SAQ 9.2

Select the most appropriate conditions from those described below [a) to e)] to maximise nitrogen fixation by:

1) Klebsiella;

2) a cyanobacterium such as *Anabaena cylindrica*.
   Indicate the reasons for your choice:

a) mineral salts medium containing glucose and nitrate under aerobic conditions in the dark;

b) mineral salts medium containing ammonia and glucose, under aerobic conditions in the dark;

c) mineral salts medium containing glucose but no combined nitrogen under aerobic conditions in the light;

d) mineral salts medium with no combined nitrogen or carbon under anaerobic conditions in the light in an atmosphere containing CO_2 and N_2.;

e) mineral salts medium containing glucose but no combined nitrogen under anaerobic conditions in the light in an atmosphere containing CO_2 and N_2.

9.2.4 How may biotechnology increase nitrogen fixation in the field?

There are many potentially useful approaches to this subject. Your instinct may have been to suggest that the production of mutant strains of nitrogen fixers with permanently de-repressed nitrogenase systems might lead to increased levels of nitrogen fixation.

∏ In practice, this is unlikely to be successful. Why?

constitutive
mutants are
likely to be
out-competed

Such mutants can, and have, been produced. However at some stage they would have to be released into the field environment. Here they have to compete with other, wild-type organisms. The permanently switched on nitrogen fixing systems means that a considerable part of the metabolism of these mutants is being continually diverted to nitrogen fixation whether or not combined nitrogen is available. In the presence of combined nitrogen these mutants are, therefore, at a disadvantage to their wild type counterparts. Indeed, they only compete on an equal basis in environments which are completely devoid of combined nitrogen. Thus seeding soil with constitutive, free-living nitrogen fixers is unlikely to be successful.

An alternative strategies could be to increase the efficacy by which symbiotic nitrogen fixers infect host plants. This might be achieved by increasing the 'virulence' of the nitrogen fixers or by increasing their host range. Alternatively, we might try to extend the range of plants that colonise the rhizosphere of host plants. In the following section, we will examine how *Rhizobium spp.* may be manipulated in the field.

9.2.5 *Rhizobium spp.* and legumes

Rhizobium spp. are known to infect a wide variety of legumes resulting in the production of nodules. There is considerable evidence that these bacteria show a high but not necessarily exclusive, degree of specificity of host plant. Those strains, for example, which infect peas, will not infect lupins. They will, however form symbiotic

specificity of
Rhizobium spp.

associations with broad bean and some other legumes. We have listed some examples of host specificities of *Rhizobium spp.* in Table 9.4.

Species	Host group
R. leguminosarum	peas (*Pisum spp.*)
R. lupin	lupins (*Lupinus spp.*)
R. phaseoli	beans (*Phaseolus*)
R. japonicum	soybeans (*Glycine spp.*)
R. trifolii	clover (*Trifolium spp.*)

Table 9.4 Examples of host specificity of *Rhizobium* species.

In the natural process, the bacteria invade the roots of their respective hosts and initiate nodule formation. They penetrate the root hairs and travel into the root via a special tube called the infection thread. This process is illustrated in Figure 9.5. The infection thread is thought to develop by invagination of the plasmalemma and proceeds to the root cortex. Here, the rhizobial cells infect root cells and stimulate them to grow and divide thereby producing nodules. The bacteria in the cells of the nodules are retained

bacteroids

within vesicles. The bacteria within these vesicles greatly enlarge and become known as bacteroids. The cells of the nodules produce leghaemoglobin which, as we noted earlier, binds O_2 and prevents the inhibition of nitrogenase.

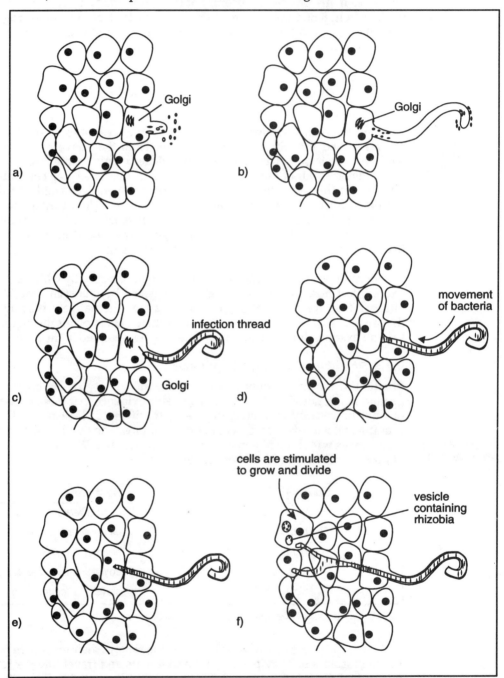

Figure 9.5 Greatly simplified representation of the processes of invasion and nodule formation in a rhizobial-legume association.

Examples of root nodules and a nodule cell packed with rhizobia are illustrated in Figure 9.6.

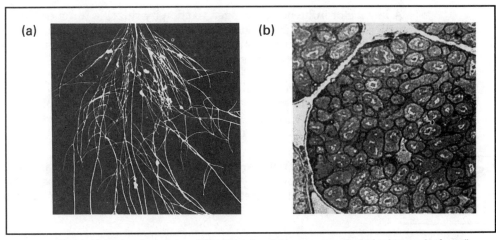

Figure 9.6 Root nodules on plant root systems, a) root nodules on a pea plant; b) micrograph of a cell within the nodule packed with nitrogen fixing bacteria. The bacteria within the nodules take on a much enlarged morphology and become known as bacteroids. They are members of the genus *Rhizobium*.

It should be noted that neither the host plants nor the *Rhizobium* cells can fix nitrogen when grown separately. The relationship is truly a symbiotic one. The hosts supply the bacteria with photosynthates as both a source of energy and carbon. In return, the bacteria supply their hosts with combined nitrogen.

nitrogen gain from legume crops

The use of legumes in crop rotations to improve yields is well established. Usually the legume is cultivated and the above ground portion harvested. The roots are usually left within the soil. There they decay releasing combined nitrogen which is made available for the subsequent crop. The extent of nitrogen gain within the soil is highly variable and depends upon the general nitrogen status of the soil, the actual process and extent of harvesting the legume and the efficacy of the nodulation process. Table 9.5 gives some typical values.

Plant	Total quantity of nitrogen bound (kg ha^{-1})	Loss or gain of nitrogen in post harvest soil (kg ha^{-1})
clover	250 to 300	+125 to +150
lupins	150	+130
peas/beans	50 to 60	-5 to -15

Table 9.5 Nitrogen accumulated annually by different crops. Data averaged from many sources.

SAQ 9.3

1) Explain why cultivation and harvesting peas and beans, although associated with nitrogen fixation, may lead to net nitrogen loss from the soil.

2) Calculate how much nitrogen is removed when peas/beans are harvested.

A key factor influencing nitrogen fixation by rhizobia in root nodules is the extent to which nodules are induced. If, for example, we have a soil with very few cells of *R. phaseoli*, then we might anticipate that very few nodules would be formed if the soil was planted with *Phaseolus spp.*

The numbers of rhizobial cells in soils will, of course, be greatly influenced by the crops that are grown. For example, under cereals such as wheat, typically as few as 10 or so *Rhizobium* cells are present in each gram of soil. The same soil, after cultivation of a legume, may contain 10^6 or more *Rhizobium* cells per gram.

| **SAQ 9.4** | Two adjacent fields are cultivated. In field A wheat is planted, in field B clover is grown. The following year both fields are planted with peas. Would the extent of nodulation of the peas be different in the two fields? Give reasons for your answer. |

natural nodulation is not assured

It is usually believed that the *Rhizobium* which infect commonly grown leguminous crops are present in sufficient numbers in soil to ensure sufficient nodulation occurs. However, this belief is not necessarily valid and the extent of nodulation can be extremely variable. An important objective of the bio-industry has been directed towards enhancing nodulation. For example, investigating the factors involved in nodulation. Of particular importance was the discovery that the legume strains used in Western European agriculture are predominantly strains that have been chosen for yield, especially for cultivation in areas where inorganic nitrogen fertilisers have been used. The selection of these strains has led to varieties of legumes which support lower symbiotic nitrogen fixing ability.

∏ See if you can explain why this is so.

energy costs of biological nitrogen fixation may reduce economic yield

Since nitrogen fixation consumes energy (photosynthates), this reduces plant yields. In a situation where combined nitrogen has been supplied artificially, then reduction of nitrogen fixing capacity has a corresponding benefit on crop yield. This of course satisfies the short term economic demands but does not help the long term environmental requirement to reduce fossil fuel consumption.

How then may we enhance the amount of nitrogen fixed by symbiotic nitrogen fixers? The key to success is to ensure that the host plants are infected by suitable, efficient nitrogen fixing rhizobia. In other words, produce suitable inocula.

9.2.6 The production and use of legume inoculants

Rhizobia grown in axenic culture may lose their ability to fix N₂

Rhizobium spp. may be cultivated on defined media within the laboratory. However, care has to be taken that the cells are still capable of fixing nitrogen and of infecting the target hosts. Bacterial cultures notoriously alter their characteristics when removed from their natural niche and cultivated in axenic culture *in vitro*. By spontaneous mutation and by repression and induction of genes, they acquire characteristics which make them more suitable for growth in culture vessels, rather than as infective agents capable of growth in plant cells. Of great importance is the retention of nitrogen fixing capabilities. For a considerable time, *Rhizobium spp.* could not be induced to grow and fix nitrogen in laboratory culture. They therefore had to be cultured in the presence of combined nitrogen. Long term culture under these conditions generated the real risk that genes necessary for nitrogen fixation might be deleted or inactivated. Under laboratory culture conditions, these deletions would not be easily detected and probably would not have deleterious effects on the cultivation of the cells. In fact, one might anticipate that such cells may be at a selective advantage. Thus we might envisage that a selective pressure might be operating that might lead to the production of ineffective inocula.

Fortunately, defined media have been developed which now allow the growth of rhizobia under nitrogen fixing conditions (for example see Barber, L E, 1979, Use of selective agents for recovery of *Rhizobium meliloti* from soil, Soil, Sa, Soc, Amer J, 43, 1145-1148). Thus it is now possible to generate inocula in which nitrogen fixing ability is retained.

∏   What other features are important in producing inocula of the appropriate quality?

The two main features you should have identified are:

• the production of inocula with the appropriate host specificity;

• the production of inocula with sufficient viable cells to be competitive in the natural environment.

selection of sources

use of acetylene reduction assay

We will examine these features each in turn. The production of inocula with appropriate host specificity depends upon the initial isolation of suitable strains and the application of criteria that allows identification of the strain. The former of these can be achieved by using a suitable source of the required organism. For example, it is logical to use soil in which peas have been grown to isolate pea specific rhizobia. Isolates from such sources can be evaluated for their nitrogen fixing capabilities. This is usually done using an acetylene reduction assay. The enzyme nitrogenase is relatively non-specific and will also reduce acetylene to ethylene. Ethylene can be readily assayed using gas chromatography. Part of the evaluation procedure will involve measurement of the efficacy of the isolates to fix atmospheric nitrogen in association with the proposed host. Suitable isolates are cultivated and stored as 'seed' cultures using conventional microbiological procedures employed to maintain suitable master cultures (for example cryopreservation). Each isolate may also be characterised in terms of morphological and biochemical characteristics and these characteristics may be used to monitor cultures in subsequent scale up processes.

inoculum viability

The large scale cultivation of suitable isolates employs the use of broth cultures. An important aspect of inocula preparation is to convert these broth cultures into a form in which cell viability is retained and which has a long shelf life. Many different approaches have been used including using the inocula simply as a broth culture, as a frozen concentrate or as a freeze dried (lyophilised) preparation. The most successful formulations include the use of a carrier. Various carriers have been used including bagasse, lignite and coal dust, but the best appears to be peat. The process of inocula preparation using peat is shown in Figure 9.7.

In this process, peat is sterilised, flash dried and milled. The pH is then adjusted to pH 6.8 using Ca(OH)$_2$ and the inoculum is sprayed on as a broth culture to give a final water content of about 40% (w/w). The mixture is then further incubated for 3 days at 28°C and is then packaged.

Produced in this way, the rhizobia continue to grow for a month or so but then viability begins to fall unless the product is kept at 4°C or lower. If stored at refrigeration temperatures, the inocula will remain viable for up to about a year.

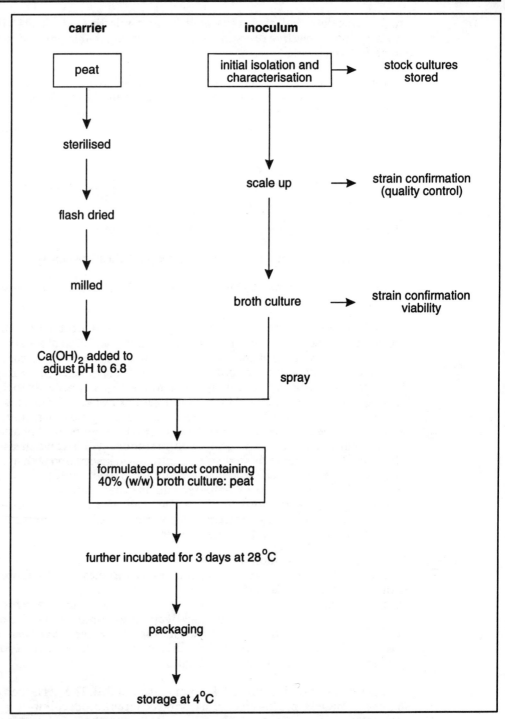

Figure 9.7 Outline of the preparation and storage of a rhizobial inoculum using peat as a carrier.

Π   How might we use such inocula?

You probably thought of two main ways. We could either add the inocula alongside the seeds during planting or pre-coat the seeds with inocula.

∏ There are advantages and disadvantages with each method. See if you can identify these.

problems arise
from chemical
protectants
coating seeds
and from the
inocula
becoming
dislodged from
seeds

The main advantages of adding the inocula alongside the seed is that more rhizobia can be added per seed. This is especially true with the larger seeded legumes. There is also less chance of the inocula being affected by chemical protectants that may be coated onto the seeds. These protectants are often broad based antimicrobial agents designed to restrict the growth of pathogens. Many, however, are also effective against rhizobia. In principle, we could also use the direct application of inocula to enhance nodulation of legumes that may already be growing. The advantages of pre-inoculating seeds is that the rhizobia are placed close to the root growth zone thereby enhancing the prospects of infection and nodulation. The disadvantages are that the rhizobia may become dislodged from the seeds during storage and handling and viability of the rhizobia may fall dramatically during seed storage. The former problem may, at least in part, be overcome by using suitable adhesives (for example gum arabic or sugars).

The evaluation of effectiveness of inoculants is extremely complex as there are many variables. We need, for example, to consider such factors as:

• inoculum density and formulation;

• nature of the soil (eg clay, sand, loam);

• the host plant variety;

• general soil fertility including the historical use of the test plots;

• prevalent climatic conditions;

• factors other than availability of nitrogen that effects yields.

It might be anticipated, therefore, that it is difficult to obtain data which shows consistent quantitative effects of inoculation on crop yields and nitrogen gains. Nevertheless in most field trials, inoculation of legumes show some improvement of yields over un-inoculated crops. We will illustrate this with some typical data (see Figure 9.8).

It should be noted however, that when clover is co-cultivated with grass, the differences are less marked (see Figure 9.9).

From the data presented in Figures 9.8 and 9.9, we can conclude that inoculation leads to improved yields. When clover is grown alone (Figure 9.8) inoculation with *Rhizobium* leads to better yields then when a combination of inoculation and chemical fertilisation is used. In contrast, in mixed cultivations of grass and clover, inoculation with *Rhizobium* improves yields compared to the control (Figure 9.9) but addition of nitrogen fertiliser improves the yield even more.

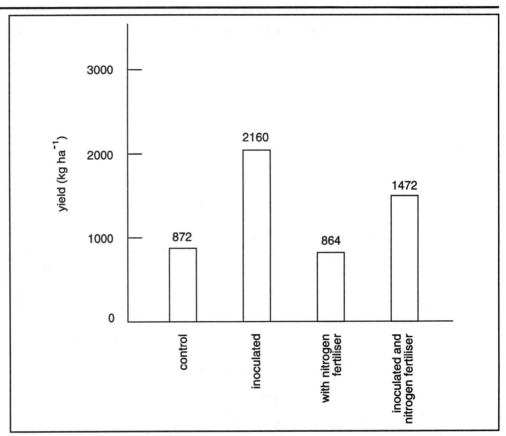

Figure 9.8 Clover yields of a Scottish site in which clover was simply seeded (control), inoculated with rhizobia, fertilised with nitrogen fertiliser or inoculated together with nitrogen fertilisation. Data from Day C A and Lisansky S G in 'Environmental Biotechnology', edited by Forster C F and Wase D A J, Ellis Horwood, Chichester, 1987, p 234-294.

Π   See if you can explain why these results were obtained.

The main reason is that inoculation with *Rhizobium* improves the growth of its host plant (clover) but did not supply much (if any) combined nitrogen to the grass. The addition of nitrogen fertilisers enhanced the growth of the grass. This may have also reduced the yield of clover since the enhanced crops of grass may have deprived the clover of other nutrients and reduced photosynthesis in the clover as a result of shading. Thus it is probable that in the mixed grass/clover crops, the plot simply treated with a rhizobial inoculum may have had a higher proportion of clover compared to the plots that had also been fertilised.

inoculation and fertilisation relationsips to crop yield

These data show the complexity of interpreting the likely effects of inoculation with and without nitrogen fertilisation especially on mixed crops such as clover and grass. Nevertheless, extensive trials using rhizobial inocula with a wide variety of legumes are continuing. Results so far indicate that improvements in crop yields and a reduction in the requirement for inorganic fertilisers can be achieved. The benefits of this approach are not solely confined to the savings of fossil fuels. The nitrogen fixed by rhizobia is either retained in the harvest or is released slowly from the breakdown of the roots left in the soil post harvest. There is, therefore, the added potential benefits of providing a continued nitrogen source for the subsequent crop and also a reduction in potential run off and leaching of nitrogenous salts which accompanies the application of single dressings of inorganic fertiliser.

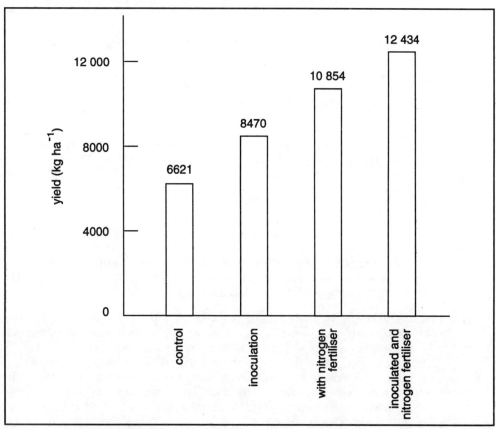

Figure 9.9 Grass/clover yields from a Scottish site using similar conditions as to those described in Figure 8.8. Data from Day C A and Lisansky S G, 1987.

9.2.7 Future prospects for biological nitrogen fixation

extension of the rhizobial inoculation strategy

In the previous sections we have indicated that the application of rhizobial inocula may be of substantial environmental benefit. The process of inoculating with rhizobia may be viewed simply as helping nature to do what is natural. We might visualise that similar processes of inoculation may be helpful in a variety of circumstances. For example, the seeding of rice paddy field with nitrogen fixing cyanobacteria such as *Anabaena spp.* may enhance rice production. Again these organisms are natural inhabitants of paddy fields. Seeding the shallow waters with selected strains of *Anabaena* have been shown to be of value.

creation of new nitrogen fixers

The transfer of nitrogen fixing genes to hitherto non-nitrogen fixing micro-organisms is a major target. The transfer of the nitrogen fixing capacity is, however, not a simple matter. Transfer of the ability to produce nitrogenase is, by itself, insufficient to achieve nitrogen fixation.

∏ Explain why this is so.

The ability to fix nitrogen requires the production of suitable electron carriers and strong reductants and the protection of nitrogenase against inactivation by oxygen. Thus the ability to fix nitrogen requires the transfer of many genes. Even if this was achieved, the release of such organisms might not result in increased levels of biological nitrogen fixation. Such engineered organisms would be at a competitive disadvantage with non-nitrogen fixing organisms unless combined nitrogen is more or less completely absent. Nevertheless the potential benefits warrant considerable efforts.

∏ See if you can cite any evidence that the transfer of nitrogen fixing ability between bacteria might have occurred in nature during evolution (Table 9.3 might help you).

Examination of Table 9.3 shows that nitrogen fixation is carried out by particular species of physiologically very diverse groups of organisms. These include both autotrophic (photo- and chemo-autotrophs) and heterotrophic organisms; aerobes as well as anaerobes. This would tend to indicate that either:

• nitrogen fixing ability evolved early in evolution and the current nitrogen fixers are evolutionary divergents from a very ancient ancestoral nitrogen fixer or;

• nitrogen fixing ability evolved relatively recently but evolved in a genetically transmissible form thereby enabling the ability to fix nitrogen to be passed to physiologically quite disparate groups.

the genes of nitrogen fixation may have been recently transferred to new species

The current consensus is that the latter explanation is probably correct. All nitrogen fixers are prokaryotic. It is well known that gene transfer, via plasmids, occur amongst these organisms and can breach the normal species-genus barriers. The failure to find nitrogen fixation in eukaryotes together with the great degree of similarities between the nitrogenases and electron carriers of nitrogen fixers may also be cited as supporting a more recent evolution of nitrogen fixing capacity and its subsequent dispersal perhaps via plasmid-based mechanisms. Care, however, must be taken in reaching such a conclusion. The great similarity of nitrogenases from different organisms might, for example, be imposed by the biochemical conditions needed for nitrogen fixation. Thus this enzyme may be highly conserved throughout a long evolutionary history rather than being of more recent origin. Nevertheless, the suggestion that nitrogen fixing ability is transmissible between organisms provides great encouragement for attempting to increase the variety of nitrogen fixing organisms that are available.

Three broad strategies to achieve this can be visualised. One is to transfer nitrogen fixing capacity to organisms that are free living but are normally associated with the rhizosphere of crop plants. The second approach is to attempt to broaden the host specificity of normal symbiotic nitrogen fixers. A third approach is to attempt to induce hitherto free living nitrogen fixing organisms to become endosymbionts within plant cells. Such associations are visualised as mimicking the origin of chloroplasts in which chloroplasts arose through prokaryotic photosynthetic organisms becoming resident within host eukaryotic cells.

∏ See if you can list the main problems that need to be overcome in order to achieve this latter objective.

There are in fact many problems but the main ones are:

- how to co-ordinate the growth of the host cells and their endosymbiont nitrogen fixers such that the endosymbiont neither outgrows the host nor is outgrown by the host;

- how to ensure that nitrogen fixation still occurs in a nitrogen rich environment (the host cell's cytoplasm).

uptake of cells by phagocytosis is not seen as a problem

control of nitrogen fixation is difficult to visualise

The actual physical introduction of the cells is not seen as a particular problem because plant protoplasts can often be induced to phagocytose (engulf) bacterial cells. Establishing the stable and balanced interaction is the critical issue. Nevertheless analogous close symbiotic relationships between cyanobacteria and fungi have been described in some lichens indicating that the desired objective is, at least biologically feasible. We must, however, anticipate that if we are able to create 'nitrogen fixing' organelles in plant cells, this is bound to have an enormous impact on the physiology of the plant. Think, for example, of the consequences of the presence of photosynthetic organelles (chloroplasts) on plant behaviour. In the case of creating nitrogen fixing organelles, these would act as substantive consumers of photosynthates. Furthermore it is, at this stage, rather difficult to visualise how the delivery of combined nitrogen would be controlled; too much of the wrong product could (would) have disastrous consequences on the plant.

transfer of N_2 fixing ability to rhizosphere organisms

It is for these reasons that the alternative strategies are seen as having more realistic prospects for success. Of particular importance are attempts to extend nitrogen fixing capacity to other currently non-nitrogen fixing micro-organisms which grow in the proximity of plant roots. Each plant type tends to attract its own special group rhizosphere organisms. Again, even if such transfers were achieved, there is no guarantee of a successful agricultural outcome. Nitrogen fixing rhizosphere organisms would have to compete with non-nitrogen fixing organisms. They appear to do so only in conditions where combined nitrogen is available in very limited amounts. Excellent examples are the associations between sand grass (*Paspalum notatum*) and the free living *Azotobacter paspali* and the tropical grass (*Digitaria decumbers*) and *Aspirillum lipoferum*.

9.2.8 Concluding remarks

the importance of economic considerations in the adoption of new technical possibilities

In this section, we have indicated the potential of extending the activities of nitrogen fixing organisms to reduce dependency on the use of inorganic nitrogen fertilisers produced by fossil fuel consuming processes. Since the combined nitrogen produced by symbiotic systems, such as in Rhizobium-legume associations, is directly assimilated by the host plants, this process reduces the prospect of leaching of the fixed nitrogen into rivers and impounded waters. This has beneficial effects in reducing the problems arising from the enrichment of these waters with combined nitrogen. The extension of the use of nitrogen fixing organisms by increasing rhizobial infection rates, increasing the capacity of rhizosphere organisms to fix nitrogen or by creating nitrogen fixing organelles, will be, however, influenced by more than environmental considerations. In the current international climate, the key factor is the question of economics. Human society and the adoption of new practices are particularly influenced by economic considerations. Farmers will consider the application of such new approaches in terms of financial costs and returns. Thus, if for example, the creation of nitrogen fixing organelle bearing plants reduces the cost of fertilisers but, at the same time, reduces the yield of the finance generating crop, then the farmer may well continue the practice of using environmentally damaging inorganic nitrogen fertilisers. We might anticipate that unless, or until, the socio-economic morality described in Chapter 2 is adopted, current farming practices will continue. It is only by the adoption of including the environmental as well as the financial costs and benefits in the evaluation of practices by the international community at large that these will become real possibilities. In the meantime, the scientific and technological advances may continue to aid the prospects of developing a more environmentally sustainable approach to soil fertility.

SAQ 9.5

In a trial experiment on a farm, plots of land were either planted with peas directly or were first fertilised using an NH_4^+ based fertiliser or treated with an inoculum of *Rhizobium*. The following results were obtained.

Treatment	Yield of seed (kg ha^{-1})
control (untreated)	11 800
fertilised using NH_4^+ based fertiliser	16 700
treated with an inoculum of *Rhizobium*	14 300

We will assume that the seed is the sole cash crop that is sold. We will ignore the use of the remaining green matter as silage for livestock feed.

It is estimated that the crop needs to yield 9000 kg seed ha^{-1} in order to cover the commercial cost of producing and harvesting the crop.

The treatment with fertiliser increases the production/harvesting costs by 25%. Similarly the use of the *Rhizobium* inoculum increases the production/harvesting costs by 15%.

Assuming that these figures are repeatable with other fields, should farmers, on strictly commercial grounds opt for the NH_4^+ based fertiliser or for the *Rhizobium* inoculum or leave their fields untreated? When you have reached your conclusion read our response carefully as it emphasises some important issues.

9.3 Biotechnological innovation in crop protection

Many biological factors may reduce the yield or value of agricultural crops. These include weeds, pests and disease. In this section we will briefly review these factors before examining, in greater depth, a particular contribution of biotechnology to the protection of crops.

9.3.1 Weeds and crop production

Weeds generally reduce crop yields by competing with crops for the limited environmental resources (water, mineral nutrients, light). The effects of weeds may, however, not be confined simply to reducing yields.

∏ See if you can list some other effects the presence of weeds may have on agriculture.

The sort of items you could have listed are:

- reduction of economic value of a crop because of contamination of the harvested crop - some examples are contamination of rye grass with the toxic black nightshade Solanum nigrum and the contamination of wheat grain by the seeds of the catchweed *Galium aparine*;

- attraction of pests and diseases in which the weeds act as alternative hosts;

- restricting agricultural practices especially harvesting operations.

Yield reduction by weeds, of course, depends upon the extent to which the crop is infested by weeds and on the nature of the weeds. It is extremely variable and can be very large. For example, over 90% of the yam and sweet potato crops in parts of Africa can be lost through the action of weeds (see Table 9.6).

Crop	Location	Loss (%)
Cereals		
maize	Ghana	55
	Kenya	34
upland rice	Ghana	84
lowland rice	Liberia	63
	Nigeria	90
	Senegal	48
Pulses		
cow pea	Ghana	67
	Nigeria	60
ground nut	Ghana	54
soybean	Ghana	53
	Nigeria	60
	Zambia	40
Roots and tubers		
cassava	Nigeria	65
sweet potato	Nigeria	91
yam	Ivory coast	91
	Nigeria	73

Table 9.6 Some examples of yield reduction in various crops in Africa. Data from Eussen, 1991 'Onknud: een onderschatte plaag' in Landbouwkundig Tijdschrift **103**, No 3, pp 11-14.

The data given in Table 9.6 are indicative of the tremendous losses in yield that may be encountered even if some limited form of weed control is operative.

use of
ploughing,
herbicides,
burning and
hoeing to
reduce weeds
Weed control aims at eradicating the weed population. Before the crop is planted this is usually achieved by ploughing (tillage) but may also be achieved by using chemicals (herbicides) or by burning. Once the crop has been planted, the removal of weeds is either mechanical (hoeing or pulling) or chemical (selective herbicides). Both these strategies consume significant amounts of fossil fuel and the chemicals used are often recalcitrant and remain in the environment for a considerable time. We will examine this aspect in more detail a little later. Now we will turn our attention to pests and diseases.

9.3.2 Pests and diseases and crop production

crops are
sources of
nutrition for
vertebrates,
invertebrates
and microbes
Within the natural order, crop plants represent a resource that may be used by a variety of biological organisms. These include wild vertebrates, such as rabbits, that may graze on the crop and invertebrates (especially arthropods such as the insects) which may also use the crop plant as a nutrient source. In this latter group, aphid infestations and locus swarms immediately spring to mind. Similarly, many crop plants may act as suitable hosts for a variety of microbial (especially fungal and bacterial) infections.

Unlike weeds, the effects of these is not limited to the production phase of the crop. They may also effect the stored product. We can, therefore, identify both pre- and post harvest losses. We might anticipate that, because of the wide variety of agents responsible for yield reduction, there might be many different mechanisms by which yield reduction occurs. In Table 9.7, we have given some examples of these mechanisms. You should realise that this list is not complete but is included to illustrate the diversity of the mechanisms by which crop yields are reduced.

Mechanism	Yield loss determining factors	Examples
tissue consumers	type of tissue eaten (root, leaf, seed), rate of consumption	birds, rabbits, beetles, lepidopteran, larvae, locusts etc
assimilate removers	rate of carbohydrate consumption, harmful excretion products	aphids, mites
assimilation rate reducers	extent and site of tissue damage	viral, bacterial and fungal infection
light stealers	covered area, type of tissue	yeasts which grow on the surface of leaves
plant destroyers	number of plants lost distribution of lost plants, ability of surviving plants to compensate for loss	damping off fungi

Table 9.7 Examples of crop yield reduction mechanisms of pests and diseases. Adapted from Rabbinge R, Ward, S A and van Laar HH. Simulation and systems management in crop protection. Simulation Monograph 32 Wageningen, 1989.

You should note that the groups listed in Table 9.7 are not mutually exclusive. For example, the tissue consumers could be regarded as plant destroyers if, as a result of their activities, plants die.

the populations of plant disease organisms and small pests may fluctuate very widely

A general feature of plant diseases and of the physically small pests (such as aphids and mites) is that their numbers may fluctuate widely from season to season. Also, if they do become established, their numbers (and therefore the damage they cause) may rise very rapidly. In some ways, the agricultural practice of growing large areas as monocultures is an invitation to generate large loses through these agents. A general feature of organisms is that the smaller they are the faster they grow and divide (compare, for example, a bacterium, an aphid and a rabbit). Thus, if a crop becomes infected by, for example, a pathogenic micro-organism, it may rapidly spread through the whole crop stand.

Much effort is therefore directed towards reducing losses of crop yields. In part this may be achieved by taking preventative measures such as:

- removal and destruction of crop residues to reduce contamination;

- crop rotation to reduce the accumulation of pathogens and other pests in a particular field;

- use of insects traps (using pheromones as attracting scents), netting (birds, some insects) and fencing (large animals);

- chemical disinfection of soils (especially for greenhouse use);

- breeding of plants to achieve some resistance.

An alternative to taking preventative measures is to use curative control. Most commonly, this involves the use of chemicals (biocides) which kill or inhibit the pathogen or pest. Post harvest treatments may also involve the use of biocides. Alternatively, the harvested crop may be protected by drying, cooling, heating or acidification.

An alternative to the use of biocides is to use biological control measures.

9.3.3 Agrochemical pesticides

In Sections 9.3.1 and 9.3.2, we have indicated that chemicals (biocides) are employed to reduce crop losses through the action of weeds, pests and diseases. The use of these chemicals is commercially attractive since they are very effective in reducing crop losses. They are also relatively cheap to apply. Thus, especially in the developed world, these reagents are used extensively. Despite this apparent success, considerable concern is expressed with respect to their continued use. In some cases the reagents used have broad specificities and kill beneficial, as well as, detrimental agents. They are also frequently recalcitrant and are not readily biodegradable. Some examples of the persistence of biocides in soils are given in Table 9.9.

many biocides are recalcitrant

Common name	Chemical formula	Persistence
Fungicides		
PCP	pentachlorophenol	>5 years
zineb	zinc ethylene-1,2-bis-dithiocarbamate	>80 days
Insecticides		
aldrin	1,2,3,4,10,10-hexachloro-1,4,4'4,5,8,8' hexahydro-endo-1,4-exo-5,8-dimethano-napthalene	>9 years
chlordane	1,2,4,5,6,7,8,8' octachloro 2,3,3',4,7,7'-hexahydro-4, 7-methanoindene	>12 years
DDT	2,2 bis (p-chlorophenyl)-1,1,1-trichlorethane	>9 years
Herbicides		
monuron	3-(p-chlorophenyl)-1,1-dimethylurea	>3 years
simazene	2-chloro-4,6-bis(ethylamino)-s-trizine	>2 years

Table 9.8 Some examples of persistence of pesticides in soils.

biomagnification

As a result of this recalcitrance, they tend to accumulate in biological materials. As we have seen from our discussions in earlier chapters (see Chapters 3 and 6), consumption of organisms containing pesticides by other organisms leads to even higher concentrations of the pesticide accumulating in the organisms of the higher trophic levels. This process, called biomagnification, can lead to detrimentally high concentrations accumulating in higher organisms. A classical example is that of DDT. The use of DDT as an insecticide, coupled with its long residence times led, by the process of biomagnification, to high levels of the pesticide in the livers of sea birds such as the Shag. Affected birds laid thin walled (fragile) eggs. This, in turn, resulted in reduction of the reproductive replacement of these birds. Despite the fact that this consequence of DDT use was established for at least two decades, DDT is still in use as an insecticide. For example, over 30 000 tonnes are still being used annually in Africa.

As we have become more conscious of such indirect and subtle effects, increasingly stringent regulations have been implemented that restrict the use of these reagents. For example, before authorisation is granted to market a product in the EC, its ecotoxicity as well as its physiological toxicity has to be evaluated. Thus, in placing a pesticide on the market, there are three main concerns:

• the protection of workers - mainly concerned with the protection of those engaged in the manufacture and application of the pesticide;

• the protection of consumers;

• the protection of the environment.

We have represented this in Figure 9.10 in which we have specified the relevant UK legislation. Analogous national legislation applies in other EC Member States.

We do not intend to examine the regulations governing the production and use of hazardous chemicals here. They are discussed in the BIOTOL text 'A Compendium of Good Practices in Biotechnology'. However, we will provide a few comments about the control of the production of these materials.

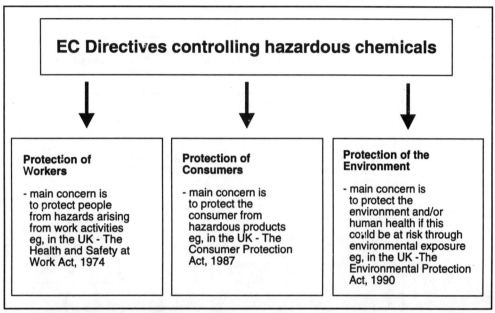

Figure 9.10 The relationship between EC Directives controlling hazardous chemicals and the protection of workers, consumers and the environment.

risk assessment

A key aspect of the control of such substances is the hazard (risk) assessment process. We have represented this process, in outline, in Figure 9.11. Note that an environmental 'dose of concern' is estimated. This is based on the dose that is suspected to give rise to environmental damage. This, coupled with the proposed use of the product and its predicted environmental concentration, enables the possible environmental consequences to be determined.

Π In your estimation, is the scheme illustrated in Figure 9.11 likely to prove satisfactory.

difficulties in determining environmental risks

Your response to this is clearly a matter of opinion. The point we would like to make, however, is that the 'reliability' of the hazard assessment depends upon the quality of the data that are available. The environmental fate of chemicals, especially recalcitrant chemicals, is very complex and it is virtually impossible to predict all of the outcomes of releasing chemicals into the environment. Processes, such as biomagnification may give rise to localised, and unpredicted, accumulations. Furthermore, the toxic effects of such compounds may be subtle and long term. Therefore, although manufacturers diligently fulfil the criteria of legislation, the requirements of the legislation are, in this case, suspect. They are, at best, based upon our best guesses. The opinion of many is that the use of pesticides, even if they fulfil the required legislative criteria, should be regarded as only a temporary measure until environmentally safer processes are developed.

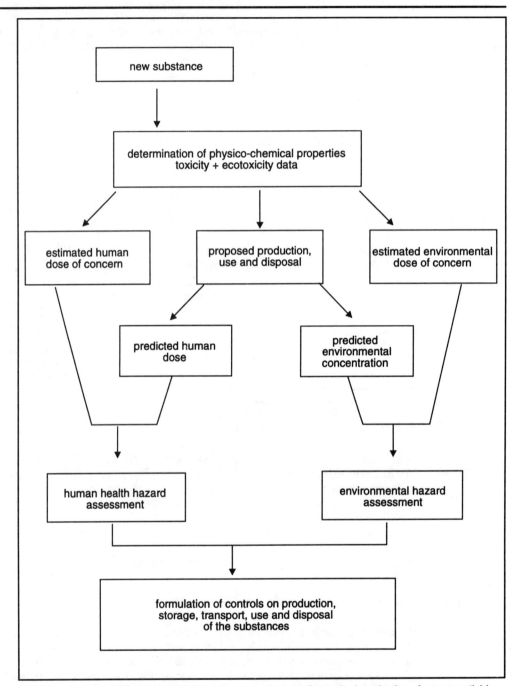

Figure 9.11 Generalised hazard assessment process which applies to the introduction of a new pesticide.

Further support for this position comes from the observation that many pests and diseases develop resistance to these reagents. Such resistance is not confined to any single group of pesticide nor to any particular group of pests.

We have reported the number of species of arthropods, with resistance to pesticides identified during the 1980s, in Table 9.9. Some of these may have been historically resistant to these pesticides. Others, through the natural processes of evolution (mutation, genetic recombination) may have developed new resistances. This situation mimics, for example, the development and diversification of resistance to antibiotics such as penicillin by micro-organisms after the introduction of these reagents.

Order	Pesticide group				
	Dieldrin/ BHC	DDT	OP	Carb	Pyr
Diptera	107	106	60	11	6
Lepidoptera	40	40	31	14	8
Coleoptera	55	24	26	9	3
Homoptera	13	13	28	9	3
Heteroptera	16	8	6	-	-
Acarina	23	21	7	2	1

Table 9.9 Number of species of arthropods with resistance to pesticides reported during the 1980s. Data from Day C and Lisansky S G (1987) in 'Environmental Biotechnology' edited by Forster C F and Wase D A J, Ellis Horwood, Chichester, p 274.

Finally, we can add to this catalogue of problems associated with the use of recalcitrant organic pesticides, the fact that the production, distribution and application of these reagents requires the consumption of fossil fuels both as a source of energy and as a source of feedstocks for their production. The key question is, therefore, whether or not environmentally friendly strategies are available to achieve the overall objective of protecting crops. We will examine this aspect in the next section.

SAQ 9.6	Apple orchards act as 'hosts' for many insects such as codling moths, wasps and honey bees. Explain, using this example, how the application of an insecticide to reduce pest insect populations in an orchard may have unwanted deleterious consequences.

SAQ 9.7	Would the inoculation of soils with micro-organisms which have the metabolic capability to degrade a particular recalcitrant pesticide likely to be successful in reducing the environmental problems associated with the use of such a pesticide?

9.3.4 Alternatives to agrochemical pesticides

We have learnt that protection of crops encompasses protection against the damage/loss of yield caused by weeds, pests and diseases. In the remaining sections, we will predominantly focus on protection against insect pests.

We can identify two broad strategies for the protection of crops which reduces the need to use chemical pesticides. These are:

• the breeding of plants so that they are resistant to the damaging agent;

• the development and use of biological pesticides.

production of
disease
resistant
varieties

Classical plant breeding programmes and the recent developments in recombinant DNA technology offers many opportunities to produce plants with new characteristics. In terms of crop protection, the greatest efforts of plant breeders are directed towards developing disease resistant varieties (that is, varieties that are resistant to microbial infection). The production of such varieties reduces the need to apply biocides (for example, fungicides) to crops. The products of breeding programmes are not necessarily directly linked to a reduced use of chemicals. For example, plant varieties have been genetically engineered to be resistant to herbicides (such as glyphosate). This, it is argued, increases the use of this herbicide by enabling farmers to use this herbicide to remove weeds from crops. The environmental arguments are complex. Herbicides belong to a wide variety of different chemical classes. Some, like the atrazines are very persistent. Others, such as phosphonitricine, are rapidly degraded in soil and are not thought to be environmentally damaging. Thus, by making the right choice, plant biotechnologists can help to change the type of herbicides used from environmentally polluting to environmentally neutral.

The application of conventional and contemporary plant breeding programmes for improved crop production as well as environmental objectives are dealt with in the BIOTOL text 'Biotechnological Innovation in Crop Production' so we will not deal with them in depth here. Instead we will now focus on the biotechnological development of biological pesticides.

9.3.5 Biological pesticides

The ideal pesticide is one that is highly targeted, does not have undesirable side effects and does not depend upon the consumption of large amounts of fossil resources. The great hope is that biological agents may fulfil this specification.

biological
control agents
should be
specific

The concept of using biological control agents is not new. For many centuries, the Chinese used ants to control granary pests and cats have been significant in controlling rodents in Europe for at least a thousand years. However, care must be taken in using biological agents as illustrated by the anecdotal story of the introduction of cats onto Ascension Island in 1815 to control the rat population. They also virtually exterminated the sea birds. This example can, however, be used to illustrate a point. We learnt that, in part, the problems of using chemical pesticides can be attributed to their lack of specificity. Cats too, are rather non-specific and will take birds as well as rats and mice. Obviously, the more specific the controlling agent is, the easier it is to predict the outcome of introducing the agent as a pesticide.

In this and the following sections, we will consider some general features of biological control strategies and deal with some biotechnological contributions in some detail. This is, however, an enormous topic about which much has been written. The material included is meant to illustrate and is not a full review. Suggestions for further reading for those who wish to follow this topic at greater depth are given at the end of the text.

∏ Write down a list of types of biological agents that might be considered as potential pesticides active against arthropods.

You could have generated quite a list, including insectivorous birds, predatory insects, fungi, bacteria and viruses.

Each of these has its own particular attractions and disadvantages. For example, birds will be relatively non-specific and are more likely to be predators of the larger arthropods. Bacteria and viruses may rapidly multiply and be relatively quick acting. They may also be more readily genetically manipulated to generate strains with enhanced characteristics.

Π Of the groups listed above, which, from your experience of biology, is likely to be most host specific?

Generally speaking, viruses are highly host specific only attacking a single or a few related species. Bacteria and fungi also show some degree of specificity. It is for these reasons that we will concentrate on these groups. We will briefly examine each in turn.

Fungi

Fungi usually attack their hosts directly and are spread by sporulation. They will therefore tend to spread throughout the host population. The persistence of spore bodies on the carcasses of their hosts means that they tend to remain in the environment for extended periods. About 400 species of fungi, mainly Deuteromycetes and Phycomycetes, are known to attack arthropods, thus there is considerable potential to use these organisms as pesticides.

There are unfortunately some disadvantages with using such fungi. They have extensive metabolisms and often have strict requirements for temperature and humidity. Under unfavourable environmental conditions, their growth might be slow and thus they may be as slow acting as pesticides. For this reason, they appear to be most effective for use in glasshouses. For both field and glasshouse applications, it is important to apply the fungus at the right time. It is also essential to apply sufficient to ensure pest control. Despite these difficulties they have been used with some success. Some examples are included in Table 9.10.

Fungus	Target arthropods	Remarks
Metarhizium anisopilae	spittle bug in pastures and sugar cane	
	rhinoceros beetle in palms	used in combination with viruses
	brown plant hopper in rice	
Beauveria bassiana	colorado beetle on potatoes	active through production of a toxin (beauvericin)
Verticilium lecanii	aphids and whitefly	mainly used in glasshouses
Hirsutella thompsonii	citrus mites	currently out of production
Nomerea rilevi	moths/butterflies	inocula difficult to produce
Entomophthorales spp.	mainly against aphids	field efficacy not yet proven

Table 9.10 Examples of fungal pesticides in commercial use.

Bacteria

Many bacteria, from different families, are known to attack insects. Some examples are listed in Table 9.11.

Bacteria	Insects attacked
Bacillaceae	
Bacillus thuringiensis	many lepidopteran
B. cereus	many lepidopteran
B. popillieae	scarab beetle (coleopterans)
B. larvae	honey bees (hymenopterans)
B. sphaericus	mosquitoes (dipterans)
Enterobacteriaceae	
Enterobacter aerogenes	mainly lepidopterans
Escherichia coli	mainly lepidopterans
Serratia marcesiens	mainly lepidopterans
Proteus spp. (eg *P. vulgaris, P. mirabilis*)	grasshoppers
Salmonenella spp. (eg *S. schottmulleri, S. enteritidis*)	some lepidopterans and hymenopterans (eg honey bees)
Lactobacillaceae	
Diplococcus spp.	some lepidopterans (silk worms, gypsy moth, greater wax moth)
Micrococcaceae	
Micrococcus spp.	various lepidopterans, dipterans (house flies), hymenopterans (saw flies)
Pseudomonadaceae	
Pseudomonas aeruginosa	grasshoppers
P. septica	various coleopterans (eg scarab beetle)

Table 9.11 Examples of bacteria pathogenic to insects.

Most of those which have found widespread practical use as biocides belong to the genus *Bacillus*. Three species in particular have found special use. These are:

- *Bacillus thuringiensis*;

- *Bacillus popilliae*;

- *Bacillus sphaericus*.

B. thuringiensis strains are toxic to a wide range of hosts

B. popilliae is used for the control of Japanese beetle whilst *B. sphaericus* appears to be effective against mosquitoes. It is, however, *B. thuringiensis* that has found most extensive use as a biocide. There are many different strains of this species usually identified serologically. Members of the group are toxic to flies (dipterans), beetles (coleopterans) and moths and butterflies (lepidopterans).

β-exotoxin

The wild type strains, first isolated from silkworms, produce a toxin (so called β-exotoxin) which is toxic to Man. However, β-exotoxin free strains have been isolated. For example, the strain HD-1 of *B. thuringiensis* Kurstaki is β-exotoxin free but remains effective against lepidopterans.

This strain has broad specificity but is not the most effective in every case. Other variants may have greater activity against specific target insects. For example strain

3A3B (HD-1) appears to be most active against caterpillars while strain H-14 (the so called Israelensis strain) is effective against mosquitoes and black fly.

Many products based on these and related strains are available. Each product consists of viable organisms together with other ingredients to improve shelf life or delivery of the product. Despite the success of these formulations, they still only represent a very small portion (about 0.1%) of the total world pesticide market. It is, however, anticipated that continued development of strains, together with increasing environmental pressures will lead to increased market penetration.

Crucial to this development will be the knowledge gained on the mechanism by which *B. thuringiensis* exerts its killing effect. We know, for example, that towards the end of their growth phase, these organisms produce resistant endospores and a large amount of a protein which crystallises to form what is known as the parasporal body. This body contains a toxin which, after digestion in the gut of the insect, is released and exerts a killing effect on the host. You will notice that, unlike the fungal pesticides, these bacteria have to be ingested to be effective. Different strains of *B. thuringiensis* produce different toxins which show different host specificities.

The gene coding for the *B. thuringiensis* toxin has been cloned into other bacterial species (for example *Escherichia coli* and *Bacillus subtilis*) opening up the prospect that the toxic factor may be introduced into micro-organisms which are ingested by other pest insects.

SAQ 9.8	The genetic information coding for the insect toxins may be cloned directly into plants. What are the advantages/disadvantages of this as a strategy for protecting plants?

use of sites specific mutagenesis to produce 'designer' proteins

The story of *B. thuringiensis* has, however, other facets. It is believed that the toxins produced by the bacteria interact with receptors for phosphatidyl choline and N-acetyl galactosamine within the insect. Molecular modelling of the toxins and their interactions with these receptors may facilitate the design of new custom-built biological insecticides. This technology is analogous to the technology which is applied to produce more effective enzymes for industrial use. By using, for example, site directed mutagenesis it becomes possible to replace specific amino acids within a protein. By doing so, proteins with new (improved) characteristics may be generated. Thus it is possible to envisage the production of a wide variety of bacterial pesticides each carrying genes which code for proteins based on the structure of *B. thuringiensis* toxin.

Viruses

viruses effective against dipteraus and lepidopteraus

A very large number of virus-insect associations have been described although only a few have been considered for application as insecticides. Many are associated with Lepidoptera and Diptera. Amongst these, the baculoviruses are considered to have enormous potential (see Table 9.12).

Insect orders/ class	Cytoplasmic polyhedrosis viruses	Entomopox viruses	Baculoviruses
Insecta			
Lepidoptera	191	24	456
Diptera	37	9	27
Hymenoptera	6	3	30
Coleoptera	2	14	9
Neuroptera	2	-	-
Trichoptera	-	-	2
Orthoptera	-	9	-
Decapoda	-	-	2
Crustacea	-	-	2

Table 9.12 Viral infections of arthropods. Data from Vlack, J M in 'Biotechnological Innovations in Crop Production', Butterworth Heinemann, Oxford, 1991.

Over 500 baculoviruses are known. They are highly specific and can attack large numbers of hosts over a short period (they are epizootic) and can reduce the insect population.

production of viruses in insect cell cultures

A feature of baculoviral infection is that there is some delay in the cessation of feeding by infected larvae and even further delay before they die (see Figure 9.12).

Furthermore, the older larvae (ie those close to pupation) tend to be insensitive to the viruses. Although much work has been done on these viruses, their use in agriculture is still limited. The main limitations are related to:

- the need to increase their speed of action, especially the cessation of feeding;

- the need to increase their virulence especially against old larvae;

- the need to extend their host range so that the same agent can be used against a range of pests. This would make them commercially more attractive;

- the need to improve their persistence in the field (they are particularly UV sensitive).

baculoviruses are targets for genetic engineering

These are, of course, all potential targets for genetic engineering. A detailed account of the genetic engineering of baculoviruses is given in the BIOTOL text 'Biotechnological Innovations in Crop Production' so we will not repeat the details of the molecular genetic manipulations here. It is, however, becoming clear that it is possible to alter the characteristics of these viruses to make them more applicable to agricultural and environmental needs and practices. For example, by introducing the toxin gene(s) from *B. thuringiensis* into the virus, the toxicity of the viral proteins may be enhanced. This may, in turn, reduce the delay in the cessation of feeding. It may also reduce the resistance of old larvae to baculoviral infections.

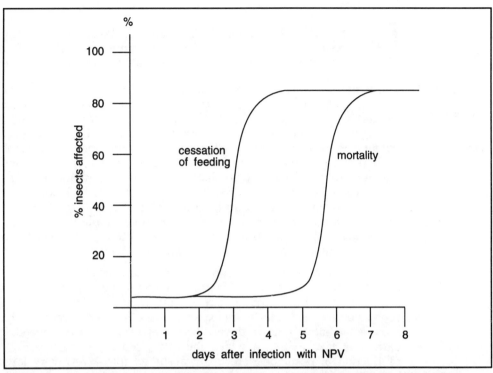

Figure 9.12 Cessation of feeding and mortality of insects versus time after baculovirus infection. (NPV stands for nuclear polyhedrosis virus - see text). Adapted from Vlack, J M in 'Biotechnogical Innovations in Crop Production', Butterworth Heinemann, Oxford, 1991.

We should not, however, confine our attention to the potential of biotechnology to genetically manipulate the biopesticide. Also important are the developments in bioprocess technology. The ability to raise larger numbers of insect cells ($1-2 \times 10^6$ cells ml^{-1}) in large reactors, means that we now have the capacity to raise large numbers of viruses (typically for baculoviruses this may be at about 2×10^7 viral particles ml^{-1}). Using airlift reactors, reactor volumes up to 1000 litres can be used. The alternative is to use cultivated insect larvae as 'bioreactors'. Usually about $1-2 \times 10^9$ viral particles may be produced per larva.

Π List the advantages and disadvantages of using insect cell cultures rather than larvae for the production of baculoviral pesticides.

The potential advantages of insect cell cultures are:

• convenient quality control;

• minimal contamination;

• convenient downstream processing;

• less cruelty.

The disadvantages are:

• the relative costs of insect cell culture media compared with raising larvae;

• the relative inexperience in using large volume bioreactors for raising insect cells;

• the relative capital costs of equipment.

∏ In the text, we have indicated that biopesticides are currently only a small portion of the total pesticide market. Explain why this is so and what factors may be cited as potentially enhancing the prospects of increased use of biopesticides.

problems of
slow actions,
impredictable
efficacy,
non-patentable
and specificity

This is another fairly open ended question. Here we will outline the main points. At the beginning of Section 9.3 we indicated some of the problems associated with the use of chemically produced pesticides. We particularly emphasised the environmentally damaging consequence of the continued and extended use of these materials and the potential for using biopesticides to reduce the environmental impact of crop protection. Such a strategy will not be implemented simply as a result of some wishful thinking on the part of a few 'environmentalists'. The main limitations to the implementation of biopesticide use is the belief that they are slow acting, have limited and unpredictable efficacy, are non-patentable and, because of their specificity, have small markets. It is, therefore, important to address these issues. For example, we described how, in the case of baculoviruses, attempts are made to broaden the spectrum of susceptible pests and the speed with which the cessation of feeding is achieved.

At the same time as improving both the image and the efficacy of biopesticides, it is also important to consider their competitors. Increasing the stringency of conditions which are applied to the authorisation to market new chemicals is making chemically produced new pesticides more expensive and thus less attractive. The evaluation and risk assessments of new products is becoming very expensive indeed. Progressively the pattern of take overs, in which the many companies historically involved in the production of pesticides, are being absorbed into a few very large companies, is indicative of the high costs of bringing new chemicals to the market.

The advent of contemporary molecular genetics, the increasing awareness of environmental issues and the development of ecological principles coincide to make for a substantial lobby in support of biopesticides. Whether or not the widespread use of biopesticides will be realised, will depend upon technical advances in biopesticide production and evaluation. Whether or not economic pressures and legal restrictions may help or hinder the transfer to biopesticides still remains debatable.

9.4 Other developments in agricultural practices which may depend upon biotechnology

In this chapter, we have discussed some of the approaches to agricultural practice that may lead to environmentally sustainable agricultural production. We have emphasised, throughout, that the adoption of such practices is dependent not only on technological advances especially arising from biotechnology, but also on the creation of a socio-economic climate that encourages their adoption. At the same time, these new practices must guarantee an appropriate yield of products to fulfil society's needs. It may, in time, be possible to modify human expectations but, inevitably, the basic requirement for food will need to be satisfied. We have also argued elsewhere in this text (see Chapter 2) that agriculture may also be called upon to satisfy energy needs and, possibly, the supply of non-food materials.

modification of
root storage
organs to
facilitate
harvesting

The two aspects covered in this chapter (the maintenance of soil fertility and the protection of crops) are not the only important issues. Our increasing ability to genetically manipulate plants, opens up other possibilities to make farm practices less environmentally damaging. For example, the natural form of the roots of sugar beet makes it quite difficult to harvest them. Considerable mechanical energy derived from fossil fuels is expended in collecting such roots which are often heavily laden with soil. This increases the energy costs (and economic costs) of transporting them to the refining plant and washing them free of soil. The large amounts of water used in this process and the production of silt from the washings are also important environmental costs. It may be possible, by manipulating the genes which specify the morphology of the sugar beet storage organ, to make these easier to collect. For example, if instead of a long tapered root, the root was near spherical and produced on, or just in, the surface, similar to a beetroot, then the energy costs of harvesting and processing may be reduced. At the same time, lighter machinery might be used to lift the beet and these may cause less damage to the soil drainage system. In other words, we might regard such a beet as a more 'environmentally friendly' crop than its traditional predecessor.

Increasingly, plant breeders are directing their attention to such factors in their breeding programmes. The traditional objectives (higher yields, more disease resistant, better storage properties) of plant breeders will, however, remain important.

creation of strains more tolerant to environmental stress

Also important in future developments is the creation of crop strains that are more tolerant to environmental stress, especially more tolerant to drought. The production of such crop varieties potentially extends the area of the Earth that can be used productively. Often the solutions to such problems are offered amongst existing plants. Most plants, when exposed to long periods of drought, die. Thus in drought susceptible areas, planting a crop is no guarantee of production. A period of drought may lead to a complete loss. Some plants (for example the so called Resurrection plants) completely dry out during periods of drought. However, on rehydration, the plant cells begin to refunction as though it had been held in suspended animation during the drought period. The ability of these plants to protect their cells from damage arising from dehydration appears to be attributable to their ability to store trehalose. The question arises, can such an attribute be transferred to other plant types? For example, if it could be transferred to wheat, the wheat might grow and, in time of drought, simply suspend growth. Upon return of the rains, the crop would then rehydrate and continue growth. Of course, such targets will not be achieved in the short term. Nevertheless, this does indicate both the excitement and the potential of biotechnology to alter agricultural perspectives. Many hopes and expectations may be unfulfilled but some, maybe many, aspirations will be achieved providing excellence in research, careful management and a long term view concerning investment are adopted.

Resurrection plants

In addition to the production processes which occur on the farm, the treatment of farm products by biotechnology may offer potential environmental benefits. We have discussed, for example, the use of agricultural by-products as an energy source elsewhere (Chapter 8) in this text. This use has some fossil fuel saving effects. We could also cite the treatment of silage to improve the quality of the silage produced by silage fermentation. It may be possible to create better silage making conditions by, for example, mixing various crop products. Alternatively special silage-making inocula may be generated.

longer shelf life products

Also generating longer shelf life products from crops has obvious environmental spin offs. Transporting tomatoes to the market place only for them to rot is not an environmentally sound practice. If (as has been achieved) strains of tomatoes are generated which will ripen to the stage at which they can be eaten but, because of a genetic block, will not develop further, the 'environmental' (and economic) savings could be substantial. Apart from reducing transport costs, the costs of storage (for example refrigeration) may also be reduced.

Summary and objectives

In this chapter, we have provided some insights into the application of biotechnology to the generation of environmentally sustainable practices in agriculture. We began by reminding readers that modern agricultural practices have become more dependent upon the consumption of fossil fuels and the use of environmentally damaging chemicals. The need to apply criteria other than parochial economics to agriculture was emphasised. By using the examples of soil fertility and crop protection, we showed how current practices may be modified or replaced using biologically mediated processes.We particularly emphasised the use of nitrogen fixing rhizobia to replace, or reduce dependency upon, chemically produced nitrogen fertilisers and the use of micro-organisms, especially *Bacillus spp.* and baculoviruses, to replace recalcitrant insecticides.

We concluded the chapter by briefly outlining how genetic engineering coupled to traditional plant breeding programmes, might be employed to generate 'environmentally friendly' crop varieties.

Now that you have completed this chapter, you should be able to:

- describe the importance of combined nitrogen to agricultural production and explain why nitrogen availability as often yield limiting in agriculture;

- describe the environmental and social consequences of being reliant upon chemically produced nitrogen fertilisers;

- explain why the production of inocula containing *Rhizobium* sp. may be valuable as agricultural reagents;

- describe the stages in the production of rhizobial inocula and explain the difficulties encountered in evaluating the effects of using such inocula;

- give examples of how biological nitrogen fixation may be extended for agricultural purposes and explain the economic barriers to the increased uses of biological nitrogen fixation in agriculture;

- describe the potential disadvantages of using recalcitrant organic materials to protect crops against weeds, pests and diseases;

- list the strategies that may be used to reduce dependency on the use of agrochemical pesticides including the potential range of organisms that may be used to control pests;

- explain the advantages and disadvantages of using bacteria and viruses to control insect pests, with special reference to the use of strains of *B. thuringiensis* and baculoviruses;

- explain, using suitable examples, how contemporary molecular genetics may lead to the reaction of 'environmentally friendly' plants;

- explain why biotechnological developments alone are insufficient to lead to the adoption of environmentally more sustainable practices in agriculture.

Application of biotechnology for mineral processing

Application of biotechnology for mineral processing

10.1 Introduction

bioleaching

Biotechnology has already played a major role in mineral resource winning and processing. A significant percentage of the copper currently in use today was obtained from source ore by a process called bioleaching. This process involves the exploitation of the activities of an interesting group of bacteria. With the rapid advances that are currently being made in biotechnology, especially with the development of genetic engineering technology, we will witness in the years to come a major 'in road' by biological systems and processes into areas of mineral processing that are served by more traditional technology. The advantages that biotechnology will offer include greater operational efficiency, safer processing and economical superiority.

In this chapter we will survey the metabolic capabilities of some micro-organisms and the properties of relevant products they synthesise and establish the potential that exists for exploiting microbial systems in mineral resource winning. Some of these technologies will have further application in pollution control and the clean-up of industrial effluents. We will discuss these towards the end of the chapter.

10.1.1 Mineral resource winning

∏ From your previous experience see if you can construct a flow diagram to represent the process of locating, extracting and utilising minerals.

mineral winning

Figure 10.1 shows a summary of the processes involved in the winning and utilisation of a mineral. A typical mineral will occur in nature in an unrefined form and in this raw state it is not suitable for direct and immediate application. Furthermore, this raw state material will most likely be unevenly distributed in nature. The first stages in the exploitation of a mineral resource will be to locate the source. Copper, for example, occurs as sulphides and oxides in subterranean rock formations. This search for the source of raw material is shown in Figure 10.1 as exploration. Having located the source, it is then necessary to extract the raw material from its environment. This may involve deep mining if the raw material, copper ore, is located at a considerable depth below ground or open cast mining if the ore is near to the surface.

The extracted ore then has to be processed ready for the refining stage. Processing and refining may take place at locations away from the extraction site (eg mine). Processing may, for example, include ore crushing. Refining is the purification of the minerals so that it is in a form that is suitable for application. With copper, refining is undertaken by smelting the crude ore at high temperatures. It is a very energy intensive process.

The purified mineral is then utilised. At the end of its useful life, the mineral may be recycled and incorporated into a new product.

∏ Examine Figure 10.1 and identify the two main 'environmental' costs of mineral winning and utilisation.

The two main costs are: the consumption of considerable amounts of energy in collecting, crushing and smelting and the production of wastes.

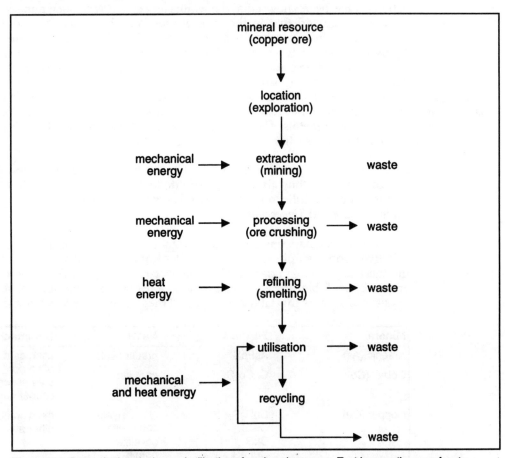

Figure 10.1 Stages in the winning and utilisation of a mineral resource. Text in parentheses refers to terminology in the mining and processing of metals such as copper.

wastes may include some minerals and be toxic

Wastes are produced at each stage of mineral resource winning production of waste material occurs. A proportion of the mineral itself will be lost as waste thus reducing the overall efficiency of the process occurring at that stage. The waste may also be toxic and so present a hazard to the environment. If this is so then containment of the waste is essential and detoxification preferable.

biological alternatives

Biological systems and processes may substitute for conventional technologies in mineral resource winning and processing. They may also be of use in the extraction of minerals from waste and in detoxification of waste prior to environmental discharge or disposal. In order to understand how we may use these systems we first need to study the biology of some microbial systems.

10.1.2 Bacterial populations in mine waters

Since Roman times when mining for coal and metals, such as copper, became a major activity it has been known that the waters associated with such mining operations can be very acidic and that in the case of metals such acidity may contribute to their solubilisation from ores such as sulphides and oxides. The origin of the acidity was thought to be solely chemical but in the early 1920's bacterial involvement was postulated as being contributory to the generation of acid. In the late 1940's a particular bacterial species was isolated from mine water and demonstrated to be of major importance in leaching of metal sulphides. This bacterium is called *Thiobacillus ferrooxidans* and it has metabolic capabilities which contribute to its ability to facilitate the leaching of metals. The characteristics of this bacterium will be discussed in more detail in the next section.

Thiobacillus ferrooxidans

In addition to *T. ferrooxidans*, a variety of other bacteria are found in the mining environment and they also may contribute to the metal leaching effect. These bacteria include thermophilic microbes, heterotrophic bacteria and members the genera *Leptospirillum* and *Thiobacillus*.

Essentially what happens is that these organisms oxidise sulphides to sulphates which effectively forms sulphuric acid (we will look at the chemistry of this a little later). Thus the these conditions under which these bacteria and sulphide mineral leaching occur are acidic. In Table 10.1 we have provided a list of the major mineral sulphides found in nature. Examine this table carefully and then attempt the following in text activities.

Mineral	Formula	Name	Comments
Arsenic (As)	$AsFeS$	arsenopyrite	used as a source of As,
Cobalt (Co)	$CuCo_2S_4$	carrolite	often contaminates other ores, associated with copper sulphide
Copper (Cu)	$CuFeS_2$ Cu_2S CuS $CuFeS$	chalcopyrite chalcocite covellite bornite	main source of Cu, often other sulphides present
Iron (Fe)	FeS_2 FeS_x	pyrite, marcasite pyrrhotite	used as a source of sulphur and iron small amounts in all sulphide ores
Lead (Pb)	PbS	galena	major source of lead, often ZnS and AgS present in small amounts
Molybdenum (Mo)	MoS_2	molybdenite	major source of molybdenum, CuS often present
Nickel (Ni)	NiS $FeNiS$	millerite pentlandite	main source of nickel
Silver (Ag)	Ags	argenite	main source of silver
Zinc (Zn)	ZnS	sphalerite	main sourve of zinc but also some associated with galena

Table 10.1 Major mineral sulphides.

Π Bearing in mind that since these organisms produce sulphuric acid they must be acid tolerant. If they are to be useful for leaching metal ores, what else must they tolerate? (Use Table 10.1 to help you).

tolerance to metals

They must be tolerant to the metals released. For example *T. ferroxidans* can tolerate 40g l^{-1} iron, 70g l^{-1} copper, 119g l^{-1} zinc and 70g l^{-1} nickel. In practice, it is possible to modify the tolerance of strains of these bacteria by slowly modifying (increasing the metal content) of the media in which they are cultured. Using this approach, it is possible to produce strains which have improved tolerance to cobalt, uranium, chromium and arsenic. However, all strains seem to have persistantly low tolerance to mercury, lead and selenium.

Π Using this information and the data in Table 10.1, see if you can identify the metals that may be leached form metal ores by sulphide oxidising bacteria.

In fact all of those listed in Table 10.1 could be leached through bacterial action, although we might expect only low concentrations of some metals (for example lead) in the leachate.

Π List the factors which will influence the concentration of metal ions in the leachate from a metal ore.

The sort of factors we anticipate you would cite include the relative concentration of the metal in the ore, the rate of water (sulphuric acid) drainage through the ore and the tolerance of the microflora present to the metal being eluted. You should note that the list of mineral sulphides given in Table 10.1 does not include examples of all of the metal sulphides. Many other metals, including tin and uranium may be leached through similar bacterial action.

SAQ 10.1

Draw a diagram showing the various stages of mineral resource winning and utilisation. Indicate where wastage may occur. Annotate you diagram as fully as possible with respect to copper exploitation.

10.2 Bacterial activity and ore leaching processes

10.2.1 Some useful terms in discussing ore leaching processes

We will begin Section 10.2 by briefly reminding you of some terms we use in describing the organisms involved in the leaching of metals from ores. Examine Table 10.2 carefully before reading on.

Acidophile	an organism that grows best in an acidic environment
Autotroph	an organism that derives its carbon for growth by fixing carbon dioxide and uses a non-organic source of energy
Chemoautotroph	an organism that derives its energy from the oxidation of inorganic substrates and its carbon for growth by fixing carbon dioxide
Chemolithotroph	an organism that derives its energy from the oxidation of inorganic substrates
Desulphurisation	The removal of sulphur (from a fossil fuel)
Heterotroph	an organism that obtains its carbon and energy for growth from organic substrates
Mesophile	an organism that grows best at temperatures 18-40°C
Mixotroph	an organism that gains its carbon from organic compounds but its energy from a non-organic source
Ore leaching	processes by which minerals locked up in insoluble ores are solubilised by liquid percolating through the ore
Psychrophile	an organism that grows best at low temperatures
Thermophile	an organism that grows best at elevated temperatures

Table 10.2 Explanation of terms used in this chapter.

Π Use the appropriate terms from Table 10.2 to describe an organism which can utilise an inorganic source of energy but obtains its carbon from simple organic molecules.

The organism could be called a chemolithotroph (not a chemoautotroph). It could also be called a mixotroph.

Π What other terms might be applied to this organism if it grows well at pH 2 and at 50°C?

It could also be called an acidophile and a thermophile. Now let us use these terms in describing the organisms involved in acid mine leaching.

10.2.2 Micro-organisms involved in ore leaching

Since the leaching of metals from sulphide ores occurs at a low pH and involves the oxidation of sulphide, you should have anticipated that the most important group of micro-organisms are the acidiophilic thiobacilli. These use the oxidation of inorganic sulphur (sulphide) to produce energy and are, therefore, chemolithotrophs. Many use CO_2 as a source of carbon (therefore they are auxotrophs or chemoauxotrophs) although some use organic sources of carbon and are, therefore, mixotrophic. Most are mesophilic (growing best at about 30-35°C). A few are, however, moderately thermophilic growing best at about 45-50°C.

general features of thiobacilli

Those strains which are most suitable for acid mineral leaching are acidophilic growing well at pH 1.5-5.0. Not all thiobacilli are able to grow at low pH. These non-acidophilic strains are not useful for sulphide mineral leaching.

Let us examine the organisms of greatest importance in leaching. It is generally regarded that *Thiobacillus ferrooxidans* is by far the most important, followed by *T. thiooxidans*, *T. acidophilus* and *T. organoparus*.

We have summarised the properties of these organisms in Table 10.3. Examine this table before reading on.

Species	Properties
Thiobacillus ferrooxidans	aerobic, acidophilic, autotrophic, rod-shaped pH range 1.5-5.0 optimum pH2, optimum temperature 32-36°C oxidises sulphides, soluble sulphur compounds, ferrous ions
T. thiooxidans	aerobic, acidophilic, autotrophic rods, pH optimum 2, range pH 1.5-6.0, temperature optimum 30-35°C
T. acidophilus	aerobic, acidophilic, mixotrophic rods, Ph optimum 3.0, range pH 1.5-5.0, temperature optimum 30-35°C, oxidises elemental sulphur
T. organoparus	aerobic acidophilic, mixotrophic rods, pH optimum 2-3, range pH 1.5-5.0, oxidise elemental sulphur
Leptospirillum ferrooxidans	aerobic, acidophilic, autotrophic, pH optimum 2-3, range pH 1.5-5.0, temperature optimum 30-32°C, oxidises ferrous ions but not sulphur compounds

Table 10.3 The major species involved in acid leaching of sulphide ores.

∏ Which of the organisms listed in Table 10.3 use carbon dioxide as their source of carbon?

You should have identified *Thiobacillus ferrooxidans*, *T. thiooxidans* and *Leptospirillum ferrooxidans* as they are described as autotrophic. The *T. acidphilus* and *T. organoporus* use organic compounds (mainly excreted by the autotrophic species) as their source of carbon. Since they use the oxidation of inorganic substrates to provide energy, they are said to be mixotrophic.

10.2.3 Metabolic activities of *Thiobacilli* and ore leaching processes

In this section, we will mainly focus on the metabolic activities of *Thiobacillus ferrooxidans*. This organism gets its energy from the oxidation of reduced sulphur compounds and by the oxidation of ferrous (Fe^{2+}) salts.

∏ How does *T. ferrooxidans* differ from *T. thiooxidans* in terms of its energy sources?

T. ferrooxidans can oxidise ferrous salts, *T. thiooxidans* only uses the oxidation of reduced sulphur compounds to produce usable energy.

solubilisation of metal sulphates

There are several chemical reactions, some biologically mediated, which will be involved in ore leaching. It is possible, for example that the bacterium *T. ferrooxidans* can oxidise directly certain metal sulphides to the corresponding metal sulphates:

$$MS + 2O_2 \rightarrow MSO_4$$

In an aqueous environment the metal sulphide dissociates:

$$MS \rightleftarrows M^{2+} + S^{2-}$$

The equilibrium position for this dissociation reaction is to the left, but as the sulphide is oxidised, more MS will dissociate:

$$MS \rightleftharpoons M^{2+} + S^{2-} \xrightarrow{O_2} SO_4^{2-}$$

Thus, an insoluble metal sulphide can be solubilised by conversion to its corresponding metal sulphate.

There are other reactions which bacterial cells may mediate and which may be important in ore leaching. Many ores contain significant traces of iron pyrite (FeS_2) and this also occurs in coal deposits in which it may be the cause of the formation of acidic mine waters since some leaching bacteria can convert iron pyrite to ferric sulphate and sulphuric acid, as shown below:

$$4FeS_2 + 15O_2 + 2H_2O \rightarrow 2Fe_2(SO_4)_3 + 2H_2SO_4$$

production of sulphuric acid

Sulphuric acid may also be generated by the direct oxidation of sulphur as a means of energy generation by bacteria such as *T. ferrooxidans*:

$$S_8 + 12O_2 + 8H_2O \rightarrow 8H_2SO_4$$

Since sulphuric acid is generated by these mechanisms, it is immediately apparent that in mines where pyrite and other sulphur compounds are present, the formation of very acidic conditions will occur. Thus acidic mine waters are produced, at least in part, through biologically mediated processes thereby contradicting the early assumptions that such acidity was solely due to chemical reactions.

Both ferric sulphate and sulphuric acid generated by the reactions shown above have an important role to play in the overall leaching process in that they can solubilise certain minerals in source ore. In addition, the presence of sulphuric acid will keep the pH low enough to maintain optimum activity of bacteria involved in the leaching processes.

The leaching of ores, therefore, involve a combination of both biologically- and chemically-mediated processes. The actual reactions contributing to the solubilisation of the mineral will depend on the nature of the ore and on other components that may be present. The biological reactions are often referred to as 'direct' processes while those mediated chemically are often designated as 'indirect'.

direct and indirect processes

Figure 10.2 shows how both direct and indirect processes may contribute in the solubilisation of a metal sulphide and shows how *T. ferrooxidans* may be involved.

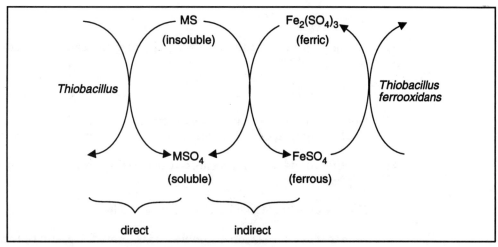

Figure 10.2 Oxidation reactions of *Thiobacillus ferrooxidans*.

On the left hand side of Figure 10.2, the reaction shown is the direct solubilisation of the metal sulphide by bacterially mediated oxidation to the sulphate. In this situation the bacteria are gaining energy by the oxidation process. The alternative indirect solubilisation of the metal sulphide is shown to the right. Here, ferric sulphate interacts with the metal sulphide. Ferric sulphate is reduced to ferrous sulphate and the insoluble metal sulphide is converted to soluble metal sulphate. The bacterium *T. ferrooxidans* can then oxidise ferrous sulphate to ferric sulphate so regenerating the leaching solution.

target ores   Ores in which bacterially mediated leaching is known to be implicated either directly or indirectly include:

iron pyrite	FeS_2
covellite	CuS
chalcocite	Cu_2S
chalcopyrite	$CuFeS_2$
uraninite	UO_2
uranium oxide	UO_3
molybdenite	MoS_2
arsenopyrite	$AsFeS$
stibnite	Sb_2S_3

10.2.4 Direct leaching processes

direct leaching   The direct involvement of bacteria in the solubilisation of minerals has been implicated in a variety of ores. Some example reactions are shown below.

1) As mentioned earlier, the conversion of iron pyrite to ferric sulphate and sulphuric acid is a direct process:

$$4FeS_2 + 15O_2 + 2H_2O \rightarrow 2Fe_2(SO_4)_3 + 2H_2SO_4$$

2) Covellite can be solubilised directly:

$$CuS + 2O_2 \rightarrow CuSO_4$$

3) In the presence of sulphuric acid, chalcopyrite can be directly solubilised to copper sulphate:

$$4CuFeS_2 + 17O_2 + 2H_2SO_4 \rightarrow 4CuSO_4 + 2Fe_2(SO_4)_3 + 2H_2O$$

Note that this reaction generates ferric sulphate which in turn may participate in solubilising further chalcopyrite, as will be shown in the next section.

4) Another copper ore, chalcocite, can also be directly solubilised in the presence of sulphuric acid:

$$2Cu_2S + O_2 + 2H_2SO_4 \rightarrow 2CuS + 2CuSO_4 + 2H_2O$$

Note that in this reaction insoluble CuS is formed. This can be further solubilised directly as indicated in reaction 2) above involving covellite.

10.2.5 Indirect leaching processes

indirect leaching

There are a number of reactions in which the ferric sulphate and sulphuric acid generated by bacterial activity can be utilised for solubilising minerals in ore deposits. When ferric sulphate is used the ferric ion is converted to the ferrous ion and so the resulting solution will contain the solubilised metal, as its sulphate, and ferrous sulphate together with, in some cases, elemental sulphur.

Covellite will react with ferric sulphate:

$$CuS + Fe_2(SO_4)_3 \rightarrow CuSO_4 + 2FeSO_4 + S\downarrow$$

Chalcopyrite can be solubilised in the presence of ferric sulphate. This reaction will occur in addition to the direct process, shown in the previous section (10.2.4) as reaction 3), which yielded ferric sulphate:

$$CuFeS_2 + 2Fe_2(SO_4)_3 \rightarrow CuSO_4 + 5FeSO_4 + 2S$$

Similarly another copper ore, chalcocite can be indirectly solubilised as copper sulphate:

$$Cu_2S + 2Fe_2(SO_4)_3 \rightarrow 2CuSO_4 + 4FeSO_4 + S$$

solubilisation of uranium ores

The oxides of uranium, UO_2 (uraninite) and UO_3, can both be indirectly solubilised to form uranyl sulphate:

$$UO_2 + Fe_2(SO_4)_3 \rightarrow UO_2SO_4 + 2FeSO_4$$

$$UO_3 + H_2SO_4 \rightarrow UO_2SO_4 + H_2O$$

SAQ 10.2	Circle the three most appropriate statements to complete the following sentence.

Thiobacillus ferrooxidans is an acidophilic, chemoautotrophic bacterium; it .....

1) Dislikes acidic environments.

2) Favours acidic environments.

3) Oxidises reduced iron compounds.

4) Oxidises reduced sulphur compounds.

5) Oxidises both reduced iron and sulphur compounds.

6) Obtains its carbon for growth by fixing CO_2.

7) Obtains its carbon for growth from glucose. |

SAQ 10.3	Decide which of the following statements are true and which are false giving reasons for you choice.

1) FeS is converted to $Fe_2(SO_4)_3$ by *Thiobacillus ferrooxidans* by a direct reaction.

2) The ferrous to ferric conversion by *Thiobacillus ferrooxidans* is by an indirect route.

3) Covellite (CuS) is a soluble compound.

4) Of the compounds FeS, $FeSO_4$, $Fe_2(SO_4)_3$, only $FeSO_4$ is soluble.

5) *Thiobacillus ferrooxidans* is a chemoautotroph.

Finally, and without reference to the text, try to balance the following equation:

$$\_\ FeS + \_\_O_2 + \_\ H_2O \rightarrow \_\ Fe_2(SO_4)_3 + \_\ H_2SO_4$$ |

10.3 Commercial ore leaching

10.3.1 Exploitation of bacterial oxidation of sulphide minerals

The principle motivations for developing processes using micro-organisms to extract minerals are mainly to:

- use less energy-intensive methods of extraction;

- use economically viable extraction methods;

- use less environmentally damaging extraction methods.

∏ Before reading on, see if you can list three or four reason why using bacterial oxidations for the extraction of metals is attractive.

There are many reasons, but the main ones we would cite are:

- it does not require high energy inputs;

- it operates at low temperatures and pressures;

- it can be used on a small or large scale;

- it can be used to extract a variety of minerals;

- it can be self-generating if soluble iron is present since the ferrous ions generated from ferric ions during extraction are re-oxidised;

- it can be used, in some instances, *in situ* and does not require initial mining. Alternatively it may be extracted close to the mine. (It does not need transporting to a smelting unit);

- it produces an aqueous effluent which can be easily neutralised and it does not produce toxic or noxious gases as by-products (compare this with smelting operation);

- it can be used either in the form of a simple percolated bed or in stirred tanks.

Of course, the disadvantages are that the micro-organisms must be kept viable and the desired metal is recovered as an aqueous solution of its salts, not as elemental metal.

The attractiveness of using bacterial oxidation of sulphide minerals means that it can be potentially be used for a variety of purposes.

∏ Again we would like you to try to think of some potential applications before reading on.

There are many potential applications but the main ones are:

- the extraction of metals from ores;

- the partial leaching of mixed mineral concentrates to remove impurities or to convert the mineral into a form suitable for processing by other means;

- to generate ferric sulphate which can be used to extract other ore types which are not suitable to support bacterial growth.

Bearing these general points in mind, we will examine some of these applications in more detail.

10.3.2 Leaching of copper ores

Although bioleaching may be applicable in a variety of mineral winning industries, by far its greatest application to date has been in the copper industry where a significant proportion of copper in current usage has been produced by this method.

commercial copper leaching

To a large extent this technology has been applied in the United States where there are massive dumps of low grade copper ore that have arisen out of conventional mining practices. The copper content of such waste ore is very low and so it is not economically viable to process such material by smelting to recover the residual copper. However, because there is so much waste as unprocessed ore, the total amount of copper 'lock up' in this material is very high and may be in the order of millions of tonnes. The total value of this unrecovered copper is considerable; at current world prices it could be worth billions of dollars (one billion equals a thousand million). What is required is a relatively inexpensive technology that can adequately recover this important reserve of copper. Bioleaching may fulfil this requirement.

A typical configuration for a leaching operation is shown in Figure 10.3.

ore dumps

The waste ore is deposited in dumps preferably on land that will be impermeable to the leaching solution. The size of the dumps may be massive; one site in the United States is reported to contain billions of tonnes of low grade waste ore. The ore is periodically sprayed with a leaching solution containing ferric ion and sulphuric acid. This slowly percolates through the ore and the leaching processes occur. Both direct and indirect processes may be involved in solubilising the copper ore.

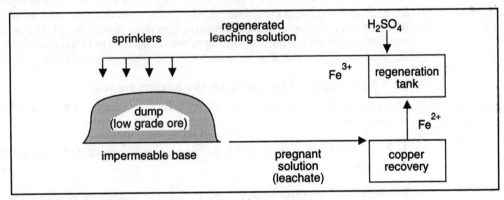

Figure 10.3 Typical configuration for a copper leaching process.

∏ Why is the ferric ion (as ferric sulphate) required and not the ferrous ion?

The answer is because ferric sulphate, having the general formula $Fe_2(SO_4)_3$, can be converted to ferrous sulphate, $FeSO_4$. Essentially the ferric ions act as an oxidant and oxidises the ore. In the process the ferric ions become reduced to ferrous ions. We remind you that the reaction is:

$$CuS + Fe_2(SO_4)_3 \rightarrow CuSO_4 + 2FeSO_4 + S\downarrow$$

pregnant
solution

Copper sulphate is formed and dissolves in the percolating leaching solution. Eventually the leaching solution reaches the bottom of the dump and is recovered ready for processing. The solution will contain copper sulphate ($CuSO_4$) and ferrous sulphate ($FeSO_4$). At this point the leachate is referred to as the pregnant solution.

There are various ways of recovering the copper from solution. One method is to use iron launders. The copper is precipitated out of solution by passing the leachate over iron; scrap iron may be used for this purpose. The following reaction occurs:

$$Fe + CuSO_4 \rightarrow Cu + FeSO_4$$

The precipitated copper is then allowed to settle out of suspension and is dried after which it is ready for utilisation. The leachate cleared of copper is then regenerated and recycled. Regeneration is achieved by aerating the solution in order to stimulate bacteria such as *T. ferrooxidans* to oxidise ferrous sulphate back to ferric sulphate. Some additional sulphuric acid may be added to ensure that the regenerated leachate is at a suitable low pH.

To give you some idea of scale, the Kennecott Chino mine in New Mexico produces about 7×10^4 tonnes of mine waste per day. This is dumped into heaps about 25-30 m high. The leachate contains 1.4g Cu l^{-1} and the daily yield of copper by this process is about 50 tonnes.

Leaching of waste copper ores in dumps is a relatively low technology operation that can be carried out on a very large scale with a minimal amount of equipment and, so long as the leachate/pregnant solution can be contained and recovered, does not present any serious environmental burden. Because of these factors it is a comparatively inexpensive technology and it will be cost effective and should facilitate the recovery of copper that is locked up in low grade waste ores.

10.3.3 Limiting factors and strain improvement

limiting factors

There are a number of factors that may influence and limit the overall effectiveness of the leaching operation.

∏ Make a list of as many factors as you can which will affect the rate of direct leaching.

The first consideration here is that leaching is mainly or wholly a bacteria-mediated process and that it is essential to maintain a healthy, viable cell population. The relevant features would include an adequate oxygen and carbon dioxide supply, correct leachate concentration, suitable environmental pH, suitable temperature and a lack of toxic metals. These features will now be discussed further.

Bacteria such as *T. ferrooxidans* which are involved in the oxidation of iron and sulphur compounds have an essential requirement for oxygen. If you refer back to the reactions shown in Section 10.2.4 (direct leaching processes) you will see that they all require oxygen. If the conditions within the dump go anaerobic then direct leaching by *T. ferrooxidans* and affiliated bacteria will cease. Such conditions may arise if the dump becomes saturated by leachate perhaps due to poor drainage or excessively fast sprinkling rate. This problem can be reduced by having alternating periods with and without sprinkling of the leaching solution. Such an alternation will also facilitate the leaching effect by allowing time for the leachate to diffuse in and out of the ore.

pH and jaroside formation

Another important parameter is pH. *T. ferrooxidans* is an acidophilic bacterium and is active in the pH range 1.5-5. Other bacteria involved in the bioloeaching process also prefer low pH. The pH optimum for the oxidation of ferrous ions to ferric ions is about 1.9-2.6 and so it is desirable to maintain the pH of the leachate within this range. Hence there may be a need to add sulphuric acid to the regenerated leachate, as described in the previous section. Maintaining a low pH will also assist in preventing the formation of Jarosite, a ferric sulphate/ferric hydroxide complex that precipitates and may cover the surface of the ore so reducing the effectiveness of the leaching process.

temperature

T. ferrooxidans, like all other life forms has an optimum temperature for growth. It is like many of the other bacteria involved in the leaching process, active in the temperature range of 25-45°C. If the temperature of the dump rises excessively above this range then the contribution of direct leaching processes will be considerably reduced leaving only the indirect reactions mediated by sulphuric acid and ferric sulphate. The temperature inside a dump may increase as a result of chemical activity thereby reducing the biological component. In such situations, the major biological contribution will be in the regeneration of the leaching solution at a point remote from the dump.

toxic metals

As we saw earlier in this chapter *T. ferrooxidans* is reasonably tolerant to many of the metals found in ores, such as copper, aluminium and zinc. However, there are metals to which it is relatively sensitive and these include: uranium, silver, mercury and molybdenum. Furthermore, the oxides of selenium, tellurium and arsenic are toxic to *T. ferrooxidans*. The sensitivity of the bacterium to these metals will restrict its potential application in ore leaching in circumstances were these toxic metals are present in the ore. This problem, can be alleviated by remote leaching systems; this will be discussed in the next section.

strain improvement

There is considerable scope for strain improvement and it is not too surprising that researchers are becoming involved in the genetics of *T. ferrooxidans* and are looking at the possibility of genetically manipulating this bacterium. It would be desirable to incorporate genes that confer resistance to mercury, uranium, arsenic and other metals to which the bacterium is sensitive. Other groups of bacteria and particularly fungi do show resistance to these substances. For example, much is known about the genetic and biochemical basis of mercury resistance. It may well be possible to use recombinant DNA technology to construct novel *T. ferrooxidans* cells with genes from other bacteria which confer resistance to these toxic metals. Furthermore, it may be feasible to develop strains of the bacterium that synthesise increased quantities of the components responsible for the oxidation of reduced sulphur and iron compounds. If successful, this would increase the rate at which regeneration of the leaching solution occurs and may also accelerate direct processes *in situ*, for example, within dumps.

thermophilic bacteria

Another area of interest is in the development of thermophilic bacteria for ore leaching. We have already seen that temperature can be a limiting factor with regards to *T. ferrooxidans* and related bacteria. If bacteria are to be used in deep subterranean mineral reserves as alternatives to conventional mining then the ability to grow at elevated temperatures will be an essential prerequisite. Thus thermophilic bacteria are being screened for capabilities beneficial to ore leaching.

SAQ 10.4

Imagine that you are the technical developments manager for a copper mining company. List the factors you think necessary to consider when deciding whether or not to introduce an ore leaching process.

SAQ 10.5

A small dump of mine waste is being acid leached. The dump was first treated with a mixed culture of thiobacilli obtained from a mine in the same region. Leachate from the dump was both acidic (pH 2.3) and rich in copper sulphate ($1.2g\ l^{-1}$). The leachate also contained significant amounts of ferric/ferrous ions ($0.25g\ l^{-1}$). The process appeared to run in a steady manner but after an accidental spillage of an aqueous phenolic solution onto the dump, the pH of the leachate rose sharply to around neutral and the leachate no longer contained much copper or iron.

After the dump had been subsequently thoroughly washed by percolating it with fresh water, it was re-seeded with a mixed culture of thiobacilli obtained from the same source as the first inoculum. Continued monitoring of the leachate indicated that its pH remained high and copper and iron were not present in significant quantities.

Explain what were the likely reasons for this failure to re-established metal leaching from this mine waste dump. (Assume that the dump had not been depleted of copper or iron).

10.3.4 Leaching of other metals

uranium leaching

Although the leaching of copper ores is currently the major commercial application of this technology there are other areas in which developments have occurred and which may become of greater economical significance in the future. One example is the leaching of uranium ores. In Section 10.2.5 it was shown that ferric sulphate and sulphuric acid can solubilise the uranium oxides UO_2 and UO_3, respectively. *T. ferrooxidans* cannot solubilise these ores directly since they are oxides but it can be used to generate the ferric sulphate and sulphuric acid. However, the bacterium is very sensitive to uranium salts in solution and it is, therefore, not possible to use the bacterium in direct contact with the ore (ie *in situ*). Instead a remote system has to be utilised where a leaching solution rich in ferric sulphate and sulphuric acid is pumped through uranium bearing ore in order to solubilise the uranium. Once the uranium has been recovered from the pregnant solution it may then be possible to use bacteria, such as *T. ferrooxidans*, to regenerate the solution to make it ready for another cycle of leaching.

It is in this type of situation where strain improvement would be greatly beneficial. For example, if *T. ferrooxidans* could be made tolerant to uranium salts then the bacterium could be used *in situ* rather than in the remote application outlined above. A further advance would be to make a thermophilic bacterium capable of uranium leaching *in situ* as this would allow the possibility of exploiting deep subterranean ore reserves where temperatures may be elevated.

gold

Gold extraction is another area where the bacterium *T. ferrooxidans* may have a role to play. In some instances, a chemical process is used to dissolve gold from ore as a means to recover it. The bacteria may assist this chemical process by solubilising iron sulphide deposits surrounding the gold particles thus allowing the reagents of the chemical process to reach gold more readily. One of the limitations in this process is that the ore may contain traces of arsenic to which *T. ferrooxidans* is sensitive. The process would thus be enhanced by strain improvement making the bacterium tolerant to arsenic.

10.4 Uptake of metals from solution

10.4.1 Microbial products and metal recovery

metal recovery
from solution

We have already discussed the need to recover metals from the pregnant solutions arising from leaching processes. With copper leaching, for example, the copper can be precipitated from solution by the addition of iron. There are other systems that could be used to recover the copper including: solvent extraction and electrolytic processes. But it is not only from leaching operations that there is a need to recover metals from solution.

pollution

heavy metals
in food chains

Many industrial processes may give rise to the discharge of heavy metals in aqueous effluents. The danger that such metals present to the environment is that they can enter and accumulate in food chains.

Π What do we call the process by which compounds accumulate in higher concentrations in higher members of a food chain?

It is called biomagnification. You should recall that we described this process with special reference to recalcitrant organic compounds at several points in this text (see for example Chapter 2). Because of biomagnification, it is essential that discharges of heavy metals do not occur. Therefore, it is necessary to develop systems to remove such metals from industrial effluents. The problem that the technologist faces may be that the heavy metals may be present in very large volumes of dilute solutions. Thus any system to remove the metals must have a high affinity for the metals and be capable of processing the large volumes of water involved.

There may be other reasons for recovering metals from industrial effluents. Some metals are precious and their loss through discharge wastes an economic resource. A good example of this is the photographic processing industry. Considerable amounts of silver are lost each year as a result of developing films. Non-microbial systems have been devised to recover silver from processing laboratory effluent. It may be that biotechnology could offer a superior silver extraction method.

silver
photographic
processing

precious
metals in sea
water

Another, but more futuristic application of the recovery of metals from solution, is the possibility of winning minerals from sea water. There are considerable quantities of precious metals such as gold dissolved in sea water. Unfortunately, the actual concentration of such metals is very low and an enormous volume of sea water would have to be processed to obtain workable quantities of metals. Furthermore, there will be difficulties due to interferences as a result of the very high concentrations of other elements such as sodium and potassium. Distillation of sea water would be prohibitively expensive on energy to be of use but other physicochemical means could be used, for example ion exchange - but this method will be prone to problems of interference.

the Dead Sea

There is interest in the recovery of precious metals from the Dead Sea. The Dead Sea is located inland on the border of Israel and Jordan and various rivers, including the Jordan, flow into it. Due to the intense heat in the region and to the fact that there is no link between the Dead Sea and the main oceans of the world there has been much evaporation resulting in the sea level being very low compared to that elsewhere. Nature has thus concentrated minerals at this location and so less effort would be required to recover minerals from the sea water in the Dead Sea.

Π   Which types of biological systems may be exploited for the winning of minerals
    from sea water and recovering toxic and precious metals from industrial effluent?

Initially one has to decide whether to use bacteria, fungi or even possibly algae. Then a
decision whether to use growing or non-growing, free or immobilised cells has to be
taken. Finally, again genetic manipulation has to be considered. Let us consider these
issues in more detail.

uptake of
metals by cells

There are several ways in which this could be done. Bacteria and fungi have an essential
requirement for certain metals and have developed specific transport systems to
facilitate the uptake of such metals. Perhaps these could be exploited using, for
example, bacterial cells immobilised on a solid support over which the industrial
effluent is passed. One problem that may be encountered in this approach is that heavy
metals (cadmium, copper, zinc and lead, for example) are toxic to bacteria and would
eventually kill the cells, thus the metal uptake would have to be selective.

Alternatively it may be possible to utilise a biological product instead of living cells.
Many organisms produce extracellular polymers, predominantly polysachharides, that
have the capability of binding metals and these could be used independently of the
producer cells.

fungal
mycelium and
metal recovery

Much interest is centering on the use of non-growing fungal mycelium as a means of
recovering metals from solution. Investigations using fungi to extract uranium from
solution have shown that the fungus *Rhizopus arrhizus* can accumulate uranium
equivalent in weight to about 18% of the dry weight of its mycelium. Toxicity will not
be a problem since the fungus is in a non-growing state. This system could prove to be
a very useful system for the recovery of a variety of metals from water.

genetic
engineering

The application of recombinant DNA technology is offering yet another approach.
Many eukaryotic organisms, including humans, produce metallothionine proteins, the
role of which is to bind metals such as toxic heavy metals within the cell thereby giving
protection against any potential toxic effects. With the advent of genetic engineering it
is now possible to produce bacteria (which normally do not synthesise these proteins)
containing metallothionine genes from higher organisms. Such novel technology could
be of great benefit in the application of biological systems for the recovery of metals
from aqueous environments.

10.4.2 Metal leaching from mine wastes and landfill sites

Π   See if you can think of examples where acid leaching of metals may pose
    commercial and environmental problems?

The most obvious example is in normal (tradiational) mining operations involving
smelting and the production of untapped mining waste dumps. Once mines containing
sulphide ores are developed, they soon become contaminated with sulphur and iron
oxidising bacteria. The same is true of sulphide-containing waste dumps. Under
favourable conditions (for example if the mine workings or dumps become wet), the
activities of the bacteria lead to acidic metal-bearing leachates. These leachates have to
be removed from the mine by pumping. The leachates are, of course, very corrosive and
cause damage to pumps and pipes and they need to be neutralised before being
discharged from the site. These features add considerable costs to the commercial
operation of mines. Untreated leachates from mine workings or from the waste dumps

pose considerable threats to the environment because of their acidity and the high concentrations of toxic metal ions.

You may also have considered that acid leaching from landfill sites may pose a similar problem. Although this is a potential problem, it is not so severe as that caused by sulphide mining operations. Remember that for the most part, landfill sites are anaerobic (see Chapter 3). This may lead to the production of sulphide (from sulphate). In the upper layers of a landfill site, this sulphide may be re-oxidised by thiobacilli to form sulphate. However, metals that are present are not predominantly in the form of sulphides. Thus although the biological production of sulphides and sulphates will cause some leaching of metal ions, these tend to be at much lower concentrations then from mining operations. Nevertheless, because of bioaccumulation and biomagnification of metal ions, the leaching of metal ions from landfill sites and their accumulation in ground water is of environmental concern.

10.5 Desulphurisation of coal

10.5.1 Sulphur content of coal

acid rain

Fossil fuels such as coal and crude oil contain sulphur compounds which, on combustion, give rise to sulphur oxides (for example SO_2 and SO_3). These oxides dissolve in atmospheric moisture and give rise to 'acid rain'. There is considerable concern about the environmental impact of acid rain. What is a particularly sensitive issue is the fact that the pollution is often produced in a country distant to where the acid rain falls. Atmospheric movements shift the clouds of sulphur oxides in the direction of the prevailing winds. Thus sulphur oxides generated by British power stations may end up as acid rain in Scandinavia. It is now realised that urgent action has to be taken to desulphurise fossil fuels or to apply some other clean-up process, particularly to power stations.

sulphur in coal

Coal contains both inorganic and organic sulphur. The inorganic fraction is mainly in the form of particles of iron pyrite (FeS_2). It is the oxidation of these by *T. ferrooxidans* and related bacteria that gives rise to acidic waters in coal mining operations. The organic sulphur complexes are diverse and include mercaptans, thiophenes and thiols and they are intimately associated with the coal matrix. The total content of sulphur varies depending on the source of coal; generally it is in the range of 0.2-8% but there are some coals with a sulphur content in excess of 11%.

precombustion method of reducing sulphur

There are non-biological ways of alleviating the sulphur problem. Firstly, of course, there is the possibility of a policy decision only to burn low sulphur coal. This is not always economically feasible. Iron pyrite can be removed from coal by chemical extraction or physical separation after coal has been pulverised to a powder. Pyrite is more dense than coal (pyrite: 5g cm^{-3}; coal: 1.2g cm^{-3}) and this can form the basis of a separation process. Such treatment is referred to as a precombustion method (ie the treatment is carried out before the coal is burnt). Organic sulphur complexes cannot be satisfactorily removed in this manner.

postcombustion scrubbing

Another non-biological approach is postcombustion scrubbing of flue gases to remove the sulphur oxides. This would only be feasible at major power stations and has been proposed as a means of reducing sulphur emissions in the United Kingdom. Postcombustion scrubbing would not be economically viable at minor plants or domestically.

10.5.2 Application of biological systems

microbial
desulphurisation
of coal

It is possible to devise microbial systems for the desulphurisation of coal and indeed in the laboratory these have proved to be technically successful. One of the principal limitations in the scale-up of this technology is that it may be relatively slow. The coal will have to be pulverised so that there is maximum surface area where the microbial process may occur.

Bacteria capable of oxidising reduced iron and sulphur compounds may be suited for solubilising the pyritic fraction. These bacteria include *T. ferrooxidans* and other *Thiobacillus spp.* However, *T. ferrooxidans* is not able to oxidise organic sulphur compounds although some other bacteria can. It may be necessary to have a multistage system whereby several different bacteria are involved in the desulphurisation process. It might also be preferable to operate the system at elevated temperatures to increase the rate of solubilisation. Thus thermophilic bacteria capable of oxidising pyrite and organic sulphur complexes would be desirable. One bacterium that has been suggested to be of potential value for this is *Sulfolobus acidocaldarius* which can grow at temperatures up to 90°C in acidic environments. This bacterium has the distinct advantage that it can oxidise pyritic sulphur and is also capable of oxidising organic sulphur compounds. A two-stage process involving the use of *S. acidocaldarius* has been proposed. In the first stage, conditions are chosen such as to maximise the solubilisation of the iron pyrite and in the second stage the process is modified to favour utilisation of organic sulphur compounds. In the view of many, however, the removal of organic S from coal is very controversial. The belief is that the organic S is built into the complex coal matrix and it may not be possible to selectively remove it.

limitations

Whether or not such processes will be commercially applied remains to be seen. There are obvious limitations including: long process times and the need to have tanks and equipment suitable for the operation of such systems. Furthermore, the coal has to be pulverised prior to the treatment process which imposes an energy cost.

SAQ 10.6

1) Which precious metal may be discharged in aqueous effluented from photographic processing laboratories? How may this metal be recovered.

2) In future it may be possible to apply biotechnological processes to the recovery of precious metals from sea water. In what part of the world would this most likely to be economically feasible, and why?

SAQ 10.7

In what forms does sulphur occur in coal? What problem does this sulphur cause when the coal is combusted, for example in the generation of electricity? List the ways in which the problem might be alleviated.

Summary and objectives

In this chapter, we have discussed the use of bacterial oxidations of sulphides and iron in mineral processing. We first outlined the processes involved in the winning and utilisation of mineral resources and then described the biological and metabolic features of microbes related to these processes. We described why the use of bacterial oxidation of sulphide minerals is a commercially and environmentally attractive option in mineral processing and we explained how both direct and indirect leaching processes may be used. We also examined the factors which may limit the leaching of metal ions from resource materials and identified some potential biotechnological targets for improving these processes. We also explained how micro-organisms and their products may be used to recover metals from a variety of sources. In the final part of the chapter, we briefly discussed the use of micro-organisms to reduce the sulphur content of coal thereby reducing the environmental problems that may arise from the combustion of high-sulphur coals.

Now that you have completed this chapter you should be able to:

- draw a flow diagram showing the various stages of mineral resource winning and demonstrate where wastage occurs;

- state how the metabolic activities of sulphur- and iron-oxidising bacteria can be exploited to solubilise metal sulphides and oxides;

- list some direct reactions mediated by bacteria;

- list some reactions indirectly mediated by bacteria;

- outline the commercial application of ore leaching in the copper industry;

- state what the limitations of the leaching process are and how these could be overcome by strain improvement;

- list the ways in which micro-organisms and microbial products can be used to remove metals from solution;

- distinguish the different forms in which sulphur occurs in soil;

- explain how bacteria may be used for the desulphurisation of coal.

Responses to SAQS

Responses to Chapter 2 SAQs

2.1   The functions of the four types of biologically relevant macromolecules can be summarised as follows:

- carbohydrates and lipids have mainly a structural role (cell wall, surface coat, biological membranes) and a role in storage and supply of energy;

- nucleic acids and proteins are both information containing biomolecules. In DNA the main accent is on storage of information, in RNA it is on conversion of information and in proteins it is on functional expression of information. In other words, DNA is the cellular databank from which information is transformed via RNA into proteins. Proteins are the cellular 'work horses' which, in addition to a structural role in for instance the cytoskeleton, are involved in all vital processes within the cell: as biocatalysts in the biochemical reactions within the cell, as receptor or messenger in intra- and intercellular communication and as immunoglobulin in the defense reactions of the organism.

2.2   A relatively small amount of biologically-usable energy is derived from the oxidation of reduced inorganic substrate (for example S^{2-} and NH_4^+ and Fe^{2+}) by chemoautotrophs. You might, therefore, argue that these are non-sunlight derived energy sources. However most of these inorganic substrates used are produced from oxidised inorganic substrates by anaerobic heterotrophs. These organisms oxidise organic molecules (ultimately derived from photosynthesis) and reduce inorganic substrates. We can represent this in the following way:

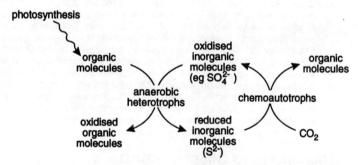

Thus we may conclude that even in these circumstances the energy obtained from the oxidation of inorganic molecules is derived from the energy input of photosynthetically derived organic molecules. We can therefore argue that the energy source of chemoautotrophs can be traced back to sunlight.

2.3   1)   0.85%.

Since each person needs 10 5000 kJ day^{-1}
= 10 500 x 365 kJ year^{-1}

Therefore 5321 x 10^6 individuals require
10 500 x 365 x 5321 x 10^6 kJ year^{-1}
= 2.04 x 10^{16} kJ year^{-1}
= 2.04 x 10^{10} GJ year

(1 GJ = 10^9 J)

Thus the proportion of the biomass that needs to be consumed $= \dfrac{2.04 \times 10^{10}}{2.4 \times 10^{12}} \times 100\% = 0.85\%$.

2)   Yes. If less then 1% of the energy stored in biomass is needed to fulfil human nutritional needs then, at least in theory, the other 99% is available for other purposes. In practice, of course, matters are not as

simple as this. There would be energy costs in collecting and manipulating the biomass to convert it into a usable form. A lot of biomass is, for example generated in inaccessible places and is not in the form that is readily utilisable as an energy source. For example, microscopic algae growing in the oceans would be very difficult to collect. Thus, although most of the energy stored in biomass could, in theory, be used as a source of energy to fulfil non-nutritional needs, in practice much of this cannot, in practice, be used. The most accessible sources are likely to be generated through agriculture. We examine the potential contribution of agriculture to the supply of energy in the next section of the text.

2.4

1) The answer is all those practices which give a ratio of energy input/energy output of greater than 1. Thus you should have included fish farming, intensive pig farming, battery hens and, perhaps milk production. You will notice that these lead mainly to high priced food products.

2) Any of the crops with low ratios of energy input/energy output look promising. Thus rice, wheat and potatoes look promising. This is because these crops provide a net energy gain. However, we must also remember that the actual crop yield is important. There is no point in cultivating a crop with a low energy input/energy output ratio if the yield of the crop ha^{-1} is extremely small.

We remind you that the data presented in the question are rather stylised. The actual ratio for a particular crop will depend on local conditions which will influence the energy costs of tilling the ground, the crop yield, the amount of fertilisers used and the transport costs.

2.5

1) A clean technology is a technology that uses energy and raw materials in a more efficient way and causes less harm to the environment. It encompasses both product and process design including the production and use of raw materials. In contrast a sustainable technology is one that aims to enhance both the current and future potential to meet human needs and aspirations and is, at the same time, capable of being used indefinitely. It should be self-evident that clean and sustainable technologies are intimately inter-related. It is difficult to visualise a sustainable technology that was not also a clean technology and it is not, therefore, surprising that the two terms are often used interchangeably.

2) Your answer to this ought to have been that the adoption of biotechnological practises and processes has the potential to be both a clean and a sustainable technology. We made the point in the text that the application of biotechnology makes it technically possible to derive the energy society needs from renewable energy sources and that the products of biotechnology are biodegradable and, therefore, less likely to lead to the accumulation of environmentally damaging factors. In essence, what biotechnology enables us to do is to enhance the natural geocycling of materials without the need to consume large amounts of non-renewable energy reserves. Whether-or-not this vision for biotechnology as the means to achieve is realised sustainability will depend not simply on technological developments but also on the political and social will of society to use this technology to replace existing environmentally damaging technologies. We also made the point in the text that the adoption of biotechnological processes will also depend upon the balance between economic and environmental criteria that are applied within society.

Responses to Chapter 3 SAQs

3.1

Contaminants	Occurrence
1) asbestos	c) building demolition site
2) copper	d) outflow from an electroplating process
3) *Salmonella typhi*	b) domestic sewage
4) methane	e) landfill site
5) soot	a) coal fired power station

3.2

Categories	Compounds
carcinogens (mutagens)	N-nitrosodimethylanine, ethylenemine
asphyxiants	hydrogen sulphide, carbon monoxide
systemic poisons	lead, mercury, cadmium, titanium, boron
pulmonary irritants	sulphur oxides, nitrogen oxide, ozone, chlorine, ammonia
fibrosis producer	asbestos

3.3

In general landfill constituents include non-noxious biological materials, xenobiotic chemicals, inorganic debris and non-biodegradable polymers derived from domestic and industrial wastes. We can divide the hazards that these present into a number of sub-groups. These include:

- flammable components in the waste and methane generated from the biodegradable constituents;

- other volatile components which may be toxic (for example H_2S, various volatile organics) and/or odorous compounds;

- soluble toxic materials which may leach from the site. These include heavy metal ions, certain anions (for example cyanide) and recalcitrant xenobiotic chemicals.

The biological debris in the landfill may contain pathogenic micro-organisms. This debris may also attract and support large populations of vectors of disease including invertebrates (for example flies) or vertebrates (for example rats).

3.4

1) 4.17 to 8.3y

Since 10^4 people produce 20-40 x 10^3 m^3 of compacted refuse y^{-1}, 10^5 people will produce 2-4 x 10^5 m^3 y^{-1}. But in using the cell emplacement strategy, about 20% of the material deposited at the site will be covering soil. Thus the total volume produced per year will be between:

$$2 \times 10^5 + \frac{20}{100} \times 2 \times 10^5 \, m^3 \, y^{-1} \text{ and } 4 \times 10^5 + \frac{20}{100} \times 4 \times 10^5 \, m^3 \, y^{-1}$$

$$= 2.4 \times 10^5 \text{ and } 4.8 \times 10^5 \, m^3 \, y^{-1}$$

The total volume of the site = $8 \times 10^5 \times 2.5 \, m^3$

$$= 2 \times 10^6 \, m^3$$

Thus the anticipated life of the site will be between:

$$\frac{2 \times 10^6}{2.4 \times 10^5} \, y \text{ and } \frac{2 \times 10^6}{4.8 \times 10^5} \, y$$

$$= 8.3 - 4.17 \, y$$

2) These figures assume that the population will not significantly change over the period nor will the pattern and quantity of refuse produced per capita. In practice, the pattern of refuse generation can be changed quite significantly. For example the introduction of local regulations restricting the use of bonfires to remove garden wastes may dramatically alter the volume of wastes that are taken to the landfill.

3.5

Methane is flammable. Thus the build up of methane in pockets within a landfill site and its release through the permeable cover presents some risk of fire and explosion. Although this has only occasionally happened, it is a factor that needs to be considered especially if the landfill site is close to habitation.

3.6

This is a difficult question because it is difficult to give a categoric answer. We can, however, give some guiding principles. If the material deposited in a landfill is predominantly biological material, it might be anticipated that the process of mineralisation would release plant nutrients and, therefore, encourage the growth of plants in much the same way as farmyard manure encourages the growth of crop plants. However, even in these conditions, the depth of the landfill and the induction of anoxia also leads to the production of methane and sulphides and other reduced materials. These are potentially inhibitory to plant growth and might prevent plants from becoming established. Fortunately, the upper layers of landfill sites remain aerobic and shallow growing plants are able to become established. In fact, these layers are enriched with plant nutrients and the surface of landfill sites often becomes covered by luxuriant plant growth. Predominant amongst these plants are those that are characteristic of rich soils such as nettles and fat hen. For these reasons, landfill is seen as a method of restoring derelict land for agricultural or horticultural use.

With deeper growing plants (trees) the presence and generation of toxic materials within the landfill make plant growth problematic. In all cases, the growth of plants will be greatly influenced by toxic materials within the landfill. As a general principle. The concentrations of these toxic materials will be progressively reduced as they slowly leach from the site.

Thus, in answer to the question, we would suggest that young landfill sites may support substantive amounts of surface (herbaceous) plant growth but that the growth of deep rooted plants may be more restricted. As the site ages we might anticipate an increasing ability to support deep rooted plants. At the same time there might be a reduction in the amount of plant nutrients in the upper layers of the site as these nutrients are lost through leaching. The chemical composition of the materials deposited in the landfill may have a profound effect. For example high levels of metal ions may substantially reduce plant growth or only allow the growth of metal tolerant varieties.

3.7

1) High.

2) High - because of low transport costs.

3) Low- filling the site with refuse will lead to the production of hazardous methane and will contaminate the area with other potential toxins. This will reduce the value of the land for building and, therefore, this will reduce the value of the site.

4) High or low. It is difficult to give a categoric answer to this. The site would be close to the source of the waste thus this would favour the use of the site for landfill (see 2). However, the social and economic consequences may outweigh this potential advantages. Questions that need to be considered relate to the proximity to residential properties, the potential reduction in the value of these properties (who wants to live next to a municipal landfill site) and the potential adverse effects of odours, fire hazards (methane) and hazards from toxins. Thus, whether or not an urban site is suitable depends on local geo-social topography.

Responses to Chapter 4 SAQs

4.1

Compounds which are degraded slowly do not contribute to the BOD value which is usually taken after 5 days. Furthermore, part of the biodegradable matter is assimilated to produce biomass and thus does not contribute to BOD.

Finally although all chemicals that can be oxidised biologically can also be oxidised chemically, not all chemically oxidisable materials can be biologically oxidised.

4.2

1) True.

2) False. Conversion of ammonia is carried out by two distinct groups of bacteria, the ammonia and nitrite oxidising bacteria (*Nitrosomonas* and *Nitrobacter*).

3) True. Allyl thiourea inhibits the nitrifying micro-organisms. Thus in the presence of this compound, no oxygen will be consumed for the oxidation of ammonia. Thus, less oxygen will be used.

4) True.

4.3

1) Fixed film processes and homogeneous aerobic processes (dispersed growth processes).

2) Tertiary (see Section 4.3.4).

3) Undesirable, as the void spaces in the filter become blocked and the operation of the bed is impeded.

4.4

1) The BOD value of the effluent will decrease.

Since $\dfrac{Li - Le}{Li} = \dfrac{1}{1 + 0.44\omega/fV^{0.5}}$

any increase in V will lower the value of $(\omega/fV)^{0.5}$ and thus increase the value of:

$$\dfrac{1}{1 + 0.44\omega/fV^{0.5}}$$

Thus: $\dfrac{Li - Le}{Li}$ will also increase.

Since Li remains constant, Le must be lower. Even without using the equation commonsense should have told you that increasing the volume of the filter would have lowered the BOD of the effluent.

2) $Le = 0.09$ kg m^{-3}.

Since $\dfrac{Li - Le}{Li} = \dfrac{1}{1 + 0.44\omega/fV^{0.5}}$

and $Li = 0.5$ kg BOD m^{-3}, $\omega = 50$ kg d^{-1}, $f = 1$ (single passage) $V = 200$ m^3.

then: $\dfrac{0.5 - Le}{0.5} = \dfrac{1}{1 + 0.44\omega/fV^{0.5}} = \dfrac{1}{1.22}$

Thus $Le = 0.5 - 0.41 = 0.09$ kg m^{-3}.

Notice that in this example we have used a very big filter (200 m^3) to treat rather a small volume of material (100 m^{-3} d^{-1}).

3) $Le = 0.08$ kg m^{-3}.

Since $f = \dfrac{1 + \alpha}{(1 + 0.1\,\alpha)^2}$ and $\alpha = 0.5$.

$f - 1.36$.

Substituting into the equation used in 2).

$$\dfrac{0.5 - Le}{0.5} = \dfrac{1}{1 + 0.44\,(50/1.36 \times 200)^{0.5}}$$

There $Le = 0.5 - 0.4205 = 0.08$ kg m^{-3}.

4) From the calculations carried out in 2) and 3), you can see that recirculation does not make much difference. Thus is because we have a very large filter and a fairly dilute waste water to treat.

4.5

1) To maintain the sludge loading rate we can in principle a) reduce the BOD of the incoming liquor (that is make the incoming liquor more dilute); b) increase the reactor volume; c) increase the reactor solids (that is return more of the sedimented microbial floc to the reactor).

These are possible because:

$$\text{sludge loading rate} = \dfrac{\text{BOD of incoming liquor} \times \text{influent flow}}{\text{reactor volume} \times \text{reactor solids}}$$

In practice, once a process has been established it is usually impractical to alter the BOD of the incoming liquor or the volume of the reactor. Thus the most obvious solution is to increase the proportion of the sedimented floc that is returned to the reactor.

2) To increase nitrification, we really need to increase the age of the sludge. This may be achieved by decreasing the organic loading rate. This may be achieved by either decreasing the flow rate of the incoming liquor or the BOD of the liquor or by increasing the volume of the reactor. In practice these are often difficult to adjust once a process has been installed. For this reason, water treatment units introduce a second aerobic treatment process specifically to improve nitrification. Thus the first aerobic unit will mainly remove BOD, the second will predominantly encourage the conversion of NH_3 (NH_4^+) to NO_3^-.

4.6   Some of the incoming nitrogenous material sediments in the primary settling tanks. Some is released as ammonia (ammonium) by heterotrophs. This may be assimilated by heterotrophs growing in the aeration tanks and, therefore, separated from the water in the settling tanks. Some of the ammonia (ammonium) may be oxidised to form nitrate. Some of this may be reduced by anaerobes to form N_2 (and other nitrogen oxides) which is volatile and will escape into the atmosphere. Thus we can describe the fate of the nitrogenous materials in waste water diagrammatically as:

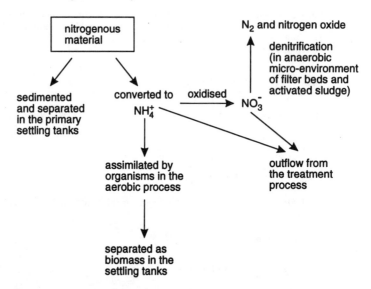

The actual proportions of the nitrogen input which ends up in the various forms shown in the diagram is highly variable. Typical figures would be for about 20% to be removed in the primary settling tanks. Perhaps about 60% as biomass produced in the aerobic processes and about 5-10% as N_2 and nitrogen oxides. The remainder remains in aqueous solution as NH_4^+ and NO_3^-.

4.7   1) Nitrification.

2) Ammonification.

3) Nitrification.

4) Nitrogen fixation.

5) Denitrification.

4.8   You might have been suggested that samples of the outflow could be inoculated into suitable media which would support the growth of human pathogens and then the pathogens identified using normal microbiological methods (for example staining characteristics, metabolic products, reaction with specific antibodies). In practice this would be extremely expensive (think of all of the different types of pathogens), it would not work with viruses (they do not grow in axenic media) and it would be dangerous. Also because the anticipated population density of any one pathogen is likely to be extremely low, each would be extremely difficult to detect.

The more usual approach is to look for *Escherichia coli* (a so called presumptive coliform count). These are present in large numbers in all human and other mammalian alimentary tracts.

Thus the presence of even a low number of *E. coli* in a water is indicative of mammalian faecal contamination. If the *E. coli* survived the water treatment process, so might pathogens from human faeces. Thus we can use *E. coli* to indicate if the treatment of the water is likely to have removed all or most of the human pathogens that are present.

4.9

1) A high NH_4^+ to NO_3^- ratio is to be anticipated. The production of H_2S in the activated sludge process is indicative of a reducing environment. In all probability, parts of the reactor are anaerobic thus encouraging the growth of *Desulphovibrios* which oxidise organic substrates and, under anaerobic conditions, reduce $SO_4^=$ to $S^=$. These are not conditions which favour the oxidation of NH_4^+ to NO_3^- (nitrification has an obligatory requirement for O_2). Thus we would anticipate most of the soluble nitrogen would remain as NH_4^+.

2) We would suggest you should use methanol. It is readily soluble (miscible) with water, it is relatively cheap and it can provide considerable amounts of reducing potential. Glucose is also a reasonable choice (metabolically) but is relatively expensive. Oxalic acid is fairly expensive and provides only a limited amount of reducing potential. Oxalic acid and phenol are both corrosive. The key to the successful use of methanol is to use a methanol-utilising denitrifier.

Responses to Chapter 5 SAQs

5.1

1) False. High levels of SO_4^{2-} in the input will result in high levels of toxic S^{2-} ions in the outflow; also noxious smelling odours would result. H_2S may be vented from the vessel. Both of these are undesirable. High levels of SO_4^{2-} will also result in less CH_4 production since the SO_4^{2-} will enable the oxidation of organic materials to CO_2 without the subsequent reduction of some carbon to CH_4.

2) False. The nitrogen in proteins is in a reduced form (that is as amines, or imines) and will not be oxidised to N_2 in the anaerobic conditions. N_2 will only be produced from oxidised nitrogenous materials (especially from NO_3^-).

3) True.

4) False. Heterotrophs use organic nutrients. The hydrogen utilising methanogenic bacteria use H_2 and CO_2 to generate methane and fix some CO_2 for cell synthesis. They are therefore autotrophs, not heterotrophs.

5) False. Although the major gases are CO_2 and CH_4, H_2 is also produced during anaerobic digestion. Therefore we might anticipate that the vented gas may also contain H_2. Also, if the input to the digester contains SO_4^{2-} we might anticipate that H_2S may also be produced (see 1) above).

5.2

The term symbiotic is used to describe situations in which two or more different types of organisms gain mutual benefit from each other. In this case, the heteroacetogens gain benefit from the methanogens which remove the metabolic products (acetic acid and H_2) of the heteroacetogens. Without this removal, metabolism of the heteroacetogens would be thermodynamically unfavourable. The methanogens, on the other hand, gain benefit from the heteroacetogens because this latter groups supply their nutrients.

5.3

1) The reduced states of the carbon atom should have provided the clue. The correct order is:

HCO_3^-; MP-COOH; MF-CHO; CoM-CH_2OH; F_{930}-CoM-CH_3; CH_4

2) All of the statements are compatible with the organism being a methanogen.

   a) Methanogens do not have conventional electron transport chains.

   b) Methanogens have a number of CO_2 fixing reactions.

   c) Hydrogen is used to reduce CO_2 and other substrates. Its oxidation provides a source of energy.

   d) Methanogens are obligate anaerobic.

   e) Methanogens use the end products of fermentative metabolism.

   f) They are resistant to cyanide because they do not contain cytochromes.

    g) Methanogens generate CH_4.

    h) Only methanogens contain methanopterin.

3) Statements g) and h) prove that the organisms is a methanogen.

5.4   1) $a = 0.2$, $x = 0.505$, $y = 0.295$, $z = -0.21$ and $w = 0.46$.

$CH_3ON_{0.5} \rightarrow 0.2\ (CH_2ON_{0.2}) + 0.505\ CO_2 + 0.295\ CH_4 - 0.21\ H_2O + 0.46\ NH_4^+$

or $CH_3ON_{0.5} + 0.21\ H_2O \rightarrow 0.2\ (CH_2ON_{0.2}) + 0.505\ CO_2 + 0.295\ CH_4 + 0.46\ NH_4$

This is how we determined this.

The C balance is:

$1 = 0.2 + x + y$ (equation i)

For O,

$1 = 0.2 + 2x + z$ (equation ii)

For H,

$3 = 0.4 + 4y + 2z + 4w$ (equation iii)

For N,

$0.5 = 0.04 + w$, therefore $w = 0.46$.

But from i)

$x = 0.8 - y$ (equation 1)

From ii)

$z = 0.8 - 2x$ (equation 2)

From iii)

$0.76 = 4y + 2z$, therefore $z = 0.38 - 2y$

Substituting equation 3) into equation 2) gives:

$0.38 - 2y = 0.8 - 2x$

Rearranging gives:

$x = 0.21 + y$ (equation 4)

Substituting equation 4) into equation 1) gives:

$0.21 + y = 0.8 - y$

Therefore $y = 0.295$

Substituting this value for y into equation 1) gives:

$x = 0.505$

Substituting this value for x into equation 2) gives:

$z = 0.21$

So the coefficients are:

$a = 0.2$

$x = 0.505$

$y = 0.295$

$z = -0.21$

$w = 0.46$

The overall equation is:

$CH_3 ON_{0.5} \rightarrow 0.2 (CH_2 ON_{0.2}) + 0.505 CO_2 + 0.295 CH_4 - 0.21 H_2O + 0.46 NH_4^+$

It would be better to write this as:

$CH_3ON_{0.5} + 0.21 H_2O \rightarrow 0.2 (CH_2ON_{0.2}) + 0.505 CO_2 + 0.295 CH_4 + 0.46 NH_4^+$

Note that water is consumed during this metabolism.

2) The ratio of CO_2 to CH_4 produced is $0.505 : 0.295$ or $1 : 0.58$.

5.5 1) $Y = 0.23$ g g^{-1}. This figure is given in Table 5.4.

2) $Y = 0.22$ mg l mg^{-1}.

Since $C_6H_{12}O_6 + 6O_2 \rightarrow 6CO_2 + 6H_2O$

180g of carbohydrate would lead to the consumption of 192g O_2.

Thus 1g of carbohydrate l^{-1} would lead to the consumption of 1.067g of O_2 l^{-1}.

In other words the BOD of the water would be 1067 mg l^{-1}.

But we know from 1) that 1g carbohydrate per litre would lead to a biomass yield of 0.23g per litre.

So 0.23g of biomass would be produced at the expense of 1067 mg l^{-1} BOD.

Thus for each mg per litre of BOD, 0.22 mg of biomass would be produced.

Thus, $Y = 0.22$ mg l mg^{-1}.

5.6 90.23h.

We use the relationship:

$$q[X]t = 2.303 \, K_s \log \frac{[S_0]}{[S_t]} + ([S_0] - [S_t])$$

Substituting in:

$$(0.1 \times 1)t = 2.303 \times 0.01 \log \frac{10}{1} + (10 - 1)$$

$0.1t = (2.303 \times 0.01 \times 1) + 9$

$0.1t = 0.02303 + 9$

$t = 90.2303$ h

5.7 1) 11000 litres

From $C = 180 P + 2000$

C = 180 x 50 + 2000 = 11 000

(The UK standard allows for a capacity of 180 litres per individual).

2) 1.83 days

Since each individual produces approximately 120 litres of waste water day^{-1}, 50 individuals will produce 6000 litre day^{-1}. Thus the mean residence time will be $\frac{11\,000}{6000}$ days = 1.83 days. Of course, in practice, there will be a distribution of residence times because the system is not mixed.

5.8

1) Crude fibre is mainly represented by fibrous plant materials which are particulate and will sediment out in the primary settling tanks. Hence the rather high levels of these in the primary sludge but rather low levels in activated sludge.

Subsequent incubation of these fibres in an anaerobic digester results in some biodegradation of these fibres. Some fibres, especially those containing lignin, are rather recalcitrant. Thus we find some fibre remaining in the products of the digester.

2) During metabolism in an anaerobic digester, much of the input organic material is mineralised (that is carbon is converted to CO_2 and CH_4, the nitrogen of amino groups is lost as $N\,H/NH_3$). This mineralisation of the organic material means that their is a loss of mass from the sludge solids. Other minerals are non-volatile and remain in the sludge. They are measured by ashing. Because of the loss of C and N by volatilisation, non-volatile minerals increase as a proportion of the total solids present.

3) Lipids are by nature not very water soluble and will aggregate in an aqueous milieu to form droplets and scums or become adsorbed onto surfaces. Thus they tend to separate out in the primary sedimentation tanks. Thus there tends to be a fairly high level of these materials in primary sludges.

Lipids are oxidisable and represent a excellent energy source under aerobic conditions and thus it is not surprising that lipids are rapidly utilised in activated sludges. In contrast, lipids are difficult to oxidise in reducing (anaerobic) environments. Thus they tend to persist in anaerobic environments. We might anticipate, therefore, that a significant proportion of the anaerobic sludge will be composed of lipids. Note that the biomass present in all 3 sludges under discussion will contain some lipid since this is an essential ingredient of bio-membranes.

5.9

This question is fairly open ended and in some instances more than one type of device would be satisfactory. Here are our thoughts on these.

1) The high organic loading of this waste is readily biodegradable and should be treated in a device with relatively short residence times. A stirred tank digester would probably be satisfactory although upflow blanket or fluidised bed digesters may also be applicable. The actual choice would depend on the volume of material to be treated and the relative capital and running costs. Methane would be a product and may be used to partially defray the costs of waste water treatment. Also important would be the availability of expertise in the company in operating these treatment processes. Anaerobic filters would appear to be unsatisfactory because the high organic loading would cause clogging of the filters by the rapidly growing biomass. It may be necessary to add nutrients (for example usable N, P and minerals) to facilitate digestion. Septic tanks would appear to be a rather inefficient way to digest such a readily degradable waste.

2) We would recommend septic tanks. These would require minimal maintenance and the relatively small volume of effluent produced would require fairly small tanks despite the long residence time required. The fact that the community is isolated means that areas would be available for discharging the effluents from the septic tanks. It would be inappropriate to install a high priced, difficult to maintain, highly engineered digester.

3) This waste would have a high organic loading and would contain a chemically complex mixture of contaminants. It is, therefore, appropriate to use a digester designed along the lines of a municipal anaerobic digester. This would be the simplest to operate. However, it might be tempting to examine the possibility of using a digester/clarifier as this would reduce the volume of vessel required and generate useful methane. The choice would depend upon the relative capital and running costs and the anticipate reliability of the device. More complex devices such as fluidised bed digesters may be appropriate but the technology is not so well established and may be both costly and unreliable. Upflow blanket digesters may also be suitable but there are uncertainties about whether the biomass produced on the waste water from piggeries will generate suitable flocs.

4) Municipal anaerobic digesters have a long tradition of successfully treating waste water from small and large urban populations. Thus it is attractive to select this tried and tested option. Anaerobic filters would tend to clog and raw sewage does not produce good flocs so that upflow blanket digesters are not reliable. The volume of waste water generated (probably of the order of 4×10^6 litres d^{-1}) would make septic tanks inappropriate as this would require very large tanks and need access to large adsorption fields.

5) Again municipal anaerobic digesters are well tried in these circumstances and would represent the easy option. However, with such a large population this would require a considerable digester capacity. It might, therefore be appropriate to examine the capabilities of the high rate (low retention time) digesters such as stirred tanks and fluidised bed digesters to handle the anticipated loading. The reduction in digester volume and land use and the production of useful methane may more then offset the capital and running costs.

5.10

1) Aerobic digestion. Aerobic oxidation yields most energy from the metabolism of substrates. Typically aerobic processes may yield 0.5 - 1.5 kg biomass for every kg BOD. Anaerobic processes yield only about 0.1 - 0.2 kg biomass for every kg BOD.

2) Aerobic digestion. The main cost is the energy required to aerate the sludge. However, bear in mind that some anaerobic digesters consume energy for mixing.

3) Anaerobic digestion. This may yield CH_4 which is an easy to handle fuel.

4) Aerobic digestion. Generally, aerobic metabolism is faster then anaerobic metabolism. The greater energy yield from aerobic metabolism means that the biomass grows and develops faster and thus there is a faster turnover of substrate into biomass and catabolic end products.

Responses to Chapter 6 SAQs

6.1

1) True - halocarbons such as halogenated aromatic pesticides are relatively stable chemically and biologically and, therefore, persist in the environment for long periods of time and can thus be termed 'recalcitrant'.

2) False - there are many naturally-occurring cyclic and aromatic compounds present in the environment, for example sterols which form part of eukaryotic cytoplasmic membranes or sugars such as glucose that are also cyclic compounds, that would never be considered as xenobiotics. Remember, xenobiotic compounds are unnatural compounds or natural ones deposited in the environment at unnaturally high concentrations due to human activities.

3) False - the presence of covalently-bound halogens in a molecule does not always result in a compound being toxic but it decreases biodegradability and hence makes the compound persistent and therefore xenobiotic. In this case the persistent nature of the compound is primarily due to its large and insoluble nature in aqueous system thus making it poorly accessible to enzymatic systems.

4) False - the toxicity of a chemical compound is indeed concentration-dependent and thus the common practice of diluting it can render it non-toxic. However, it cannot be considered as a safe method since the compound may become re-concentrated by biological systems in food chains by the process of biomagnification. An example of this latter effect has been observed for the insecticide DDT.

5) False - the statement as written is true **except** for the word 'unsubstituted'. The materials described are toxic primarily because they contain polychlorobiphenyls which are halogen-**substituted** aromatic compounds.

6.2

This SAQ tests your understanding and knowledge of the sequence of events involved in the metabolism of aromatic compounds such as benzene which form the basic structure of many xenobiotics. The order of events should be: 5), 4), 7), 2), 9) because, as indicated in the text is degraded by the *meta* ring cleavage route.

Let us briefly consider each option:

1) Acetyl CoA and succinate are end-products of the *ortho* cleavage pathway for benzene metabolism.

2) The formation of an enoate, a 2-oxo-pent-4-enoate, is an intermediate of the *meta* cleavage pathway formed by the decarboxylation of 2-hydroxymuconic semialdehyde.

3) A 1,2 dioxygenase catalyses the *ortho* cleavage of catechol to yield cis,cis-muconate.

4) A 2,3 dioxygenase catalyses the *meta* cleavage of catechol to yield 2-hydroxymuconic semialdehyde.

5) The initial step in the metabolism of benzene is the dihydroxylation of the ring to yield catechol.

6) The formation of an enol lactone is an intermediate, (3-oxoadipate enol lactone) of the *ortho* cleavage pathway of benzene metabolism.

7) Hydroxymuconic semialdehyde is the initial product of the *meta* cleavage of catechol catalysed by catechol 2,3 dioxygenase.

8) Cis,cis muconate is the initial product of the *ortho* cleavage of catechol catalysed by catechol 1,2 dioxygenase.

9) Aldehyde and pyruvate are end-products of the *meta* cleavage pathway for benzene metabolism.

6.3

1) Compound 'b' is the most recalcitrant because substitution of hydrogen atoms with chlorine atoms makes the compound less recognisable by enzymatic systems and therefore potentially more toxic.

2) Compound 'a' is the most recalcitrant because the presence of a nitro ($-NO_2$) group on an aromatic ring impedes the oxygenation reactions to form the catechol more than would the presence of a carboxyl group.

3) Compound 'b' is the most recalcitrant because the presence of the chlorine atom in the *meta* (C-3) position on the ring hinders the hydroxylation of the ring in the C-2 position (in this case, between the hydroxyl and the chlorine groups) more than does the presence of a chlorine atom in the *para* (C-4) position.

4) Compound 'b' is the most recalcitrant. 2,4-dichlorophenozyacetate (2,4-D) is biodegraded within days; 2,4,5-trichlorophenozyacetate (2,4,5-T) differs by only one additional chlorine substitution in the *meta*, yet this compound persists for months. The additional chlorine interferes with the hydroxylation and cleavage of the aromatic ring.

5) Compound 'a' is the most recalcitrant. 1,1,1-trichloro-2,2-bis-(*p*-methoxyphenyl) -ethane (DDT), since the *p*-methoxy groups are susceptible to dealkylation while the *p*-chloro substitution endows DDT with increased chemical and biological stability.

6.4

Reaction 1 is a substitution (hydroxylation) reaction since it shows the dechlorination of pentachlorophenol by substitution of a chlorine atom with an hydroxyl (-OH) group at the C-4 position.

Reaction 2 is also a substitution (hydroxylation) reaction since it shows dechlorination by substitution of a chlorine atom with a hydroxyl (-OH) group at the C-2 position.

Reaction 3 is a substitution (reductive) reaction since it shows dechlorination by substitution of chlorine atoms with hydrogen atoms at positions C-3, 5 and 6 of the aromatic ring to yield 1,2,4-trihydroxybenzene.

6.5

The answers to the statements are: 1) D; 2) B; 3) F; 4) C; 5) E; 6) A.

Let us consider some of the indicative results to look for:

1) A high % removal (for example 98%) indicates that the compound, PCB or 4-chlorobenzoate, is being completely metabolised;

2) A zero % removal is indicative that the organism or pairs of organisms, do not have the metabolic capabilities for the removal of the compound;

3) A low removal of 40, 42 or 58% is indicative of the organism or pairs or organisms having the metabolic capabilities for removal of the compound but that the metabolic pathway is eventually inhibited by the accumulation of a product. This product may be 4-chlorobenzoate or 5-chloro-2-hydroxymuconic semialdehyde which is the toxic product of the *meta* cleavage pathway of 4-chlorobenzoate.

Let us now look at each organism in turn:

Organism A - alone it exhibits high (98%) removal of both PCB and 4-chlorobenzoate via the *ortho* cleavage pathway.

Organism B - alone it shows 0% removal of PCB and high % removal of 4-chlorobenzoate. It, therefore, does not have the metabolic capability for PCB removal but is capable of metabolising 4-chlorobenzoate via the *ortho* cleavage pathway.

Organism C - alone it shows 0% removal of PCB and a low % (42) removal of 4-chlorobenzoate. This organism, therefore, does not have the metabolic capabilities for PCB metabolism but is capable of partially metabolising 4-chlorobenzoate via the *meta* cleavage pathway.

Organism D - alone it shows partial (40%) removal of PCB and 0% removal of 4-chlorobenzoate. This organism is, therefore, capable of metabolising PCB but the reaction is prematurely terminated due to the accumulation of 4-chlorobenzoate which it is not capable of metabolising. In combination with organism B (an organism that is unable to remove PCB but can completely remove 4-chlorobenzoate) the PCB is completely removed and this is an example of sequential metabolism.

Organism E - alone it shows 0% removal for both PCB and 4-chlorobenzoate. Therefore, this organism is not capable of metabolising either PCB or 4-chlorobenzoate.

Organism F - alone it shows 0% removal of PCB and high % removal of 4-chlorobenzoate. Therefore, organism F cannot apparently remove PCB but can metabolise 4-chlorobenzoate via the *ortho* pathway. It does, however, show a high % removal of PCB in the presence of organism E which is itself apparently incapable of PCB removal. Thus either organism E or F has the capabilities for PCB metabolism but that this capability is only exerted in the presence of specific nutrient provided by the other organism. Organism E cannot metabolise 4-chlorobenzoate so the answer to statement 3 has to be organism F.

6.6 The method considered most unsuitable would be Method 9), it would be at a competitive disadvantage to the other organisms present and would probably soon be lost from the microbial flora. We also need to consider the regulations regarding the use of genetically engineered micro-organisms.

Not all of the other methods are of practical value for increasing the degradation rate of a halogenated compound so let us consider each of the methods in turn.

Method 1) - one can consider each cell as a catalytic unit capable of speeding up reaction rates so increasing the number of cells present should increase the rate of reaction and hence the degradation rate. How one can increase the number of cells present is another problem but one which can be solved.

Method 2) - the external pH surrounding microbial cells will affect their growth and metabolic rates so the implementation of a pH-control system to maintain the pH to the optimal value for halocarbon degradation would be of value. One may, however, have to consider the optimal pH value for any other processes taking place such as the degradation of any other organic compound.

Method 3) - generally degradation rates of compounds are higher under aerobic conditions so the implementation of an aeration system may help to prevent the system from going anaerobic and so maintain high degradation rates. If, however, a degradative system has been developed under anaerobic conditions then it is likely that the metabolic pathway involved would not operate under aerobic conditions.

Method 4) - nitrogen, phosphorus and sulphur are essential elements for microbial growth and metabolism so their addition to an environment in which they were poorly available would increase degradation rates.

Methods 5) and 6) - degradation rates are substrate concentration-dependent at the two extremes (low and high). Too high a concentration of a halogenated compound may result in it becoming toxic towards the microbial cells with consequential inhibition of metabolism and/or cell death. Too low a concentration may result in degradation rates being below optimal since, at low concentrations, the reaction rate will be proportional to the concentration of the compound of interest. Thus either method may result in increased or decreased degradation rates depending on what the initial concentration of the compound was.

Method 7) - certain compounds do not induce the enzymes required for the metabolism of that compound. The pathway can, however, be induced by the addition of a similar compound such as benzoate for the aromatic degradation pathways. Alternatively benzoate may provide the source of carbon, energy, and or reducing equivalents required by a cell for metabolism of the halogenated compound.

Method 8) - reaction rates are temperature-dependent so increasing temperature could increase degradation rates as long as the temperature does not exceed the optimum for the particular cells involved.

Method 9) - such an organism could indeed increase degradation rates but this method can be considered as the most unsuitable because of the legal regulations governing the release of genetically-engineered micro-organisms due to their unknown potential danger.

6.7   All of the effects listed may occur.

1) Halogenated aromatic compounds are potentially toxic and could thus inhibit microbial activity or result in cell death.

2) It would be unlikely that an efficient population of cells would exist in the treatment plant capable of removing the compounds.

3) The effects described in 1) would lower the efficiency of the system for the oxidation and removal of other 'normal' organic molecules and so increased levels of these compounds would be present in the discharge water and hence increase the COD (Chemical Oxygen Demand) value which is a measure of oxidisable matter.

4) Because of the same reasons as for 3) except that BOD (Biological Oxygen Demand) is an estimation of biologically oxidisable matter.

6.8   The proposals are considered below.

1) This is a traditional approach of water treatment organisations and one often carried out. Environmentally it is not a good approach because the potential toxic problems of the waste still exist, albeit in a different place, and there would always be the danger that these toxic compounds could leach out of the burial site into surrounding land and water courses.

2) This is a good approach as the compounds could be detoxified by microbial mineralisation or transformation prior to discharge to a water course or to a water treatment plant treatment.

3) This is a possible approach as this would prevent shock loadings (sudden high concentrations) of the material entering the water treatment plant and thus allow for a microbial population capable of degrading the waste to establish and maintain itself. One would have to first establish that the continuous addition of the compound would result in the establishment of a microbial population capable of completely degrading the compound of concern and that the concentration of the compound did not ever reach toxic levels.

4) This is an unlikely solution to the problem since the water treatment system will already have an aerated system (activated sludge tank or similar) and the additional amount of organic material entering the system as a result of the halogenated aromatic discharge would be negligible and would, therefore, not increase the oxygen demand of the system. If the amount of halogenated aromatics entering the system was not negligible then it is highly likely that such a concentration of halogenated aromatics would toxify the treatment system.

5) We could follow a similar reasoning as for 4) in that the sewage treatment system will already have a pH-control system and the biological oxidation of negligible amounts of halogenated aromatics would not significantly alter the pH. If the effluent from the factory was highly acids or alkaline due to the addition of acid or alkali during manufacture then this would indeed have to be neutralised but this would be done prior to its discharge to the sewage works.

Responses to Chapter 7 SAQs

7.1   1) False - Member States are only obliged to set standards if a parameter is given Maximum Admissible Concentration or Minimum Required Concentration. If only a Guide Level value is given in the Directive, Member States may use their discretion as to whether or not to set a standard.

2) False. Although the absence of coliforms indicates that the water is free of contamination from excreta (sewage), it could, for example, contain high levels of toxic materials from other sources.

3) True - It should be self evident that a Guide Level (GL) is a level that it is hoped can be achieved. The Maximum Admissible Concentration (MAC) is the level that must not be exceeded. Thus the GL must never exceed (and is usually lower than) the MAC.

7.2   Examination of the data reveals that waters used to raise coarse fish may have greater levels of chromium, lead, zinc, phosphate, ammonia and BOD than the waters used for salmonid fish. These indicates that coarse fish are generally more tolerant to both inorganic and organic materials than are salmonoids.

7.3   1)   You should have chosen enzyme A. From the data given, this enzyme appears to be more specific than enzyme B. Note however it is not completely specific - enzymes rarely are.

If you were interested in producing a biosensor capable of detecting a range of compounds related to $X_{reduced}$, then enzyme B might be a better choice.

2)   The other parameters should include:

- availability of the enzyme (influences costs);

- stability of the enzyme (influences reliability and costs);

- affinity for the substrate (influences sensitivity);

- activity versus pH (and temperature) profiles (influence operational factors).

7.4   There are several possible ways to tackle this. Here we will deal with a genetic solution. The basic problem can be represented diagrammatically as:

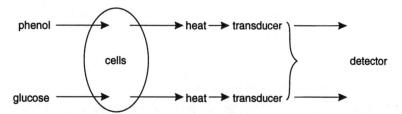

What needs to be done to make this device successful is to allow discrimination of heat generation from phenol metabolism from that from glucose metabolism. The most obvious approach would be to produce mutants which were blocked in glucose metabolism. Thus cells could be treated with a mutagen and plated out onto phenol-containing medium. Then they could be replica-plated onto glucose-containing medium. Those colonies which did not grow on glucose but did so on phenol (ie phe^+ glu^-) would be appropriate to study further. It might be that even with the phe^+ glu^- cells, the glucose repressed/inhibited phenol metabolism. If so, a further round of mutations would be undertaken to isolate cells in which glucose no longer inhibited phenol metabolism. Cells with glucose-sensitive phenol metabolism would not grow well in phenol media containing glucose whilst glucose-insensitive cells would grow quite well under such conditions. Such cells would be suitable as the bio-component of the proposed phenol sensor.

You may have thought of alternative approaches. For example, you might have considered transferring the phenol-metabolising plasmids to host cells that do not metabolise glucose. You may have considered adding inhibitors that blocked either glucose uptake or metabolism. All of these may be valid. The point we are making is that there is usually more than one route to solving a problem. The actual route chosen will depend upon the circumstances and upon the knowledge, skill and judgement of the designer.

Responses to Chapter 8 SAQs

8.1   1)   The answer is a) CH_3OH.

There are two basic ways to do this question. One is to use the equations shown in Table 8.1. The other is to use the following argument. Methanogenesis using compounds composed entirely of C, H and O leads to the complete dissimilation of C as either CH_4 or CO_2. The more reduced the substrate, the greater the proportion of C is dissimilated as CH_4. Essentially what is happening is that the C, H and O atoms are rearranged to form CH_4, CO_2 and, in some cases, H_2O.

$$(CHO) \longrightarrow \begin{array}{l} CH_4 \\ CO_2 \\ (H_2O) \end{array}$$

Using the equations given in Table 8.1, we could calculate the actual amounts of methane that could be generated from these substrates.

Thus:

$$4CH_3OH \rightarrow 3CH_4 + CO_2 + 2H_2O$$
$$4 \times 32 \text{ kg} \rightarrow 3 \times 16 \text{ kg}$$
128 kg → 48 kg   Therefore 1 kg would produce 0.375 kg CH_4

Similarly for:

$$4HCOOH \rightarrow CH_4 + 3CO_2$$
$$4 \times 46 \rightarrow 16$$
184 kg → 16 kg   Therefore 1 kg would produce 0.0714 kg CH_4

and

$$CH_3COOH \rightarrow CH_4 + CO_2$$
60 kg → 16 kg   Therefore 1 kg would produce 0.26 kg CH_4

2) The proportion of CH_4 to CO_2 in the gases produced from:

   a) CH_3OH wouldbe 3:1.

   b) $HCOOH$ would be 1:3.

   c) CH_3COOH would be 1:1.

   (see Table 8.1 if you did not come to these conclusions).

8.2

1) False.

2) False. The sludge is retained for a long time but there is little methane production.

3) True.

4) False. The nature and size of the flocs produced in digestions is dependent upon the composition the material being treated.

5) True.

8.3

You should remember that to attempt to harvest and utilise the biogas produced there will have to be an initial high investment to build equipment for collection, storage and distribution of the gas.

1) The site must be large enough to exploit - it is generally reckoned that a site of several thousand tonnes of organic waste will be needed. This suggests that the proximity of one or more large towns or cities is essential to provide material and to keep down transport costs.

2) The industrialised (Western) world is a more likely place because such communities tend to produce more waste per head as they are less resourceful at recycling than underdeveloped nations.

3) A suitable geographical location is required, preferably some sort of small valley or man-made quarry so that the site can be confined.

4) There needs to be a nearby, constant market for the biogas produced, preferably close to the landfill site. An ideal customer is one who is only partially reliant on biogas who will, therefore, generally be able to utilise total production thereby avoiding the need for storage facilities.

8.4

1) False. There is little hydrogen in biogas. Any hydrogen evolved during metabolism would be combined with the carbon dioxide to yield more methane.

2) False. Due to the 40% carbon dioxide it contains, biogas has a much lower energy value than natural gas which is virtually pure methane.

3) True. Only a very few bacterial species have the property to generate methanol.

4) True. The major source of material will be decaying plant material, particularly cellulose and xylans.

5) False. Although decaying plant material could be the predominant substrate, the likelihood is of a very mixed substrate source containing varied but limited amounts of plant waste.

6) True. If you re-examine Figure 8.2 you will see that all of the end products except for carbon dioxide and hydrogen are simple acids and alcohols.

Responses to Chapter 9 SAQs

9.1

The key point is that industrial nitrogen fixation leads to products which are directly applied to agricultural land. Much of the biological nitrogen fixation occurs in uncultivated land and in the ocean and seas. Nearly half of the world's biological nitrogen fixation occurs in the ocean. Although this will ultimately contribute to the total combined nitrogen available for plant growth, its production and distribution is not targeted for agricultural use.

It is important to note however, that biological nitrogen fixation associated with legumes contributes significantly to the nitrogen input to plants. In the UK this contributes about 40% of the total nitrogen input into crops. In the USA, it contributes about 60% of the total nitrogen input. Free living nitrogen fixers contribute less than 5% of the total nitrogen input to crops in the UK (10% in the USA). The remainder of the nitrogen input (approximately 60% in the UK, 40% in the USA) comes from nitrogen fertilisers. Thus although industrially fixed nitrogen makes a significant contribution to maintaining soil fertility, there are still some important relics of the traditional reliance on natural processes.

9.2

1) Klebsiella is a heterotrophic facultative anaerobe. It only fixes nitrogen under anaerobic conditions (see Table 9.3). Thus it needs an organic carbon and energy source (eg glucose) and anaerobic conditions. The presence of combined nitrogen will repress its nitrogenase. Therefore the most suitable conditions for nitrogen fixation by members of this genus is given by e). Light has little effect on the growth of these organisms.

2) Anabaera cylindrica is an obligate photoautotroph. It requires light as a source of energy and uses CO_2 as a source of carbon. It has a requirement for O_2 but this can be supplied via its own photosynthesis. Its nitrogenase is repressed by combined nitrogen. Thus, the conditions most appropriate for nitrogen fixation in cultures of this organism would be those described in c) and d). However the conditions described in e) might also be appropriate. A. cylindrica will grow in the presence of glucose providing CO_2 and light are also available.

9.3

1) The conclusion you should have reached is that although 50-60 kg N ha^{-1} are fixed, more than this amount of nitrogen is removed when the crop(s) is harvested. In other words, although nitrogen is fixed in the nodules of the peas and beans, combined nitrogen is also taken up from the soil. The calculation shown in 2) gives quantitative data on this.

2) 55 to 75 kg ha^{-1}. Since 50 to 60 kg N are fixed ha^{-1}, but there is a net loss of 5 to 15 kg N ha^{-1} from the soil when the crop has been harvested, then (50 to 60) + (5 to 15) kg N ha^{-1} has been removed.

9.4

The answer is 'probably not'. The growth of clover in field B would have increased the number of clover-infecting rhizobia (R. trifolii). These, however, will not infect peas (Pisum) thus we might anticipate that there would be little difference in the extent of nodulation of peas grown in the two fields. In practice, however, the situation is never as predictable as this. The growth of wheat in field A and clover in field B will result in substantial differences in the soil microfloras of these two fields. Without a great deal of further study, it would be difficult to predict the effects of this on the numbers of R. leguminosarum in the two fields. Therefore, it is difficult to make confident predictions about the extent of nodulation. Nevertheless substantive studies have been, and continue to be, undertaken to examine the factors which influence the extent of nodulation and we might anticipate that some general 'rules' will emerge.

9.5

We need to calculate the profitability of the three options: the untreated (control) crop; the NH_4^+ fertilised crop and the inoculated crop. The cash crop for peas is mainly through the sale of seeds. Thus we can use the yield of seeds as a measure of profitability.

For the untreated crop, the profit margin is represented by 11 800 - 9000 kg seed ha^{-1} = 2800 kg ha^{-1}.

For the NH_4^+ treated crop, the costs are 25% higher. In other words 11 250 kg seed ha^{-1} would be required to break even. The profitability of this treatment would therefore be represented by 16 700 - 11 250 kg seed ha^{-1} = 5450 kg ha^{-1}.

For the inoculated crop, the costs are 15% higher. 10 350 kg seed ha^{-1} would be required to break even. The profitability of this treatment would, therefore, be represented by 14 300 - 10 350 kg seed ha^{-1} = 3950 kg ha^{-1}.

We may, therefore, rank the potential profit for the three types of production as:

untreated $\equiv$ 2800 kg ha^{-1};

inoculum treated $\equiv$ 3950 kg ha^{-1};

NH_4^+ treated $\equiv$ 5450 kg ha^{-1}.

It would, therefore, appear that the ammonium treated crop would be most profitable in terms of saleable crop.

In practice the decision might be influenced by a number of factors. For example, if the farmer did not have the money necessary to purchase the NH_4^+ fertiliser, this option may not be available to him. He might borrow such money as was needed but then he would incur additional interest payments thereby reducing his profit margin. The inoculum treatment requires less financing, but still incurs some additional costs. A poor farmer could probably not be able to afford either treatment and would, therefore, be forced into leaving his fields untreated.

From what we have described in the text, there are additional (environmental) costs especially associated with the NH_4^+ treatment which, currently, are unlikely to figure significantly in the individual farmer's considerations. It is, however, only by including these 'costs' that the true cost of a particular strategy can be evaluated. Increasingly, attention is being drawn to the discrepancy between commercial costs and 'total' (including environmental) costs. Both national and international legislative bodies are working towards formulating strategies which encompass a holistic approach to agricultural practices.

You will notice also that, just by applying commercial criteria, poor farmers are likely to remain poor since they cannot afford the treatments that would improve their profit margins. There are therefore also important sociological consequences that arise by simply applying commercial consideration.

Of course, in a paper exercises of this type, we have simplified the issues. They are, indeed, very complex. Nevertheless, this SAQ may have helped you appreciate that applying stark economic criteria may not be necessarily environmentally nor socially good. The challenge to the international community is how to achieve the successful introduction of the environmentally good practices that arise through biotechnology.

9.6 This is a rather open ended question, but it will serve to illustrate the point we are trying to make, that the environmental consequences of applying chemical pesticides may have far reaching effects.

The application of the pesticide may well reduce the population of the pest species. It will, at the same time, increase the selective pressure for the development of pesticide resistant variants of the pest. Thus, in the longer term, may eventually result in the pesticide being rendered ineffective.

Many so called pesticides are relatively non-specific. In the case described in the question, it is likely that the honey bees will also be affected by the applied pesticide. This will, of course, have consequences on the pollination of flowers in the orchard and the surrounding areas. It will also lead to reduction in honey yields. Thus, although the use of the pesticide may reduce the loss of yield caused by the pest(s), it may reduce the yield by reducing pollination.

Furthermore, the pesticide may accumulate in the apples and honey products of the orchard. The apples, either eaten directly or fermented to produce cider, will provide the vehicle for the pesticide to be passed onto humans. The presence of the pesticide may also have an effect on the fermentation process.

Within the orchard, birds feeding on contaminated, but not killed, insects would also accumulate the pesticide. This may have long term effects on the bird population (remember the case of DDT and the Shag described in the text). It may ultimately reduce the degree of predation of these pests by birds.

If the pesticide was recalcitrant, it may eventually leach out of the soil into ground water and into rivers and may, through biomagnification, cause problems elsewhere.

You may have thought of other consequences of using the pesticide in the situation we described in the question. The points we are emphasising is that the application of pesticides may have many undesirable consequences and that it is difficult to identify and predict all of these consequences.

9.7   The answer is probably no. Although organisms with such metabolic capabilities are either known or could be generated using contemporary biotechnological techniques, their application to soils is unlikely to be successful. Even after a recent application of a pesticide, the soil concentration on a w/w basis is still extremely low. If we inoculated such a soil with a pesticide metaboliser, the pesticide would represent only a tiny fraction of the total metabolic energy resource within the soil. Thus other organisms would be present in large numbers. The pesticide metaboliser would itself be a source of metabolic energy (see Chapter 2). Thus we might anticipate that such an organism would soon be out-competed and, probably, eradicated from the soil. In all probability the effect on the amount of pesticide remaining in the soil would be very marginal unless, of course, a very large inoculum was used. Even then, success could not be guaranteed. A more detailed discussion of the application of micro-organisms capable of metabolising recalcitrants is given in Chapter 6.

9.8   The process described in the question has, for example, been achieved, using contemporary genetic engineering techniques, with tobacco. The gene for the toxic protein has been transferred into tobacco cells *in vitro*. Whole plants have been regenerated from these transgenic cells and the plants have been shown to express the toxin gene in every cell. The plants have been shown to be 'resistant' to insects since the caterpillars die soon after they begin to eat the plant tissue. Thus the process is seen to be feasible. However, if the plant is to be consumed by humans, important questions arise concerning the possible effects of the protein on the consumers. Thus, although this strategy may eliminate or reduce the pest, it may do the same for humans!

Furthermore, the consequences of this gene transfer in terms of biomass production and harvestable crop yields is also uncertain. In a world in which economics largely govern decisions, this remains an important issue.

Responses to Chapter 10 SAQs

10.1   You should compare your response to Figure 10.1 given in the text.

10.2   We have learned that *Thiobacillus ferrooxidans* requires a pH of 1.5 to 5.0 to allow it to grow therefore 1) must be incorrect and 2) must be correct.

The bacterium can reduce both iron and sulphur compounds therefore response 5) is indicated rather than 3) or 4).

The bacterium is a chemoautotroph, thus by definition it obtains carbon for growth by fixing of carbon dioxide and it does not utilise glucose as carbon source. Therefore, statement 6) and not 7) must be correct.

10.3
1) This statement is true. As we learned in the text, iron pyrite can be converted to ferric sulphate by a direct reaction.

2) This statement is false. *Thiobacillus ferrooxidans* can covert ferrous to ferric sulphate easily by a direct process.

3) This statement is false. Covellite, as mentioned earlier is insolubilised though it can be converted to the soluble copper sulphate.

4) This statement is false. In fact both of the sulphates are soluble, it is the iron pyrite which is insoluble.

5) This statement is true. *Thiobacillus ferrooxidans* is certainly a chemoautotrophic bacterium which derives its energy from the oxidation of inorganic compounds and utilises carbon dioxide as sole carbon source.

Finally the correctly balanced equation is:

$$4\ FeS_2 + 15O_2 + 2\ H_2O \rightarrow 2\ Fe_2(SO_4)_3 + 2\ H_2SO_4$$

10.4   The most important feature is a commerical one. Presumably there is insufficient copper to extract by smelting but there has to be sufficient in terms of concentration and total amount to justify the venture economically.

The site must be geologically suitable in that the dumps must allow permeation of the leaching solution but the surrounding land must be impermeable such that the pregnant solution can be contained.

Environmental conditions must be correct to allow growth of essential bacteria such as *T. ferrooxidans* to grow and participate in the leaching. For example, a site near to either of the poles would be too cold.

The dump must not be overcontaminated by other, toxic metals.

10.5   There are several possible explanations for the observations given in this question. The one we tend to favour is as follows. The accident spillage of a phenolic solution probably resulted in a large portion of the microflora of the dump being killed (phenols are well known disinfectants). This would, of course, curtail acid production and thus the pH rises. In effect, this means that the concentration of OH^- ions increases. These in the presence of ferric ions form insoluble complexes similar to those formed in the processes of rusting. In fact, in this case, since the ferric ions are present mainly in the form of ferric sulphate, the high pH favours the formation of a complex of ferric, sulphate and hydroxide called Jarosite. This is insoluble and forms an impervious layer over the surface of the ore. Thus, despite being re-innoculated, the thiobacilli find it difficult to re-establish themselves in the dump. As a consequence, the pH remains relatively high and the metal ion concentrations are low.

10.6   1) Silver is released when films are developed. Currently chemical systems are available to recover the silver. However, the facts that silver is a precious metal, is in relatively high concentration in photographic waste and that only a relatively small volume of waste is produced renders the extraction suitable for microbial exploitation.

2) The Dead Sea, because the sea water has evaporated down considerably and, therefore, minerals in it have been concentrated. This would make a mineral winning process more likely to be economically viable.

10.7   The sulphur in coal occurs as inorganic iron pyrite and as organic sulphur complexes. When combusted, the sulphur is emitted in flue gases as sulphur oxides (including SO_2 and SO_3). These combine with moisture in the atmosphere leading eventually to acid rain and subsequent deleterious effects on the environment.

The problem can be alleviated by:

*   burning only low sulphur coal in the power stations;

*   apply postcombustion scrubbing systems to remove the sulphur dioxides from flue gases;

*   removal of the sulphur compounds from coal by precombustion treatment. Either physicochemically or, possibly, biologically.

Suggestions for further reading

BIOTOL texts

Many BIOTOL texts provide background to the techniques and applications relevant to this text. The main ones are:

Bioprocess Technology: Modelling and Transport Phenomena, ISBN 07506 1507 9

Bioreactor Design and Product Yield, ISBN 07506 15095

Operational Modes of Bioreactors, ISBN 07506 1508 7

Biotechnological Innovations in Animal Productivity, ISBN 07506 1511 7

Biotechnological Innovations in Crop Production, ISBN 07506 1512 5

Biotechnological Innovations in Chemical Synthesis, ISBN 0 7506 0561 8

Energy Sources for Cells, ISBN 0 7506 1505 2

In Vitro Cultivation of Plant Cells, ISBN 0 7506 0554 5

In Vitro Cultivation of Micro-organisms, ISBN 0 7506 0507 3

Techniques for Engineering Genes, ISBN 0 7506 0556 1

Strategies for Engineering Organisms, ISBN 0 7506 0559 6

Crop Physiology, ISBN 0 7506 0560 X

Crop Productivity, ISBN 0 7506 0562 6

Other texts

In addition to the texts cited in the main chapters, there are many excellent texts available. Below are a few examples.

Barrow C.J., Land Degradation, Cambridge University Press, Cambs, 1994, ISBN 05214 6615 6

Bates J.H., Water and Drainage Law, Sweet and Maxwell London, 1990

Benarde M.A., Global Warning - Global Warming, J. Wiley & Son, Chichester, 1992, ISBN 0 4715 1323 7

310

Boon, P.J., Calow P. and Petts, G.E., River Conservation and Management, J. Wiley & Son, Chichester, 1992, ISBN 0 4719 2946 8

Botkin, et al, Changing the Global Environment: Perspective on Human Involvement, ISBN 0 1211 8731 4

Bourriau J., Understanding Catastrophe, Cambridge University Press, Cambs, 1994, ISBN 05214 6615 6

CAMLAB: Design of Municipal Waste Water Treatment Plants (MS/M008), CAMLAB, Cambridge, 1991

CAMLAB: Operation of Municipal Waste Water Treatment Plants (MS/M008), CAMLAB, Cambridge, 1991

Chambers J.E. and Levi, P.E., Organophosphates: Chemistry, Fate and Effects, Academic Press, London, 1992, ISBN 0 1216 7345 6

Christensen, N., Sanitary Landfilling, Process, Technology and Environmental Impact, Academic Press, London, 1989, ISBN 0 1217 4255 5

Forster C.F. & Wase D.A.J., (Editors) Environmental Biotechnology, Ellis Horwood, 1987, ISBN 0 85312-838-3

Greenberg A.E., Clescerl, L.S. and Eaton, A.D., Standard Methods for the Examination of Water and Waste Water, CAMLAB, Cambridge, UK (MS/S0040)

Gilpin, A., Environmental Impact Assessment, Cambridge University Press, Cambs, 1994, ISBN 0 5214 2976 6

Holmes, G., Theodore, L. and Siagh, B., Handbook of Environmental Management and Technology, J. Wiley & Son, Chichester, 1993, ISBN 0 4715 8584 X

Lerche, I., Oil Exploration, Academic Press, London, 1989, ISBN 0 1244 4175 0

Mannion, A.M. & Bowby S.R., Environmental Issues in the 1990s, J. Wiley & Son, Chichester, 1992, ISBN 0 4719 3366 0

Nriagu J.O., Gaseous Pollutants: Characterisation and Cycling, J. Wiley & Son, Chichester, 1992, ISBN 0 4715 4898 7

Purdon, P.W., Environment Health, Academic Press, London, 1990, ISBN 0 1217 4299 5

United Nations Environmental Programme 'Environmental Data Report 1991/92', Blackwell, ISBN 0 631 18083 4

Index